The Self Beyond Itself

The Self
Beyond Itself

___

An Alternative History of Ethics, the New Brain
Sciences, and the Myth of Free Will

Heidi M. Ravven

Requests for permission to reproduce selections from this book should be mailed to:
Permissions Department, The New Press, 38 Greene Street, New York, NY 10013.

Published in the United States by The New Press, New York, 2013
Distributed by Perseus Distribution

LIBRARY OF CONGRESS CATALOGING-IN-PUBLICATION DATA

Ravven, Heidi M., 1952–
  The self beyond itself : an alternative history of ethics, the new brain sciences,
and the myth of free will / Heidi M. Ravven.
      p. cm.
  Includes bibliographical references and index.
  Summary: "A critique of 'free will' that draws on neuroscience, philosophy,
and religion"—Provided by publisher.
  ISBN 978-1-59558-537-0 (alk. paper)
  ISBN 978-1-59558-800-5 (e-book)
  1. Ethics.   2. Neurosciences.   3. Free will and determinism.   I. Title.
  BJ1012.R349 2013
  170—dc23
                                                                 2012031769

Now in its twentieth year, The New Press publishes books that promote and enrich public
discussion and understanding of the issues vital to our democracy and to a more equitable
world. These books are made possible by the enthusiasm of our readers; the support of a
committed group of donors, large and small; the collaboration of our many partners in the
independent media and the not-for-profit sector; booksellers, who often hand-sell
New Press books; librarians; and above all by our authors.

www.thenewpress.com

Composition by Westchester Book Composition
This book was set in Adobe Caslon

Printed in the United States of America

2   4   6   8   10   9   7   5   3   1

To the memory of my parents, z"l
Avi mori, Robert Maurice Ravven, M.D., Ph.D.,
"A free man thinks of death least of all things, and his wisdom is a
meditation on life, not of death" (Baruch Spinoza)
and
Lucille Morrison Ravven, M.A.,
"All things noble are as difficult as they are rare" (Baruch Spinoza)
and to the memory of my exalted teacher, z"l
Professor Alexander Altmann,
"Intellectual thought in constantly loving Him should be aimed at"
(Moses Maimonides)

"That thing is said to be free (*liber*) which exists solely from the necessity of its own nature."

—BARUCH SPINOZA, *Ethics*, part 1, definition 7

"Freedom of mind . . . is a private virtue."

—BARUCH SPINOZA, *Tractatus Politicus*, chapter I, section 6

"The supreme mystery of despotism, its prop and stay, is to keep men in a state of deception, and . . . cloak the fear by which they must be held in check, so that they will fight for their servitude as if for salvation."

—BARUCH SPINOZA, *Theological-Political Treatise*

"The human heart is conquered not by arms but by love and nobility."

—BARUCH SPINOZA, *Ethics*, E IV, appendix 11

Contents

Acknowledgments

The germ of the idea for this book was planted by Constance H. Buchanan, senior program officer in religion at the Ford Foundation. Unexpectedly but fatefully, in January 2004 I received a call from Connie inviting me to submit a proposal for a project that would rethink ethics. That call set me on the path that, nine years later, has resulted in this book. I am grateful beyond measure to Connie and the Ford Foundation for making it possible for me, through their extraordinary grant of $500,000 over a five-year period, to think more broadly and in a multidisciplinary way about why people are ethical, why they are not, and how to get them to be more ethical. For five years the Ford Foundation's Progressive Religion and Values group met twice a year. I thank the other members, a fascinating and diverse group of scholars and practitioners of religion, for their willingness to listen to my work in progress, their ideas and critiques, and their encouragement. Connie's gentle encouragement, sharp intellect, and critical reading carried me through the stages of the development of the project.

I have been blessed with astute, thoughtful, rigorous, and kind readers. I cannot adequately express the gratitude I owe so many, of whom I will name but a few. Marc Favreau, editorial director of The New Press, is the best reader and editor one could ever imagine having. Warren Zeev Harvey of the Hebrew University read version after version of crucial chapters. Bernadette Brooten of Brandeis University, Jaak Panksepp of Washington State University, John McCumber of the University of California at Los Angeles, Joe Keith Green of Eastern Tennessee University, Wendell Wallach of the Yale Center for Bioethics, Karen L. King of the Harvard Divinity School,

Sheila Greeve Davaney of the Iliff School of Theology and subsequently of the Ford Foundation, James Wetzel of Villanova University, and Bonnie Kent of the University of California at Irvine offered crucial avenues to pursue and thoughtful advice. My Hamilton College colleagues Marianne Janack, Bonnie Urciouli, Richard Werner, Henry Rutz, Doris Rutz, and Richard Seager read chapters and discussed ideas with me on multiple occasions. Many colleagues in philosophy and its history and related disciplines also generously gave of their time and shared ideas, including Richard C. Taylor of Marquette University, David Burrell of Notre Dame, Daniel Boyarin of the University of California at Berkeley, Lee C. Rice of Marquette University, Vance Maxwell of Memorial University of Newfoundland, David Novak of the University of Toronto, Norbert Samuelson of Arizona State University, Hava Tirosh-Samuelson of Arizona State University, Anne Klein of Rice University, Ebrahim Moosa of Duke University, Edwin Winckler of Columbia University, Patricia Longstaff of Syracuse University, Warren Montag of Occidental College, and Allen Manning of the State University of New York at Oswego.

Because of the multidisciplinary nature of the project, I met with a wide range of scholars and practitioners. I wish to thank Don Denetdeal, Herbert Benally, James McNeley, and Mark C. Bauer of Dine (Navajo Community) College; Oswald Werner of Northwestern University; Duncan Ryuken Williams of the University of California at Berkeley; Reverend Masao Kodani of the Senshin Buddhist Temple, Los Angeles; Elizabeth Napper of the Tibetan Buddhist Nuns Project in Dharamsala, India; Dr. Maher Hathout of the Muslim Public Affairs Council of Los Angeles; Thomas Lickona of the State University of New York at Cortland; Merle Schwartz of the Character Education Partnership in Washington, D.C.; and Max Malikow of LeMoyne College in Syracuse.

I am deeply grateful to Hamilton College; its academic deans David Paris, Joseph Urgo, and Patrick Reynolds; and the chair of my department, Richard Seager, for having been wonderfully supportive and protective of me and of the project, granting me a total of three and a half years of leave in the course of it and also granting me the flexibility to focus on its demands while trying to keep up with teaching and service. I also thank my students in my course on Spinoza's *Ethics* for their lively engagement with, wonderful insights into, and passionate love of the text.

On a more personal note, I cannot thank my family enough for their patience and support throughout this lengthy project. It has consumed me at times, sometimes appearing as if it were a mission impossible or perhaps without end. There have been within my immediate family illness, marriage, divorce, the birth of a granddaughter (the beautiful and delightful Lucy Morris-Ravven), and my remarriage in the years the writing of this book has spanned. I am grateful to the late Dr. Robert Seidenberg for helping me through these life changes and keeping me on track. My daughter, Simha Ravven, MD, along with my work, are the sources of my greatest joy and pride. To the extent that Simi had to defer to the book, especially at the end and at a crucial time in her life, I am regretful. I am grateful to my life partner, Eric J. Evans, for his appreciation of the project and for his understanding and tolerating its demands upon me. Finally, I have lived for almost four decades with the knowledge of the faith that my eminent professor Alexander Altmann placed in me. I hope I have at long last somewhat redeemed that faith—and that of my parents. May their memories be blessed.

The Self Beyond Itself

1

Searching for Ethics:
How Do People Become Good (and Bad)?

Human Moral Nature

Why are some people ethical and others unethical? How do people become
ethical or unethical? Do people sometimes act in ethical ways and at other
times act in quite unethical ways? How can that happen? Are there situations and times when people *tend* to act in ethical ways and other times when
they *tend* to act unethically? How can we get people to be more ethical and
more consistently ethical? How can we get ourselves to be better people and
act more ethically more of the time? These are the questions I address in this
book. Philosophers refer to these questions and related ones as the problem
of "moral agency." This book is about moral agency. I look at the problem of
moral agency—how we become moral (and immoral) and why we act
morally and immorally—from many different perspectives. I circle around
it, exploring different ways of thinking and rethinking what our *experience*
of being ethical is all about—especially where our ethical capacity comes
from, how it develops, and finally how to strengthen it and put it to best use.

One fairly popular idea among some scientists and philosophers looking
to discover where our moral sense comes from is to search for an ethical
module in the human brain. These brain scientists set out to discover and
locate a special innate ethical capacity in the brain. They conjecture that
some of us inherit a more effective ethical brain than others—that is, some
of us are born with a strong moral brain capacity and others with a weak one.
Other scientists and philosophers conjecture that perhaps some of us use
and develop our ethical capacity better than others. These folks ask what

1

certain people do to become better at being ethical than others. So some scientists and philosophers regard the variation as primarily between individuals because of an innate difference, while others chalk up the difference to how people are brought up. Still others raise the question of the effects upon moral agency of present context and situation, proposing that our moral capacity may be more about context and group behavior than about individuals.

The most common assumptions in the United States and the West more generally about the human moral capacity differ from the innate moral module view (nature) and also from the individual or social training view (nurture) just outlined. The view most prevalent among people all around us (and also nearly universally held by philosophers till very recently) is that we have free will. The free will view goes like this: we might have a brain that has certain biological tendencies toward good or bad, and we might have a biography replete with all kinds of terrible moral models and have suffered painful and harsh conditions and even abuse, and we might be in fairly coercive political and social situations and institutions, yet we all know what doing the right thing is, and we can and ought to do the right thing no matter what. We can rise above both our nature and our nurture and even our situation to be good people and choose to act ethically. This capacity to choose our actions—to rise above our genetic inheritance whatever it might be, above our upbringing no matter how terrible it was, and above our present situation despite its social pressures—is what we mean by "free will." On this account, we are all capable of being good, and we are all *equally* capable of it because we are all human. Being human means that we can freely choose the good over the bad no matter what hand nature or nurture has dealt us. *The choice is completely our own.* Our actions have no other origin, no other ultimate causes, than ourselves as free agents. Even if we are somewhat shaped by our hardships, by our luck, or even by our brains, nevertheless we still have a sacrosanct core of free will that we can use to rise above all of that and be moral beings. *We do moral acts for moral reasons, for no other reasons, and out of no other fully determining causes*—such as brain modules, group pressures, or upbringing. And that is why we can be and ought to be held morally responsible for what we do and for what we fail to do. This free will story, about how and why we are moral and also at times fail to be moral, is everywhere around us. It probably seems and feels absolutely obvious and obviously true as you read my account of it here. But the evidence from the new brain sciences is amassing that the free will account of the nature and origin of our ethical capacity, of our moral agency, is in fact false, or at least highly unlikely; at best it may work that way in some rare individuals, who are probably philosophers.

In this book I argue that it is not obvious that human beings have free will, as we like to believe, in the way that it is obvious we have hands and feet and noses; instead, free will is a cultural assumption. And it is an assumption that turns out to be false. I make the case that, rather than serving as a description of human beings in general, free will is a particularly American and Western way of conceiving human nature. Even though it *feels* natural to us, the belief in free will is actually conventional and provincial. While we generally believe that this way of thinking about our moral nature is universally human, an account of human nature—everyone knows that we have "free will," that all human beings experience this inner freedom and lay their claim to moral virtue or sin and to the right to praise or blame upon the basis of that freedom—it turns out that most other cultures have no notion of free will. They base their understandings of human moral nature on different cultural assumptions. They conceive both human nature and the human place in the universe quite differently from the way we do. The belief in free will is actually part of a larger story, a story we take for granted or have even forgotten. Other cultures have different stories. We are as culturally provincial as they are, for ours is just one way among many of thinking about the human moral capacity and human nature generally. One of the aims of this book is to expose the free will account of moral agency as a mere cultural assumption and inheritance. I argue that when we interpret our moral agency in terms of having freedom of the will, we are not discovering in our inner experience of ourselves something *all* human beings share, but instead are discovering cultural assumptions that deeply and implicitly shape the ways we envision our place in the universe. The notion of free will is based on a theological story whose religious origin and meaning we often tend to be unaware of and which some of us even explicitly reject. Nonetheless, the standard Western theological vision of the human place in the universe still has an implicit and quite pervasive hold over us. The belief in free will, I recount at considerable length in Chapter Four, has a unique history that more or less began at one time—in early Latin Christianity—and was widely disseminated through authoritative thinkers who worked to make it sacrosanct and to delegitimize and even outlaw other points of view advocated by other individuals and groups. The presupposition of free will has been embodied in our institutions, practices, and laws and transmitted for hundreds of years by systems of education. These practices and institutions, with their implicit notion of human moral agency, still govern our lives to a great extent in the West and especially in the United States. And that is why they feel natural and universal when they are really, instead, the products of a particular cultural point of view and hence peculiar to ourselves.

Once we have uncovered our own standard and ubiquitous cultural presuppositions about our moral capacity, we can begin to discover where they come from. We can also question their validity by looking at the new brain sciences to see if they are borne out. And we can turn to explore other ideas from other cultures to open our minds to different ways of thinking about why people act ethically and why they don't, and why and when they think they can hold both themselves and others morally responsible. Can we learn anything from other cultures? How can we revise our own cultural conception of moral agency to reflect new and better understandings of how the brain works? Our first aim here, in this chapter, is to expose our deep presuppositions about how and why we come to act ethically and unethically. Then in the next chapter we shall turn to test cases, those of perpetrators and rescuers in the Nazi Holocaust, to determine whether the standard assumptions we hold about free will moral agency can explain either the evil of the perpetrators, the virtue of the rescuers, or the passivity of the bystanders.

In order to tease out our standard beliefs about moral agency, I begin, in this first chapter, with an investigation of moral education in America from colonial times to the present. I chose this starting point for my research on moral agency because I thought that how we as a society teach our children to be moral will expose our basic assumptions about our moral capacity, how we generally believe we can get our kids to become good people. Here we have our own cultural answer to Socrates's famous question in the *Meno*: can virtue be taught? Americans have always believed that virtue *can* be taught, and taught in school as well as in church and at home in the family. I discovered that from our early beginnings to today, the ubiquitous assumption is that our moral capacity rests on free will, albeit a free will that needs some training in the classroom and in the home. I began with the present. The widespread introduction of (moral) character education into public schools since the 1980s makes it the predominant contemporary form in which children are instructed in ethics in the United States. I met with several of the leading proponents of the movement; I read lots of the books and articles pertaining to this movement; and, with the help of professional advice, I selected several elementary, middle, and high schools to go to so that I could observe their programs in character education. What I discovered was fascinating.

Character Education: How (We Think) We Teach Our Children to Be Ethical

I set out on my journey to meet with prominent thinkers in the character education movement and also to observe teachers and schools nationally known for their successful implementation of moral character education programs. I went first to visit Fillmore Elementary School in an outlying suburb of a medium-size American city, which I'll call here Park Center.[1] Park Center is 97 percent white and its population is by and large neither affluent nor poor. It covers a large geographical area of parts of three counties and includes rural areas, semiurban village centers, and a growing summer resort area. Fillmore Elementary School is an award winner, a National School of Character, one of ten across the nation so designated each year by the Character Education Partnership (CEP) in Washington, D.C. The Character Education Partnership defines itself as "a national advocate and leader for the character education movement." On its website it says that it is "an umbrella organization for character education, serving as the leading resource for people and organizations that are integrating character education into their schools and communities." Each year since 1998 the CEP has given out awards to schools and districts. When it refers to "character," the meaning is *moral* character, for the website defines its mission as "developing young people of good character who become responsible and caring citizens." Here we have an encounter with the teaching of ethics that's not about teaching philosophy in the college classroom or even in special ethics classes for budding professionals in business, law, and medical schools. This is where ethics is being formulated and transmitted in ways that affect all of us because this is the moral education that is being introduced to our kids in schools. In addition to the family and the church or synagogue, mosque, or temple, here we are on the front lines of ethical training. This is no arcane theoretical enterprise of professors teaching Plato and Wittgenstein but a major site of the moral education of our children.

So my attention is rapt and I am soaking up the dedication of Fillmore Elementary to teaching character across the curriculum. I have just been talking with Mrs. Finch, the current principal, who developed the character education program more than a decade ago, initially without national or professional guidance or connections, she tells me. Character education is at the center of the school's mission and is not a separate curriculum but integrated into all the activities and programs, including gym. The designation National School of Character is just below the name of the school on the outside of the building over the front door. Mrs. Finch takes me to the main entrance of the school, where she shows me a large colorful ceramic mural that all the kids and teachers contributed to a couple of years ago. Its purpose is to convey the mission that is written in ceramic letters at the top, "Building Character." Embedded in the mural are the names of the values that the school stands for. Each month one of these values becomes the focus of teaching and activities throughout the school. All the monthly values also contribute to the overall moral theme of the year, which this year is "respect." Mrs. Finch points to each value word embedded in the mural and also to a small white ceramic building in the center of the mural that looks like a columned Greek temple. "These columns represent the values that are the pillars of our community," she says. "The values that we honor and teach in our school are fairness, respect, responsibility, perseverance, honesty, helpfulness, patience, good manners," and the like. "Each month we choose a moral value as the special one and all month we learn and think about that value. We plan activities around it, read stories about it, and practice it in our daily work and school life. Children who excel at it are given special public recognition, too."[2] The whole month is dedicated to transmitting that particular value, she tells me, and each month begins with an assembly where the value is introduced and a skit illustrating it presented. The school librarian's job is to find a storybook that expresses the character trait of the month, and she reads that story to every class in the course of the month. Another short story illustrating the month's value is photocopied and sent home with the kids to be read together with their parents. There are questions at the bottom of the sheet that the parents and children are asked to discuss together and then answer. Some examples of these questionnaires hang on a wall outside a second-grade classroom. During the month the children who best exemplify or articulate the value are given public recognition and awards, both in writing and over the loudspeaker. The award certificates are taken home to show parents.

As I walk down the school corridors I see walls covered in three-by-five cards with children's names on them and graphs. The character traits they have received awards for are written at the top, while below that, on the graphs, are colored stars marking their progress in reading and arithmetic. Some few children are chosen for an even greater recognition of their work on the monthly character trait, and these kids are given leadership roles in handing out awards during that month or the next. Teachers are also given awards by Mrs. Finch for exemplary service. Mrs. Finch has me walk with her and Joel, the current character award winner in the second grade, as we go from classroom to classroom handing out award certificates to students and teachers alike. A number of teachers are receiving awards for coming in on a Saturday to plant flower beds. Children's awards are to be taken home to show parents, but a teacher's awards are fastened to the door of her class-room. Also, on the door are signed pledges by the teachers, the children, and the children's parents to abide by a set of school moral principles. It's called the Fillmore Elementary School Pledge and it is as follows:

To Be Careful and Happy: I pledge not to hurt others inside or out.
To Learn: I pledge to always do my best and help others do their best.
To Be a Good Citizen: I pledge to respect myself, other people, and my school.[3]

In a few glass cases I see handwritten statements by parents, along with a picture of their child. At the beginning of the school year, the parents were asked to write a paragraph about which moral value or values they think their child particularly exemplifies. One mother writes about her son Phil, who is helpful with his younger brother. Another tells how her daughter perseveres in her homework even when it's hard. And a third tells about her little boy's good manners at home. As we walk down a corridor looking at the school pledge cards, the award certificates, and the special glass cabinet displaying the parents' paragraphs praising their children's virtues, we see an athletic-looking woman in her thirties with a pageboy hairdo walking toward us. Mrs. Finch introduces me to Sally Laury, a mother of both a second grader and a kindergartner, who volunteers in the school on a part-time basis. Mrs. Laury tells me how much she loves what the school does for her kids. She says that kids this age need to be told clearly the difference between right and wrong, and the moral lessons they learn in the school help with parenting at home. She can follow up on those lessons and use some of the same techniques at home. I ask Mrs. Laury what happens if a child disobeys or in some other way violates a character trait. How is that dealt with in the school? She tells me that the child is taken aside and asked, "What did you do? What character trait did you disobey? How did your action violate that moral value? How are you going to act the next time to uphold that value?" And then an appropriate punishment is meted out and recompense decided upon. That year's overarching virtue, respect, and the additional monthly virtues provide a ready-made, clear framework for discipline both at school and at home. They set up clear, non-negotiable rules and unswayable lines of authority, Mrs. Laury tells me.

Mrs. Laury mentions that there is an after-school Character Club for kids that her daughter, Emily, goes to. I ask her if the club is involved in service projects in the community. She tells me that they do a little of that at Christmas, making decorations to sell and giving the proceeds to the needy. "We all love the moral character songs that the teachers and parents on the PTA Character Education Steering Committee make up," she tells me. They are set to the tunes of songs everyone knows, such as "Here We Go Round the Mulberry Bush" and "Baa, Baa, Black Sheep," and they list all the character traits for the year. "The whole school sings them at the assemblies, and the kids in the Character Club sing them each day in the after-school program." The school song is sung to the tune of "Row, Row, Row Your Boat" and has the following lyrics:[4]

Kindness, sharing, and respect,
Following safety rules,
We're proud to show how much we care
All around our school.

We give our best at Fillmore School
Every single day,
Working hard and learning well
To play and obey.

Each month one clear moral message is being communicated to kids in a variety of contexts and settings and through a variety of media and activities. Throughout the school the message at any given time is the same. And the overarching moral character message that each monthly value fits into is everywhere—it's on the mural, in songs, on the walls, in pledge sheets. The school's complete devotion to teaching morals is evident everywhere, and the almost complete absence of any display on the corridor walls of academic work unrelated to a moral message tells all.

Mrs. Finch now walks me down to a kindergarten classroom to meet and observe the school's most highly regarded teacher and moral character educator, Mrs. Danvers. Mrs. Danvers is perhaps in her late fifties and is clearly an experienced teacher. She perches in a large, comfortable armchair above the eighteen or so kindergartners sitting in a wide circle at her feet. Mrs. Danvers reads a book to the class about how to control anger. It begins, "If you're angry and you know it and you really want to show it, stomp your feet."[5] The following pages offer different possible responses: if you're angry and you know it, take a deep breath, bang on a drum, walk away, or talk to a friend. It can be sung to the tune of the children's classic "If You're Happy and You Know It Clap Your Hands," and the singing follows the reading. The children seem to like the singing. Upon the heels of this there follows a lesson in the moral value of the month, self-control. Mrs. Danvers gets up and comes back to her chair with a wicker basket in which there are a number of spongy Nerf balls of various colors and textures. Mrs. Danvers gently tosses these yummy balls into the center of the circle of seated children. Not one child reaches out a hand to touch a ball or even leans forward. They sit like statues glued to their spots. They have clearly been through this before and have learned the proper moral lesson, self-control. The balls remain in the center of the circle, untouched, unreached for, inert, as Mrs. Danvers now turns to me to say that she is going to tell me a secret. It is a secret, she says, that the class knows but has been sworn to silence on, and at this some of the kids nod. Mrs. Danvers explains that other teachers aren't to know this secret, nor are other children, but she and the class share this special secret: Mrs. Danvers requested from the principal, Mrs. Finch, and was granted a classroom of the kindergartners who best exemplify moral character and whose parents are most involved in character education in the school. Their special status is a secret from other children and perhaps from other teachers within the school but not from the principal or from me. As an outsider doing a study of character education, I have been brought to the exemplary classroom to observe the exemplary teacher and children.

At this point, one of the kindergartners, Danielle, raises her hand and asks Mrs. Danvers if she may collect the Nerf balls and put them away.

Mrs. Danvers grants her permission and the tempting balls are gathered together and returned to the basket, never having been played with or even touched. Temptation has been successfully resisted. The children in this class have thoroughly learned their lesson. Somewhat sullenly they return to their desks; circle time has ended.

Next Mrs. Finch suggests I visit a second-grade classroom. I choose a seat at a table with some children doing an assignment in their workbooks about caterpillars. The classroom has a terrarium with two caterpillars, and the caterpillars are in the early throes of sloughing off their cocoons. Every child I talk to mentions the chrysalis; perhaps it is a word of the day. The kids love to watch the caterpillars, and when they mention them their faces light up and they all want to bring me over to look at what's happening in the terrarium. They draw wonderful pictures of the chrysalis in their note-books and put captions below. I try to gently introduce the topic of the moral value of the current month, self-control, and that of last month, helpfulness. "What does self-control mean?" I ask. "Can you give me an example?" Again and again I hear the same answer: "Don't talk unless you raise your hand." When I ask for another example, there is silence, and I feel the tension mount; the joy brought on by the caterpillars is gone.

I ask about helpfulness. "How are you helpful?" I ask. Most say they wash the dishes. I ask what else could be helpful. More silence. One boy starts to tell me a story about his brother who is not really his full brother but lives with another father and was mean to him. He tells me about this brother "getting what was coming to him." A sad, distraught little girl comes up to me and blurts out that her mother is in jail because she stole something. The children come alive when they tell their own stories. Their anguish and con-fusion is palpable. They reach out to me to listen; they want to share with me something they think is important and which troubles them. They want my understanding and perhaps even my advice or intervention. They clearly yearn for my help, or someone's help. I respond to their emotions, emotions that are immediately rehidden when I ask them what I'm there to ask. Then they feed me what they think I want to hear, mimicking a standard defini-tion of self-control or the same tired examples of helpfulness. Their school mask is securely back in place and a wall is erected between us. The danger-ous moment of self-disclosure and need has passed. Their own personal moral dilemmas and confusions are left hanging and unaddressed. Real help is not on the way, so the kids go underground again. The children in this school are here to learn the mask of obedience, the outer neutrality of self-control. They look generally subdued, with momentary flashes of joy (the caterpillars) or anguish and sadness (some personal tales) revealed to a receptive stranger. Most of these kids have already learned to put on a happy face, or at least to do what Archie Bunker used to yell at Edith: to "stifle."

I return to the school office to thank the principal, Mrs. Finch, and say good-bye. Mrs. Finch hands me a packet of materials as I leave: sample moral pledge cards, examples of how moral values can be introduced into both the standard social studies and literature curriculums. I promise to return to the school in a couple of weeks to attend a PTA meeting where parents will offer ideas and plans for next year's after-school Character Club.

The Recent History of Character Education in American Public Schools

The current character education movement had its origins in the 1960s and 1970s, according to B. Edward McClellan, whose *Moral Education in America: Schools and the Shaping of Character from Colonial Times to the Present* remains the most complete history of moral education in the United States.[6] The impetus for the recent movement for moral character education came from two quarters: those who had supported the character education movement of the early twentieth century, which had become eclipsed by midcentury, and a number of politically conservative intellectuals alarmed by what they regarded as the moral decline of youth.[7] The movement had its initial headquarters in San Antonio, Texas, in the American Institute of Character Education (AICE) and was organized and funded primarily outside of mainstream educational circles but, despite that, has had a significant impact on public schools.[8] By the late 1980s roughly eighteen thousand elementary classrooms in forty-four states had adopted AICE's Character Education Curriculum. That curriculum was made available to schools in kits that included books, films, story wheels, transparencies, and teachers' manuals to guide discussion, role-playing, and stories introducing virtues.[9] According to McClellan, AICE generally avoided educational organizations and university departments of education, preferring to go directly to teachers and principals to further the adoption of its aims and materials.[10] A substantial amount of the institute's funding has come from the Lilly Endowment, which has as its mission "to support the causes of religion, education and community development."[11] Lilly defines its religious mission in two ways: first, to "deepen and enrich the religious lives of American Christians," and second, to "support projects that strengthen the contributions which religious ideas, practices, values and institutions make to the common good of our society."[12] Moral character education began to take hold in the 1980s and 1990s when a number of public intellectuals called for the development of programs to (re)introduce into American schools the explicit defense and transmission of a set of virtues. Since the Columbine, Colorado, high school massacre in April 1999 by two alienated, disgruntled students, character education has really taken off in American schools at all levels.[13] Columbine Elementary School became one of ten National Schools of Character in 2000. Character education now receives financial support from all levels of government, with most of its funding coming from the federal government.[14]

"Virtues-centered" moral education had been a movement and then a commonplace of American schooling early in the twentieth century but was discredited in the 1930s and remained marginalized until recent decades, when a call for its revitalization could be heard especially from conservative quarters. In the words of B. Edward McClellan, who is not only the movement's historian but a clear sympathizer, "a newly alarmed group of elite intellectuals and educational leaders" were "appalled by the growing 'amorality' of the school and blamed it in part for the soaring rates of social pathology among youth in the modern era . . . alarming rates of teenage suicide, crime, drug use, and unwed pregnancies."[15] They claimed that American schools had given up teaching moral values and were especially guilty of failing to convey a notion of individual moral responsibility and insisting upon its practice.[16] The most prominent of the conservative intellectuals calling for a revival of character education to combat what they regarded as the moral decline of American youth were William J. Bennett, director of the National Endowment for the Humanities early in the Reagan administration and later secretary of education; Bill Honig, superintendent of public instruction in California; and several prominent academics, some associated with conservative think tanks (among them the American Enterprise Institute): Andrew Oldenquist (professor of philosophy, now emeritus, at Ohio State University), Kevin Ryan (founding director, now emeritus, of the Center for the Advancement of Ethics and Character at Boston University), James Q. Wilson (expert on crime and public policy, UCLA and Harvard, emeritus),

and Edward Wynne (a leading theorist of the character education movement and of Catholic religious education).[17]

The public debate about the explicit teaching of moral values in the schools
was initiated within the context of an expressed alarm over general societal
moral decline and especially the waywardness of youth and within a conservative ideological framework.

> Today in America we have far too may twelve-year-olds pushing drugs,
> fourteen-year-olds having babies, sixteen-year-olds killing each other; and
> kids of all ages admit to lying, cheating, and stealing. We have crime and vio-
> lence everywhere and unethical behavior in business, the professions, and
> government. In other words, *we have a crisis of character all across America* that is
> threatening to destroy the goodness that . . . is the very foundation of our
> greatness. . . . We need to dramatically *uplift the character of the nation.* [Em-
> phasis added.]

So states Sanford N. McDonnell, chairman of the board of the Character
Education Partnership and chairman emeritus of McDonnell Douglas, in
the foreword to Kevin Ryan and Karen Bohlin's *Building Character in
Schools.*[18] In response to a claim of a precipitous decline of morals in America, those at the forefront of the character education movement called for
what McClellan describes as an "educational counterrevolution . . . to restore both educational and behavioral standards they believed had been destroyed by the disruptions of the [nineteen] sixties and seventies."[19]

To get a fuller idea of the public debate in which the movement to introduce or reintroduce explicit and directive moral character education into
American schools began to be heard, I need quote only a few more passages
from some of the widely discussed and loudly trumpeted books by several of
its main advocates. Their overriding presupposition was that American
society was in a precipitous moral decline and that American youth had lost
its moral compass and was out of control. The overall assessment was that
things were going to hell in a handbasket and youth were the victims of a
wider societal moral decline—the moral bankruptcy of families, neighbor-
hoods, and other aspects of the social fabric. In addition, the argument
went, American children were the victims of morally misguided and even
pernicious recent policies of the public schools themselves. Together these
two factors were thought to amount to an alarming feedback loop in which

this sorry state of social decline, rather than being mitigated by the schools, was in fact being aggravated by public schools' neglect and even abandonment of the teaching of morals. This purported failure was, in the view of these conservative intellectuals, to be chalked up in part to the dominance in education since the 1960s of liberal models of ethics that amounted to nothing short of a moral relativism in which anything goes. Another factor, they believed, was educators' reluctance to bring religion into the classroom, which was a result of Supreme Court decisions about school prayer and other decisions that strengthened both the separation between church and state and the rights of children and minorities in schools. Finally, they argued that the decline was due to the increasing focus of education on technical and (what they regarded as morally) neutral knowledge and skills.[20]

William Bennett remains the most prominent advocate of the view that there is a decline in American social institutions and in the public schools and that the two declines reinforce each other. Bennett also believes that introducing moral education—both in schools and outside them—can turn the situation around. In *The Book of Virtues*, Bennett set out to provide children with the means to "moral literacy" through the telling of moral tales, a practice that he believes has been lost.[21] He indicts not only the schools but also what he calls the decline of the American family. (He titled his 2001 book *The Broken Hearth: Reversing the Moral Collapse of the American Family*.) So character education through morality tales taught both in and out of school is the hoped-for remedy to America's social problems. The purpose of these books and many others like them is to shape children's individual personal behavior, or character, by offering a catalogue of virtues and corresponding examples of virtuous behavior. Such tales are held to be capable of playing a major role in reversing the alleged state of social disintegration that Bennett attributes to a widespread and catastrophic failure of individual morality. In *The Broken Hearth*, Bennett argues that individuals across society have *chosen* to ignore the standard and universal knowledge of right and wrong in "a vast social experiment" to reinvent family life, in turn leading to the "collapse" of "the" American family, a collapse that can be seen everywhere, he says, in high divorce rates, children born outside of marriage, the widespread neglect of children, and sexual immorality, including the acceptance of homosexuality. All this in turn gave rise to a further loss of morality in the next generations in an ongoing vicious circle. Bennett proposes that what we are witnessing is a social experiment with devastating effects, and he quotes the late senator Daniel Patrick Moynihan as holding that "the family structure has come apart all over the North Atlantic world."[22]

"The dissolution of the family is the fundamental crisis of our time," Bennett argues, whereas "the ideal of the nuclear family [is] the essential foundation of society."[23] "Marriage and family are cultural universals," he goes on, and "throughout history they have been viewed as the standard to which humans should aspire."[24] Bennett points to "cohabitation, illegitimacy, fatherlessness, homosexual unions, and divorce" as "the most important contemporary challenges to marriage and the modern family."[25] And the underlying cause of this social catastrophe is unwise and ultimately immoral individual decisions on a society-wide scale: "To put it simply: We could not have experienced the scale of marital breakdown we have witnessed since 1960 unless huge numbers of our fellow citizens—conservative and liberal, believers and non-believers alike—had *willingly* detached themselves from once-solemn *commitments* made to spouses and children."[26] (Emphasis added.)

"One reason so many American families are dissolving or never forming," Bennett goes on, "is that many of us have forgotten *why* we believe—and why we *should* believe—in the family."[27] It is not the case, he assures us, that conditions of poverty and of economic dislocation and transition drive social breakdown, because "the decline of marriage and the American family happened during one of the greatest periods of economic *expansion* ever seen on earth." In harder times, "the black family was relatively stable" and "the vast majority of black children lived in two-parent homes," while at present "eighty percent of black women will be heads of family at some point in their childbearing years."[28] So we are led to conclude that the problem is not economic conditions but individual moral failure, especially of black women but also of liberals and others, on a vast societal scale. And the proposed remedy, in *The Book of Virtues*, is moral literacy as the basis for transforming individual moral decision making. We have, as individuals, failed to learn right from wrong, and we are suffering the consequences. But the "we" here, we now realize, does not really apply equally to all of us. Only a large dose of moral education can save us from ourselves—or, by insinuation, save Us (prosperous whites, especially men) from Them (blacks and women, and particularly black women, homosexuals), and the country as a whole. But the situation is dire and the need immense.

James Q. Wilson goes even further than William Bennett and makes the explicit claim that poverty is the *result* of personal immoral decision making.

Poverty is, in Wilson's estimation, a kind of karmic punishment for bad individual moral decisions. Hence the poor (by implication) are getting what they deserve. Nevertheless, they are bringing the country (Us) down with them, and that is where the problem lies. This blame game is insinuated in Bennett's analysis but is made explicit by Wilson, a prolific writer on crime and punishment—for example, in his essay "The Rediscovery of Character." Commenting on a 1985 essay by economist Glenn C. Loury, Wilson says:

> The very title of Loury's essay suggested how times had changed [since Daniel Patrick Moynihan's 1965 report on the problems of the black family]: whereas leaders once spoke of welfare reform as if it were a problem of finding the most cost-effective way to distribute aid to needy families, Loury was now prepared to speak of it as "the moral quandary of the black community."
>
> Two decades that could have been devoted to thought and experimentation had been frittered away. We were no closer in 1985 than we were in 1965 to understanding why *black children are usually raised by one parent* rather than by two or exactly what *consequences, beyond the obvious fact that such families are very likely to be poor, follow from this pattern of family life.*[29] [Emphasis added.]

Here's the argument: "the black family" (all lumped together as one entity) has made the fateful immoral choice of single parenthood, and hence is poor and the children immoral. A negatively stereotyped black family is demonized here in a way reminiscent of Ronald Reagan's excoriation of "welfare queens" as the model of immorality, an immorality consisting of bad personal decisions and choices. By implication, "these people" are to blame and they get what they deserve. Nevertheless, we need to morally educate them. Enter moral character education.

A similar alarmist message pervades William Kilpatrick's *Why Johnny Can't Tell Right from Wrong*, another influential book in this genre.[30] Kilpatrick, a professor of education at Boston College, begins his book with a look at "The Crisis in Moral Education." "The core problem facing our schools is a moral one. All the other problems derive from it," he insists.

> If students don't *learn* self-discipline and respect for others, they will continue to exploit each other sexually. . . .

If they don't *learn* habits of courage and justice, curriculums designed to improve their self-esteem won't stop the epidemic of extortion, bullying, and violence. . . .

If . . . schools were to make the formation of good character a primary goal, . . . hitherto unsolvable problems such as violence, vandalism, drug use, teen pregnancies, unruly classrooms, and academic deterioration would prove less intractable than presently imagined.[31] [Emphasis added.]

The title of Kilpatrick's book recalls a 1955 book by Rudolf Flesch, *Why Johnny Can't Read*, which was an attack on the whole-language technique of teaching reading as a disastrous fad that replaced an earlier focus on phonetics and precise skills.[32] (Most schools now use a combination of whole language and phonetics to teach reading; the claim that the whole-language method of teaching reading is disastrous is not well founded, but that need not concern us here.) Kilpatrick argues that "the failure of moral education in the schools parallels the failure of the schools to teach reading." Not only are "students . . . being taught by the wrong method, a method that looks more and more like a fad that won't go away," he says, but that method both "fails to encourage virtuous behavior" and "seems to actively undermine it."[33] Kilpatrick's objection to the whole-language method seems to be that it does not consist of rote rules transmitted in an authoritarian way by teachers to be memorized by students and then applied to concrete, specific situations. Analogously, Kilpatrick believes that moral education should consist of a principle to be memorized and then a freely willed decision to apply that principle—a moral choice for which each of us is individually responsible. It is this approach, of directive training in identifying the relevant moral principle and then applying it through choosing the actions that accord with it, that Kilpatrick identifies as "character education." "All the various attempts at school reform are unlikely to succeed," Kilpatrick warns, "unless character education is put at the top of the agenda."

Kilpatrick uses the analogy to reading education to indict liberal models of moral education as disastrous for American children and for society as a whole. The underlying message seems to be that a newfangled, nondirective, nonauthoritarian method derived from some crazy theory is being imposed on us normal people and on our innocent children by elite intellectuals removed from normal life in isolated ivory towers, and this is ruining the country. There is an appeal to what is purported to be (but of course isn't) pure common sense—to what "everyone knows" is really the right way of doing things—and there's more than a hint of conspiracy theory in the various alarmist claims that follow:

> In addition to the fact that Johnny still can't read, we are now faced with the more serious problem that he can't tell right from wrong.
>
> Not every Johnny, of course, but enough to cause alarm. An estimated 525,000 attacks, shakedowns, and robberies occur in public high schools each month. Each year nearly three million crimes are committed on or near school property—16,000 per school per day. About 135,000 students carry guns to school daily; one fifth of all students report carrying a gun of some type. Twenty-one percent of all secondary school students avoid using the rest rooms out of fear of being harmed or intimidated. . . .
>
> The situation is no better outside of school. Suicides among young people have risen by 300 percent over the last thirty years. . . . Drug and alcohol use is widespread. Teenage sexual activity seems to be at an all-time high. . . . Forty percent of today's fourteen-year-old girls will become pregnant by the time they are nineteen.[34]

Leaving aside the factual validity of the claims, how do we know that this situation is due to the failure of moral education? Kilpatrick makes his case by claiming that "many youngsters have a difficult time seeing any moral dimension to their actions; getting drunk and having sex are just things to do," he says. And not only that, but "police say that juveniles are often found laughing and playing at homicide scenes."[35] Now that we are thoroughly alarmed, he thrusts home: "One natural response to these grim statistics might be to ask, 'Why aren't they teaching values in the schools?'" Though Kilpatrick admits that moral values programs have been present in the schools for more than twenty-five years and that more research is being conducted on moral education than ever before, he claims that "these attempts at moral education have been a resounding failure."[36] "The same educators and experts who still cling to the look-say [whole-language] method [of teaching reading] want desperately to hold on to this failed philosophy of moral education," a philosophy he now identifies as a "moral reasoning" and "values clarification" approach to moral education, lumping the two together as a "decision-making" approach. He contrasts this approach, one that he regards as disastrous and as "leaving children morally confused and adrift," to one that "like phonics . . . has been tried and proven."[37] This "tried-and-true" method of character education was used "in school and society in the past" and "seemed to serve our culture well over a long period of time."[38] Character education, Kilpatrick tells us, "is based on the ideas that there are traits of character children ought to know, that they learn by example, and once they know them, they need to practice them until they become second nature."[39] Kilpatrick, like Bennett, identifies both the problem and its remedy as matters of "learning"—really, of training. Virtues must be identified and transmitted in an authoritarian way so that each individual child adopts them and then chooses to act upon them. The underlying message is that if any child fails to act morally he or she is to blame, and "we" can wash our hands of social problems.

The successful institutional model for the teaching of moral values that Kilpatrick wishes schools to emulate, he tells us, is the military.

> What the military has that so many schools do not is an ethos of pride, loyalty, and discipline. It is called esprit de corps. . . .
>
> How does the military manage to create such a strong ethos?
>
> First, by conveying a vision of high purpose: not only the defense of one's own or other nations against unjust aggression but also the provision of humanitarian relief and reconstruction . . . Second, by creating a sense of pride and specialness . . . Third, by providing the kind of rigorous training . . . that results in real achievement . . . Fourth, by being a hierarchical, authoritarian, and undemocratic institution which believes in its mission and is unapologetic about its training programs.
>
> Schools can learn a lot from the Army. . . .
>
> In the past, schools *were* run on similar lines. They had a vision of high purpose. . . . There was a sense of pride in one's school. . . . Schools were serious about their academic mission. . . . Finally, schools were unapologetically authoritarian. They weren't interested in being democratic institutions themselves but in encouraging the virtues students would need for eventual participation in democratic institutions.[40]

The implicit model has become explicit, and the battle lines are now drawn: moral character education means the authoritative transmission of principles or values that the child (and later the adult) adopts and then applies to situations through an act of personal decision or will. The adversaries identified here are liberal versus conservative models of moral education; for Bennett and others, the former amounts to the absence of any real moral education at all. Kilpatrick clarifies his position as a conflict over authoritarian versus democratic teaching methods, which is to say, over an authoritarian transmission of values as the basis of personal moral commitment versus a more open-ended evaluation of values as the basis of such commitment.[41]

William Kilpatrick, inadvertently and perhaps against his own intentions, has exposed something important: liberal and conservative models of moral education are more similar than different. They share a basic set of assumptions about what morals are and what the moral life consists in. Both hold that morals are explicit principles about values or virtues that individuals (freely) choose and commit themselves to and then apply to specific situations by choosing the actions that accord with them, for which they are then responsible. Ethics is about individual choices of action and about holding individuals responsible for those choices and actions, assigning the individuals praise and blame. That understanding underlies both versions. What differs is largely the mode of transmission: hierarchical and authoritarian (with a good dose of fear and a punitive orientation) versus more egalitarian and open-ended (with much more leeway and encouragement). The point I am making is, in a sense, an anthropological one: there is a deep structure to this notion of moral agency, what it means to be moral and act ethically, that pervades and underlies all (or perhaps most) of the various versions we see around us, liberal as well as conservative. In their analysis of the character education movement, Robert W. Howard, Marvin W. Berkowitz, and Esther Schaeffer have defined what practitioners and theorists mean by character education as "an attempt to prepare individuals to make ethical judgments and to act on them, that is, to do what one thinks ought to be done." It is a "process of defining what is the ethically correct action and having the integrity, or character, to do the right thing."[42] When it comes to our children, we think it's all about free will and choice.

The Blame Game: Social Problems Are About Lots and Lots of Individual Moral Failures to Make Good Choices

Thomas Lickona, a professor of education at the State University of New York College at Cortland and the founder and director of its Center for the 4th and 5th Rs (Respect and Responsibility), is a former president of the Association for Moral Education.[43] He is one of a small number of educators who have worked to bring Bennett and Kilpatrick's vision to practical application in American education by translating it into specifics on the ground.[44] Lickona has developed programs and curriculums to implement the teaching of moral character in schools, and he also gives workshops to teachers and educators all over the country. His version of the moral character curriculum is captured in the portrait of Fillmore Elementary School with which I began this chapter. He proposes the inculcation of specific virtues month by month; the use of areas of the curriculum, especially history and literature, to teach morality tales; a focus bringing individual behavior into conformity with a personal commitment to obedience (through choice or free will) to authoritatively transmitted values and principles. Lickona, in his widely influential *Educating for Character: How Our Schools Can Teach Respect and Responsibility*, prominently quotes Kilpatrick's assertion that "the problem facing our schools [is] a moral one."[45] He, too, argues that "escalating moral problems in society—ranging from greed and dishonesty to violent crime to self-destructive behaviors such as drug abuse and suicide" are resulting in a "summons to the schools" "from all across the country, from private citizens and public organizations, from liberals and conservatives alike" to "take up the role of moral teachers of our children."[46] For "schools cannot be ethical bystanders at a time when our society is in deep moral trouble," he warns.[47] Lickona sounds the alarm of a society in moral danger and points to moral character education as the needed remedy:

> The premise of the character education movement is that the disturbing behaviors that bombard us daily—violence, greed, corruption, incivility, drug abuse, sexual immorality, and a poor work ethic—have a common core: the absence of good character. Educating for character, unlike piecemeal reforms, goes beneath the symptoms to the root of these problems. It therefore offers the best hope of improvement in all of these areas.[48]

Lickona's evidence for the moral deterioration of American society is of the same type and just as weak as Bennett's, Kilpatrick's, Ryan's, and Wilson's: it consists of carefully picked examples, broad and unverifiable claims, and correlations with no causal evidence. His tone is equally histrionic: "There is today a widespread, deeply unsettling sense that children are changing—in ways that tell us much about ourselves as a society."[49] "Children with the most glaring deficiencies in moral values almost always come, their teachers say, from troubled families. Indeed, poor parenting looms as one of the major reasons why schools now feel compelled to get involved in values education."[50]

Moreover, "young people growing up in mass media culture are stunted in moral judgment." The result is that "the most basic kinds of moral knowledge . . . seem to be disappearing from our common culture." "Educators" now speak of the "ethical illiteracy" of young people.[51] So schools must step up to the plate and "do what they can to contribute to the character of the young and the moral health of the nation."[52] The judgment of pervasive societal decline and social fragmentation is attributed to the cause of a society-wide epidemic of individual moral deficiency, itself chalked up to bad parenting, in turn caused by families with bad values. The solution is to intervene in the schools by inculcating moral values to children no longer able to tell right from wrong because of their moral neglect.

> Most of us would be likely to agree that our contemporary society faces serious social-moral problems and that these problems have deep roots and require systemic solutions. Many of us are now coming to realize the link between public life and private character—that it is not possible to develop a virtuous society unless we develop virtue in the hearts, minds, and souls of individual human beings. . . . [T]he health of our nation in the century ahead depends on how seriously all of us commit to this calling.[53]

Lickona calls what he is advocating the coming together of "head, heart, and hand": the head proposes (a value or principle is learned and known), the heart adopts it (one wills or commits to do it), and the hand performs it (behavior or action follows).[54] Moral character, Lickona writes, consists of three components: moral knowing, moral feeling, and moral action.[55] It involves the training through direction, inculcation, and practice of each individual child's decisions and choices, that is, of his or her will, in order to address what he calls a nationwide social emergency. While subscribing to the standard conservative analysis and developing a largely authoritarian model of direct inculcation, Lickona has nevertheless been supportive of several forms and kinds of moral education in schools, in "a spirit of accommodation and cooperation," he says.[56] Most recently both the debate and the model of moral education advocated and instituted generally have broadened and become more varied, and its proponents are no longer just those of a narrow ideological sector.[57]

Thomas Lickona is one of the few theorists of moral education in schools who grasp that liberals and conservatives share certain basic assumptions in their models of moral education and that these shared assumptions are of Christian origin.[58] In keeping with his own conservatism, he calls for the reintroduction of religion into the public schools as the basis for their moral education programs.[59] Lickona claims that "there is an emerging consensus that the exclusion of religion from the public school curriculum is neither intellectually honest nor in the public interest." So he has developed "seven ways that [public school] educators can constitutionally incorporate religion into character education."[60] Although I do not echo his call for more explicitly religious moral education in the public schools, I believe that Lickona is right in claiming that it is dishonest of moral educators to maintain that the moral education offered in our schools is culturally and religiously neutral and universal. Moral education in early America was simply Christian education, and I argue below that today it remains true to its Christian origins in important respects. As a result, we Americans are particularly vulnerable to diagnosing societal ills as due to individual moral failure of choice and decision making, and hence to instituting the moral training of children as the remedy. American moral education took shape as the transmission and inculcation of Protestant Christianity to children from colonial times to the present. That shape, I'll now argue, has not substantially changed.

The History of Moral Education in America: The Transmission of Protestant Christianity

In early colonial times in America, moral education consisted primarily of Protestant versions of the Christian catechism. "It was Protestants from northern Europe, especially from Great Britain, who did the most to give moral education its character in the thirteen colonies," McClellan writes in *Moral Education in America*.[61] The Puritans of New England were "deeply committed to moral education and extraordinarily fearful that their children would drift away from the faith and culture," despite their understanding of their own journey to the New World as a special mission to establish a model Christian commonwealth. Religious and moral education were inextricable, and the family was the locus of both.[62] "The most devout families among the Puritans . . . conducted family devotions at the beginning of each day; . . . drilled [their children] in the church catechism; and exercised a careful, sometimes severe, discipline."[63] Children heard Bible reading, psalm singing, lessons, and stories of piety and moral instruction at the dinner table and, as soon as they could, participated in prayers. McClellan quotes the following telling passage from *The Diary of Cotton Mather*:

I began betimes to entertain them [the children] with delightful Stories, especially *scriptural* ones. And still conclude with some *Lesson* of Piety; bidding
them to learn that *Lesson* from the *Story*.

And thus, every Day at the *Table*, I have used myself to tell a Story before
I rise; and make the *Story* useful to the *Olive Plants about the Table*.

When the Children at any time accidentally come in my way, it is my
custome to lett fall some *Sentence* or other, that may be monitory and profitable
to them.[64]

In most Puritan families, McClellan goes on, the catechism, the basic
instruction in the denominational doctrines of the Christian faith, was "the
single most important element in formal moral instruction"; it was even more
important than reading the Bible, he says.[65] The short version of the West-
minster Catechism, which was the standard in early New England, stated
that "man's chief end is to glorify God," whose nature is "a spirit, infinite,
eternal, unchangeable in his being, wisdom, power, holiness, justice, goodness, and truth." Hence *"the duty which God requires of man"* is *"obedience to his
revealed will."*[66] Although the family was the primary locus of religious moral
instruction, what schools there were in colonial New England reinforced
the same message in primers and hornbooks.[67] Religious instruction was not
primarily the job of churches but a task of the home and, to a lesser extent,
of school and apprenticeship. Even higher education was theological, including classical, Greek, and Latin texts in its curriculum. And of course
Puritan New England was known for its public scrutiny of private life.[68]

What about the rest of early America? We know that both traditions and
conditions in the southern and middle colonies, although not as thoroughly
documented, were different from those of New England. The Anglicans of
Virginia, for example, while concerned about the religious and moral education of their children, had come to the New World for economic reasons rather
than religious ones. Nevertheless, they passed laws mandating the teaching
of the catechism to youth, servants, and apprentices in the family, in church,
and in the workplace. In Virginia, however, the harsh conditions of life, the
very high mortality rate, the much lower literacy rate, and the pattern of
settlement of scattered plantations and farms distant from one another did
not permit the development of the kind of community and educational and
religious institutions common in New England, with their focus on religious
moral instruction.[69] The Quakers of Pennsylvania, while emphasizing liter-

acy and Bible reading less than the Puritans of New England, nevertheless taught children the Quaker catechism and basic Quaker values and exercised firm discipline. Southern society generally followed the Virginia pattern, whereas the middle colonies followed the example of New England. There were common assumptions and common patterns in all regions: the family was thought to be the major transmitter of moral values, while school, apprenticeship, and church were supplementary. Religion and morality were seen as intertwined and even indistinguishable, and the catechism was the main document transmitting both.[70] After the Revolution and up to the 1820s, growing prosperity and social stability and hierarchy brought a loosening of the severity of the moral instruction of the earlier colonial period.[71] McClellan points to a new gentler tone in child rearing and "a more affectionate and egalitarian structure challeng[ing] the rigid patriarchal forms" of earlier times.[72] A prolonging of childhood and the allowance of play came to be accepted. McClellan characterizes this change as toward moderation and also toward a divergence in the roles of mother and father in child rearing, the mother now taking the primary role as moral educator in response to the new Victorian conception of women as the moralizing force of society.[73] Churches and schools began to have an enhanced role in the moral education of children. Following the vision of Thomas Jefferson, Noah Webster, and Benjamin Rush to teach the virtues appropriate to the new nation as a republic, a system of public education began to develop.[74] At the universities and colleges, the Enlightenment and its vision and values softened the older sternness and rigidity of early colonial Christianity while nevertheless maintaining "a general Christian framework."[75] Also, at this time, religious revival and intensity were taking hold in certain areas, especially along the frontier. Methodists and Baptists, other evangelicals, and newly arriving settlers of pietistic sensibilities emphasized strict and harsh methods of child rearing and moral education.[76]

Moral Education in America from 1820 to 1900: Pan-Protestantism as a Civic Religion for American Schools

The transformation of moral education in America took place in the context of great social, economic, demographic, and political changes. These included the end of the rule of the propertied elite and the beginning of popular rule;

the weakening of the family economy that accompanied the decline of the family farm and small business; the growth of suffrage to roughly all white male adults; the opening of the West to settlement; and the growth of urban centers and large commercial and manufacturing enterprises. Mobility and opportunity replaced stable hierarchical social arrangements as the norm.[77] These were the years of Jacksonian democracy, a revival of evangelical Protestantism, crusades for various moral reforms, a new utopian literature, and a new form of moralistic sentimental literature.[78] Perhaps not as paradoxically as one might expect, in morals and personal behavior Americans at this time abandoned the relaxed moral style of the earlier period and moved toward more institutional restraints and "an insistence on rigid self-restraint, rigorous moral purity, and a precise cultural conformity."[79] McClellan suggests that "a distinctly evangelical temperament pervaded the society."[80] The new freedoms bred a concomitant moral rigidity, perhaps because of both the fears and the opportunities let loose by the new freedoms and mobility and the loosening of structures and hierarchies. An intense focus on self-restraint accompanied the new freedoms, especially in the years 1820 to 1865. Parents, keenly aware that the new opportunities offered their children might soon distance their offspring from them forever and also bring their children into urban areas or the western frontier, where lawlessness, danger, and "alluring evils" ruled, turned with a new urgency to giving children a firm and direct foundation in moral instruction.[81]

At this time education and especially moral education began to be institutionalized and systematized in ways previously unknown, and there was also an emphasis on beginning such education early. The family and the school were to be focused on providing this urgent and early moral education, as one educator put it, in "prepar[ing the child] for this transition to freedom by effective training in self-control and self-guidance, and to this end, the will must be disciplined by an increasing use of motives that quicken the sense of right and make the conscience regal."[82] As another wrote, "Having ordained that man should receive his character from education, it was ordained that early instruction should exert a decisive influence on *character*."[83] This was moral character training through a prism of normative Christian free will and conscience, and it was incorporated into the new nineteenth-century view of a "special role for mothers" and the new social demand that women and especially mothers "exhibit . . . a constant Christian virtue in their own lives" and take up the primary responsibility in moral education "through daily readings and exhortations to children designed to increase piety and teach proper conduct."[84] Mothers were less focused on the catechism's doctrinal specificities than on inculcating simple moral values, now readily available in a vast new popular literature of moral instruction for children.[85] Mothers were seen as in charge of shaping the moral character of their children and could leave their academic education to the schools. The preference for women teachers, especially in the early grades, was indicative of the focus of public education on the moral shaping of children, for it was thought that only, or especially, women could provide that moral instruction.[86] As McClellan writes, "The primary task of the female teacher in the classroom was to exercise a strong moral influence on the child, reinforcing the lessons of the mother both by serving as a model and by eliciting proper behavior from the child."[87] Both Sunday schools (a new institution gaining popularity as a result of early nineteenth-century evangelicalism, which brought a wider constituency than the urban poor, who were the initial target of reformers in their hope to "civilize" the urban poor) and daily schools were seen as continuing the moral instruction begun in the home.[88] A vast new public school system was just being developed in the 1830s–1860s as education became universal in America and all white children were to be educated together in the public schools.

The middle and upper classes envisioned the introduction of universal public education as serving to integrate, Americanize, and morally educate the children of the growing immigrant population. Moral education was particularly directed at the children of poor immigrants. As McClellan puts it, "Leaders of the movement to create public schools sought to use moral education . . . as an instrument for remedial moral instruction" of "the poor and the immigrants, not to mention blacks and Native Americans." "What was critical was that all children learn self-restraint through a common moral code."[89] McClellan writes that the very aim of the classroom "was to win student assent to certain values, to cultivate in the young minds a love of virtue, and to develop moral commitments that would last a lifetime."[90] Moral lessons in the form of maxims and morality tales were pervasive in nineteenth-century textbooks such as the McGuffey readers, and the values were "a blend of Protestant morality and nineteenth century conceptions of good citizenship": thrift, honesty, hard work, love of God, love of country, duty to parents, the dangers of drunkenness and pride and deception, and the rewards sure to follow from courage, honesty, and respect.[91] The "evangelistic and moralistic tenor of the times" even led colleges and universities, which had drifted toward a more critical and freethinking perspective in the eighteenth century, to return to a devotion to moral concerns. Hand in hand with a new emphasis on Christian pieties, many colleges put in place a course in moral philosophy as the culmination of a curriculum in which morality was seen as a "a matter of bringing the will into conformity with absolute and universal moral rules." Universities thus distanced themselves from the more utilitarian perspective on morals of the previous century.[92] "In the antebellum years," McClellan suggests, "the college experience only reënforced the basic values that children had first learned at their mothers' knees."[93] "The centrality of moral education remained an article of faith from the creation of the public school system in the 1830s until the last decade of the century," McClellan writes, while, at the same time, the denominational allegiances were becoming blurred and to some extent eliminated.[94] The aim of non-sectarian public schooling "was not to forbid religion in the classroom but rather to teach a nonsectarian Christianity at public expense," "to teach children universal moral values and a generalized Protestant religion in the public schools."[95] The movement to establish nonsectarian public schools "was a thoroughly Protestant campaign," an attempt "to turn their particular worldview into a kind of civic religion."[96] It drew "heavily on Protestant social thought and Protestant modes of organization, and it recruited a disproportionate number of its leaders from the Protestant clergy." Protestants openly proclaimed their hope to "put the stamp of their own values on the entire society" and insisted on "the connections between morality and religion."[97]

The notion of a common, broadly Protestant Christian culture as the basis for the social harmony and unity of the American polity came out of the Second Great Awakening of the first decades of the nineteenth century. The active "soldiers" in its various "crusades" "began to think of themselves not simply as Presbyterians or Methodists but also as a part of a great pan-Protestant moral empire, an empire they found it increasingly easy to identify with America itself."[98] A telling example of this nondenominational Protestant approach was the introduction into public schools of Bible reading but without any interpretation following. "The presence of the Bible in the schools became a powerful symbol of the connections between religion and morality, and Protestants resisted any effort to remove it."[99] While Catholics eschewed the unvarnished anti-Catholic and anti-immigrant biases of the public schools and set up their own system, Jews, especially Reform Jews, of the eighteenth and nineteenth centuries largely embraced the opportunities public schools afforded, attending them in large numbers. There were also Protestant Nativist efforts, especially in the 1920s, to restrict Catholic parochial education so as to force Catholic children into public education.[100] As the wall between church and state became higher as the twentieth century progressed, however, the openly Protestant character of public schools lessened in favor of a heightened official non-sectarianism.[101]

Moral Education in America from the 1890s to the 1940s: Early Character Education and the Quest to Return to Nineteenth-Century (Christian) Moralism

In the first decades of the twentieth century a call for character education began as a reaction against modernity and as a way to recapture an older moral framework. Its purpose was to "develop educational mechanisms to stem the erosion of moral training and preserve traditional values." Those who rallied around the banner of "character education" favored a number of educational programs that sought to make the teaching of a list of specific virtues and the cultivation of good traits of character central to the curriculum.[102] At the same time a new progressive movement in education was emerging whose aim was to match values to new and changing times. By the mid-1920s the progressive movement had gained strength.[103] Character education's advocates feared the new freedoms and also the growing specialization of knowledge and occupation. They advocated a strong effort to preserve traditional values against a progressivist tide. Many traditionalists lamented that modernism in religion and the decline of the importance of religious authority in favor of the authority of science had diminished the "fear of eternal punishment," leading to a loss of "some of its power to divert men and women from pleasures that were increasingly available and alluring."[104] McClellan quotes a teachers' manual of the day as declaring that "the day of science has taken away from mankind most of the fears that once censored his conduct."[105]

By the mid-1910s a movement to develop and implement codes of conduct took wide hold. "It was the use of character codes that most clearly set these reformers off from the progressives." These codes were essentially lists of virtues in the form of laws or pledges. The most famous of these was the result of a competition initiated by Milton Fairchild, who in 1911 founded the Character Education Association. The winning entry was published in 1917 as the "Children's Morality Code." Its substance was "Ten Laws of

Right Living: self control, good health, kindness, sportsmanship, self-reliance, duty, reliability, truth, good workmanship, and teamwork."[106] Many school systems adopted the code, while frequently modifying it according to their needs. The Boston schools, for example, added to it a "law of obedience" and made that law the center of their moral education program.[107] In Birmingham, Alabama, schools focused on one virtue per year. The code became the subject of posters in classrooms and hallways and the focus of extracurricular activities and especially of student clubs, some of which formed around particular virtues.[108] In Boston there were, among others, the Courtesy Club, the Prompt Club, and the Thrift Club. Girls and boys often had different clubs, with girls' clubs devoted to such domestic virtues as sexual purity, gentleness, and meekness, while boys' clubs were focused on the complexities of the modern workplace and on athletics.[109] The content of the moral education promoted in these clubs "was derived primarily from nineteenth-century morality." Their overarching purpose was "to use every means available to them to ingrain good habits and to strengthen the will of students against the temptations of the day."[110] Their general approach "subordinated ethical reasoning to an emphasis on training the will."[111]

By the late 1920s, however, the broad appeal and pervasiveness of the first round of moral character education had begun a slow decline. The decline was due to an exhaustive study that exposed its ineffectiveness. In its place a progressive approach to moral education began to take center stage.

The Failure of Moral Character Education Exposed

The decline and eclipse of the first round of character education were precipitated by a broad study whose findings seemed to demonstrate definitively that the direct and didactic approach of moral education programs such as character education was ineffective. Research into the effectiveness of the character education widely implemented in schools in the early twentieth century was proposed by the Religious Education Association at its 1922 meeting. The question that was to be addressed by an empirical study was: "How is religion being taught to young people and with what effect?"[112] The Religious Education Association requested that the Institute of Social and Religious Research fund a study of the question, and the institute agreed to do so in 1924. Hugh Hartshorne (University of Southern California) and

Mark May (Syracuse University) were chosen to serve as co-directors, and later the research project was housed at Columbia University's Teachers College and supervised by Edward L. Thorndike. The first phase was a three-year "inquiry into character education with particular reference to religious education."[113] In 1927 the research funds were extended for another two years so that the project could continue. Hartshorne and May broadened the original intent of the study and widened the research to include a primary study and seven ancillary ones, only three of which ended up included in the study results, which in the end came to three volumes (1,782 pages) of data and interpretation.

> The primary study focused on the development of a large body of standardized test materials for use in the field of moral and religious education. Tests were to be developed in the areas of knowledge and skills, attitude, opinion and motive, conduct and self-control. Student character was assessed through innovative classroom tests of honesty (deceit) and altruism or prosocial behavior (service).[114]

The three secondary studies addressed the problems of character traits: how behavior, on the one hand, and knowledge and attitudes, on the other, affect each other; the biological, social, and cognitive causes of behavior; and the effectiveness of current techniques to develop moral character and habits.[115] "The findings of the study represented a potential body blow to the enterprise of character and religious education."[116] The primary finding was that moral character traits were not attributable across contexts; instead, "character was found to be situationally specific."[117]

Hugh Hartshorne, writing later in his *Character in Human Relations*, doubted whether moral character was really a viable description of human beings at all:

> If, for example, honesty is a unified character trait, and if all children either have it or do not have it, then we would expect to find children who are honest in one situation to be honest in all other situations, and, *vice versa*, to find dishonest children to be deceptive in all situations. What we actually observe is that the honesty or dishonesty of a child in one situation is related to his honesty or dishonesty in another situation mainly to the degree that the situations have factors in common.[118]

The authors' conclusion was devastating: "The mere urging of honest behavior by teachers or the discussion of standards and ideals . . . has no specific relation to conduct. . . . The prevailing ways of inculcating ideals probably do little good and may do some harm." Hartshorne and May recommended a shift away from direct methods of teaching moral character traits and virtues toward "indirect methods such as the creation of a positive school climate and service oriented activities for students."[119] Subsequent research over the years has only strengthened the evidence that "children cannot be sorted cleanly into behavioral types on the basis of presumptive traits, habits, or dispositions."[120]

A Brief Interlude: John Dewey and the Progressive Education Movement of the Early Twentieth Century

The progressive education movement associated especially with the American pragmatist philosopher John Dewey introduced what McClellan calls "a radically different approach to moral education," which gained more and more supporters beginning in the 1920s, especially among "liberal Protestant clergy, intellectual leaders, professional elites, and educators associated with major universities and large urban and suburban school systems."[121] Dewey rejected the notion that ethics carved out a particular sphere or segment of life, instead arguing that the social and the moral coincide: morals are embedded in contexts and necessarily articulate those contexts. All education is ethical or moral (or unethical or immoral) insofar as it expresses the values and structure of the community. Dewey regarded schools as mini-societies and, furthermore, considered them potentially ideal ones that could exemplify even more perfectly the values and arrangements of the larger American society, namely, its democratic values, institutions, and practices. "The controlled environment" of the school was the chief means for initiating society-wide social reform.[122] He believed that thoughtful and considered moral deliberation took place within social contexts that deeply inform ethical decision making yet did not completely determine its outcomes.

So moral education was to model flexible and innovative responses to a changing world and the unprecedented and unique situations and experiences therein. As Dewey scholar and philosopher Richard Bernstein remarks, "It should . . . be clear that ethics conceived of in this manner blends into social philosophy," for Dewey conceived "the entire universe as consisting in a multifarious variety of natural transactions" in which the person "is at once continuous with the rest of nature" while also "exhibit[ing] distinctive patterns of behavior that distinguish him from the rest of nature."[123] Bernstein characterizes Dewey as "a robust naturalist or a humanistic naturalist." On this model, justice and ethics are intertwined and not clearly distinguishable; as a result, children's concerns for the justice of the school and for the justice of the larger society were inseparable. Ideally, the school would look outward as well as inward. There was no such thing as a politically neutral ethics, in Dewey's conception, for ethics was inseparable from justice, the justice of institutions near as well as far. John Dewey's notion of moral education marks a momentary glimpse of ethics as a social phenomenon, implemented largely structurally and institutionally, from which individual personal decisions and choices emerge and are shaped.

The History of Moral Education in American Public Schools Since World War II

In 1951 the National Education Policies Commission of the National Education Association and the American Association of School Administrators published a report, *Moral and Spiritual Values in the Public Schools*, in an attempt to define moral education for the post–World War II era. The report emphasized both the teaching of certain central values as well as a degree of moral flexibility in the face of a fast-changing world. Its authors identified a number of "essential" values that schools had the responsibility to inculcate. These included "respect for the individual personality, devotion to truth, commitment to brotherhood, and acceptance of individual moral responsibility."[124] They also urged teachers to respect and encourage children's spiritual and religious values and expressions.[125] There was a good deal of consensus about this report for its moderate tone and recommendations, but some vocal dissenters outside the mainstream called for the revival of an older form of character education that emphasized the development of specific character traits and the adoption of formal codes of moral conduct.[126] At issue was a progressive notion of the evolution of values versus a notion of the unchanging character of values—hence the latter's call for "the direct teaching of the eternal verities."[127] The dissenters, harking back to an even earlier vision of moral education, were largely affiliated with, or had their origins in, religious institutions, and some prominent advocates received generous funding from the Lilly Endowment.[128] Generally the 1940s and 1950s witnessed a new emphasis on cognitive development and skills, and the sidelining of the moral, due to the needs of a more technologically oriented and quickly expanding economy.[129] Perhaps less emphasis on moral education in schools was also due to the concentration of moral focus on anti-Communism above all else.[130] A judicial atmosphere that encouraged the separation of church and state and the enhancement of the rights of children also contributed to a retrenchment, a treading lightly, when it came to moral instruction in public schools.[131] Beginning in the 1960s, however, the quest for moral education in schools was renewed with three new individualist, free choice approaches with a liberal tenor: values clarification, cognitive developmentalism, and a caring approach to morals.[132]

Values Clarification

The immediate appeal of values clarification when it emerged in the mid-1960s was that it offered a ready alternative to traditional inculcating approaches to moral education. Louis E. Raths, Merrill Harmin, and Sydney B. Simon, in their *Values and Teaching*, set out a clear and easy program with curricula and exercises for teachers to follow. It met with great success and was widely adopted in schools in the 1960s and 1970s.[133] The approach focused on individual moral decision making and emphasized the situational character of moral decisions. Children, it was held, needed to learn how to think about values and make value judgments in changing situations. McClellan points out that "what made the matter especially pressing to these reformers was their sense that the troubles of youth in modern America stemmed . . . from the difficulty of choosing values." Hence they emphasized the "personal and individual nature of valuing."[134] The teacher's role was to help students engage in a process of discovering, discussing, developing, and then freely choosing values. These values were chosen from among alternatives after thoughtful consideration and class discussion, and then each student was to act by choosing the actions that fit with the principles he or she had chosen.[135] The role of the teacher in the discussion was as facilitator rather than as moral authority. The teacher, like a therapist, was to help students determine their own values, to help them "find a personal path in a bewildering world."[136] Conservative critics charged values education with promoting moral relativism, on one hand, or with presuming controversial and politically charged moral commitments such as to the environment, on the other.[137] The philosopher Andrew Oldenquist criticized values clarification for presupposing too blithely that human beings are essentially good and just needed help in clarifying how to direct their basic goodness. With these criticisms, enthusiasm for the program waned and values clarification increasingly fell out of use.[138]

Kohlberg and Moral Reasoning: Cognitive Developmentalism

Concurrent with the popularity of values clarification was a movement to help children develop moral reasoning and judgment. This was the cognitive moral development approach of the widely influential Harvard psychologist Lawrence Kohlberg. "From the mid-1960s to the present," McClellan writes, "Kohlberg's theories have occupied a central place in the discourse about moral education."[139] Kohlberg wished to counter conservative charges of moral relativism. He focused narrowly on the cognitive component of moral growth and proposed a universally human six-stage schema of moral reasoning, from the earliest, most primitive to the fully mature. The burden of the teacher was to help children rise in their cognitive moral development to more mature levels. Kohlberg's concern, like that of the values clarificationists, was the process of individual moral decision making.[140] Despite various modifications over the years, Kohlberg's schema retained an overall structure of discrete developmental stages that every person had to move through sequentially. Moral development began with primitive selfishness (egoism) and developed by stages to an ultimate stage of personal commitment to and application of universal principles. Not everyone reached the final, ideal stage, but children could be helped to improve their moral reasoning by teacher-led discussion of moral case studies carefully prepared for that purpose. Later on, after working in both prisons and troubled schools, Kohlberg turned to a model of restructuring schools democratically as essential to transmitting a notion of justice. With the Just Community Schools model Kohlberg transformed his notion of how moral development occurs—it is not just as an internal psychological developmental process within reasoning resulting in personal choice and commitment but also needs to takes place within a specific kind of social context. The teaching of moral thinking was transformed from the abstract, hypothetical stories he first developed to the lived community of the school itself. Free rational choice and commitment to act on principle, Kohlberg's stage six, was now set within and supported by democratic structures, practices, and social arrangements, a nod to Dewey. Nevertheless, the ultimate goal of individual choice of action on principle, through personal independent rational decision making, remained intact.

Caring-Based Moral Choices and Decision Making

A variant of the moral cognitive developmental approach to moral education in schools emerged in the 1980s out of a critique of Kohlberg, especially by his student Carol Gilligan. Gilligan criticized Kohlberg for what she held was his masculinist bias in thinking that ethics was about justice and rights rather than about caring. Caring, she held, was a feminine way of looking at morals, whereas justice was biased toward the masculine. Women, unlike men, conceive ethics in terms of personal relationships rather than as an impersonal arena of justice, she argued. The difference was not due to women's occupying a lower stage of moral development as Kohlberg's schema would suggest, Gilligan said, but due instead to a different way of being a moral person. Women, she argued, had a more relational and emotional approach to morals than the emotionally detached and impersonal and impartial stance of men.[141] Caring, rather than justice, informed women's moral sensibility and should inform the various stages of an account of female moral developmental and how it differs from male moral development. Gilligan developed an alternative version of Kohlberg's stage theory of moral development that traversed an initial stage of caring for self, then an intermediate stage of caring for dependents (mothering), and culminated in caring for all, insofar as all human beings are within interconnected webs of relation.[142] Nel Noddings, Jane Roland Martin, and others joined Gilligan in calling for a moral education in schools that includes what they regarded as female as well as male moral orientations and voices. The intimate relations of women's domestic life were seen as offering a moral orientation of special importance, equal to the male public arena of justice. Practice in caring ought to inform moral life within the classroom and school, and Noddings proposed that the school curriculum could be reorganized around the theme of caring—"caring for self, for intimate others, for strangers and global others, for the natural world and its non-human creatures, for the human-made world, and for ideas."[143] Noddings envisioned the moral model of caring as relation-centered.[144] Virtues are defined "situationally and relationally" rather than abstractly.

Care theorists focused on virtues, "put[ting] far greater emphasis on the 'social' virtues . . . [for example] of congeniality, amiability, good humor, emotional sensitivity, good manners, and the like," than the standard character educators do.[145] Noddings criticized the out-of-context nature of the teaching of virtues in standard character education. Parents, she suggested, most often introduce a moral lesson in context rather than as a lesson in general principles: for example, they often say such things as, "You must not hit your little brother; be nice."[146] That kind of intervention with a direct and directive moral lesson is preferable because it is immediate and relevant rather than theoretical and distanced, Noddings proposed.[147] Nevertheless, she applauded the use of cultural narratives and stories, suggesting that these can be from many cultures and offer differing moral points of view, bringing up legitimate moral conflicts and, hence, moral options to *choose* from. Noddings and the other care theorists rejected the impersonality of principles and opted for appealing to individual emotional relationships as the proper guide for individual moral decision making. We may no longer be rational choosers, on this model, but we are still choosers, individual subjects who engage in decision making according to our emotional commitments. Noddings highlights and wishes to encourage personal emotional responsiveness as the basis for an ethics of free choice. It is these emotionally rich choices that are to bring the (potentially isolated) individual into a world of relationships.[148]

My Brief Encounter with the Ethics of Caring

I became acquainted with the caring approach to character education when I attended a workshop for educators, "The Ethics of Caring," at the Boston University Center for the Advancement of Ethics and Character, part of the School of Education, in April 2007. The introductory session of the workshop began with the showing of the 1983 German film *The White Rose*, which tells the story of the group of five Munich University students and their philosophy professor, who came together to form a secret, nonviolent resistance organization in defiance of the Nazis. The group took their name, the White Rose, from a Spanish novel about peasant resistance in Mexico and also gave that name to the underground newsletter they published opposing the Nazis and calling for general resistance to the Nazi regime. The newsletter called attention to the mass murder of Jews in the east and also opposed Nazi militarism and tyranny. The group wrote and distributed six issues of the newsletter clandestinely over a period of eight months in 1942–43, mailing them from distant cities and also distributing them by courier runs to various locations. The Gestapo led a concerted search to find the source of the publication and distribution, and the group was eventually betrayed to the Gestapo by a university custodian who witnessed members hastily doing drops of leaflets on campus in February 1943. The six were tried by a court devoted to political offenses against the Nazi state, found guilty of treason, and sentenced to death. Three were executed by guillotine on the same day as their verdict, and the other three some months later. Several of those who assisted the White Rose Six in publication and distribution and in collecting funds for their surviving family members were sentenced to long prison terms.[149] After the showing of the film, Bernice Lerner, director of the Boston University center, offered the participants in the workshop an interpretation of the film that was strangely apolitical: it was to be understood as depicting a quintessential illustration and model of caring. The film was not seen as portraying grassroots political resistance to injustice but instead as an example of personal attachments being the source of choosing courageous actions in response to feelings of caring. The lesson to be taken away from the film concerned the personalizing and individualizing of the social-political moral arena in terms of personal decision making. This interpretation of *The White Rose* alerted me to some of the underlying assumptions of the caring approach to moral education: its depoliticizing of the moral domain and its de-emphasis of, perhaps even blindness to, issues conceived in terms of social justice and injustice (which are regarded as masculinist and impersonal). Thus social structure and the distribution of power, especially of political power, are taken off the table in this liberal model of ethics, as is done also in conservative character education.

While eschewing an approach to ethics focused exclusively on rational discourse and decision making according to abstract universal principles divorced from emotion, the caring approach to moral education is nevertheless just as individualist and personalist in its understanding of moral agency. If anything, caring heightens the underlying notion of morals as emergent from individual, freely chosen personal commitments. Both Kohlberg's stage of cognitive moral development and Gilligan's alternative notion of stages originate in the individual and move outward to connect to others and finally to the larger human and natural worlds. Gilligan's model of moral development makes that movement completely explicit. The individual does not begin as relational; instead, relationality is the moral goal and achievement to be taught and learned. Children are to be brought from self-centeredness to community. They begin as infants and small children, having what could be an isolating self-focus if carried on into adulthood, but ideally they end up in family and community, in a full life of personal emotional relations of caring. The bonds of community are believed to depend on individual, personal choice.

The Contemporary Harking Back to the Gilded Age: The Radicalizing of Personal Choice and the Privatization of Moral Life

The current focus on caring and character exposes the privatization of moral life, and also has parallels and resonances with the moral perspective of the Gilded Age. It was in the Gilded Age that the privatization of the family and a new idealization of private life took hold.[150] It was only at that time that "the nuclear family was made the sole repository for standards of decency, duty, and altruism."[151] "Middle-class Americans," writes Stephanie Coontz, a historian of the American family, "elevated family values and private rectitude into the defining features of Gilded Age morality," with the concomitant mission to try to get the poor to adopt the private virtues whose lack, it was claimed then as now, was the cause of their poverty.[152] This moral outlook coincided with a withdrawal from public life associated with the end of Reconstruction after the Civil War and with the rise of great disparities in wealth and of the tremendous power of capital. "The notion that enhancing private family morality could substitute for forging public values and societal bonds," Coontz says, "developed comparatively late in American history." The privatization of the family, "far from being a source of social commitment and responsibility," instead "helped erode" them.[153] "In the Jeffersonian tradition, public engagement was considered the primary badge of personal character; *honor* and *virtue* were political words, not sexual ones."[154] In her influential study of the American family from colonial times to the present, *The Way We Never Were: American Families and the Nostalgia Trap*, Coontz argues that ethical sensibilities were different in different eras in America. Today, she says, we are undergoing a revival of an ethical outlook of the late nineteenth-century Gilded Age: a withdrawal of moral concern from the public arena to the family seen as a private domain. The family in America was not always seen as "private," and in fact rarely was. She writes, "The idea that private values and family affections form the heart of public life is not at all traditional. It represents a sharp break with Enlightenment thought and the republican tradition, which held that public values . . . were qualitatively different from and superior to private values of love and personal nurturance."[155] Even in the Victorian era, the unique "womanly" moral character of mothers, their "domesticity," was not seen as contributing exclusively or even predominantly to the family but was instead, in the first half of the nineteenth century, thought to be a vital public contribution that helped curb the "masculine" competitiveness and ruthlessness of the public arena.[156] Women's public service and reform organizations abounded and had tremendous influence on society, from the temperance movement and the abolitionist movement to caring for the indigent.

The Progressive Era, Coontz says, marked a change toward a renewed public engagement from a Victorian focus on the family. Parallel to the Progressive Era was the tremendous public moral engagement of the civil rights era of the 1950s through the mid-1970s. The current retreat into private life and the concurrent growth of income disparity and the power of corporations are equal only to the similar phenomena of the Gilded Age. "The Gilded Age of the mid-1870s to mid-1890s resembles the period since the mid-1970s in most intriguing ways. After the intense idealism of the 1860s, most middle-class individuals entered a phase of political disengagement and economic reorientation that required them to disavow old alliances and beliefs. Turning away from social activism, many people focused on their personal lives and material ambitions."[157] Even Gilded Age political discourse succumbed to the shrinking of the moral domain to the privatized family. What we now think of as the recent "American practice of selling candidates' sincerity and family values instead of their positions on issues began during the first Gilded Age."[158] All this marked "a retreat of the middle class from previous involvement in social reform."[159] The turn toward the private arena was also reflected in the religious concerns and ideologies of those times, focused as they were on a social message of law and order and an ethical appeal to repentance for private vice.[160] It was not until the late 1890s that "the middle class participate[d] in a revival of mass action around women's suffrage, ma[d]e new alliances with workers and immigrants, and beg[a]n to move in the direction of Progressive Era reform."[161] Moral education, while focused on individual choice, nevertheless was not as disconnected from civic engagement, commitment, and identification with the public arena in those more public-spirited times as it now is.

A Brief Analysis

From the beginning, moral education in America has taken an individualist free will and personal commitment perspective, emerging from the memorization of, recital of, and commitment to the Christian catechism. That Protestant model of moral education has been the bedrock to which all other experiments have eventually returned. It was everywhere in the schools and classrooms I visited, it was evident in the moral educators I talked with, and it still dominates current popular books on the subject. The central model buried deep in the American psyche revolves around individual moral choice: using one's free will to act in conformity with principles and virtues. Even if we consider liberal models of moral education since the 1960s (values clarification, Kohlbergian cognitive moral developmentalism, and Gilligan's caring alternative to Kohlberg), we still find that, like character education, the focus is on individual choices and decision making. All four models seem to presuppose that social problems are moral problems and are due to aggregates of bad individual choices and personal decisions. As a result, social problems are to be attacked and remedied by changing personal decision-making processes, and hence individuals' choices. The set of shared assumptions about what it means to be moral, how a moral person develops, why a moral person acts morally, and how moral weakness can be changed to moral goodness and strength share implicit presuppositions about how social problems arise and can be remedied. In the American context, the term *character* points to individual morality and behavior driven by personal free choices and decisions. Insofar as social problems are seen as signaling the failure of what is variously termed character, caring, or rational choice, they are chalked up to failures of personal will, the remedy for which involves changing that will, person by person.

Both standard character education and its seeming opposite, the ethic of caring, use stories to sway individuals' emotions—although the stories in character education are focused on illustrating and inculcating various virtues, while the focus in the ethic of caring is to illustrate and recommend cases of caring behavior. In both, however, the final stage is presented as making the right decision. Children are instructed to make a choice according to a learned and adopted value or principle illustrated in the story. From that choice or decision a moral commitment and obligation emerge. While both character and caring moral education involve both emotions and thinking, Kohlbergian moral education is strictly cognitivist or about rational choice. Nevertheless, the notion of personal commitment, which is to say some form of free will, pervades all the models as the sine qua non of ethics, for the individual will is brought into conformity through commitment (whether affectively tinged or not) with larger virtues or principles and with other wills. It is choice and decision that are thought to bring the individual into the moral arena, whether that domain is thought of as universal moral truths or, alternatively, as relation and community (Gilligan, Noddings). Both liberal and conservative versions take the form of free commitment to moral demands even though both character education and also caring moral education, in contrast to Kohlberg and values clarification, use the language of virtues and habits rather than free decision making and choice.

Aristotle Versus Kant

Daniel K. Lapsley and Darcia Narvaez in the authoritative *Handbook of Child Psychology*, seventh edition, make the point that all the current kinds of moral education programs in schools today are fundamentally about individual free will, choice, and responsibility, and that makes them Kantian. The major model in the West of what it means to be ethical is Immanuel Kant's: (1) the identification of universal moral principles of right action, (2) the discernment of how these principles can be applied in actual situations, and (3) the commitment and resolve of the free will to act upon those principles when such practical situations arise.[162] Kantian ethics is about moral obligation to principles that set out as a duty the performance of certain kinds of actions on those principles in relevant situations. It depends on a kind of cognitive skill to determine the relevant principle in a situation, and then a resolve of the will to act upon it. Classical Greek philosophical ethics, in contrast, generally was based on the Delphic maxim "Know thyself." For the Greek philosophers the fullest understanding possible of human nature, political institutions, and the nature of the cosmos was the basis of the ethical life. From a Kantian perspective, however, ethics is more narrowly about action: how discrete actions are chosen and what standards they conform to. The kind of knowledge involved is, therefore, narrow and focused rather than broad and open-ended. Knowledge in the modern, Kantian perspective offers precise answers, whereas for Aristotle, and for Plato before him, it is a way of life, an open-ended engagement in a quest for understanding the world and the human place within it, which is the virtuous life itself. For moderns, ethics is properly about doing, whereas for the Greek philosophical tradition, it was about the transformation of the self through gaining wisdom about what it means to be human, within the biological and cosmological natural order, within the social arena (for Aristotle), and within an underlying mathematized scientific universe and within the political community (for Plato). For the Greeks, all knowledge, not just some discrete arena of moral education, was thought to be contributory to ethics. In the Kantian notion of ethics, ethics and broad knowledge of the world seem to have been torn apart in ways that would be completely anathema to the ancient Greeks.

The Kantian notion of ethics expresses a Latin Christian approach to our human moral nature—albeit secularized on the surface. Perhaps it is ironic that the three liberal versions of moral education so criticized by the main revivers of traditional character education share with moral character education a perspective based on free will and personal decision making, for both liberal and conservative varieties are Kantian and Christian. And this is despite the fact that character education on the face of it seems to be not about freedom of the will but instead about training the personality in habits of virtue—an ancient Greek, Aristotelian view of moral agency. Lapsley and Narvaez recognize that the term *character education* is, however, a misnomer, for the American movement of character education is in fact not at all the transformation of character in the Aristotelian sense as the name suggests and as it purports to be.[163] Ironically, perhaps, character education programs do very little to train kids in particular situations; instead they present hypothetical stories and ask kids to identify the correct virtue that was chosen and acted upon in the story.[164] They repeatedly instruct children in the identification of abstract moral principles and in action that conforms to those principles, with much rewarding of the correct answers to questions, but they do not provide much in the way of the behaviorist-style training of action that one would expect from the name. Instead, they tell and retell stories and reshape historical events and personalities so that they exhibit clear black-and-white moral principles that pupils then are supposed to identify and be inspired to commit themselves to choose and act upon accordingly.

The idea of habits, in contrast to decisions and free choices, implies an automatic and even unconscious way of acting, rather than the reflective and self-aware independent choices and decisions of the free will. The training of character in habits of virtue goes back to Aristotle's ethics. The advocates of moral character education are explicitly aligning themselves with an Aristotelian notion of moral psychology and agency. But in the American context, the Aristotelian notion of personal character has been reshaped through the lens of free will, that is, through a Kantian lens. And that Kantian lens is fundamentally Christian, not Aristotelian. It owes a great deal to the Christian appropriation and transformation of Aristotle, especially by Thomas Aquinas in the thirteenth century. (I discuss the Christian origins of free will and the Christianization of Aristotle at some length in Chapter Four.) The use of the term *character education* muddies our understanding of how character education programs actually attempt to educate children's morals. In practice, character educators use a (Kantian) model of choosing actions that accord with principles (or virtues) to which children have freely committed themselves, rather than a model that involves training behavior. Why, then, use this term, which is a misdescription, rather than a term that more accurately describes the process?

Character educators appear less committed to the individualist model of ethics than liberals do. But when it comes to looking at what they actually do, they are in fact more committed than are liberals to a model of society based on individual free decision making and individual responsibility. What the moral character educators seem to be signaling by their use of the term is their disdain for nonauthoritarian models of moral education, in contrast with their own more authoritarian one. But in terms of the dependence on individual free will and responsibility, both liberals and conservatives are playing the same game—a Western, Christian, Kantian game. The character educators no less than the liberal moral educators have little use for an Aristotelian conception of ethics. They do not train children in moral habits, nor, as Lapsley and Narvaez point out, do they raise the question of what kind of life brings overall fulfillment. In contrast to the model used by character educators, a (true) ethics of virtuous character (virtue ethics) has two outstanding features: (1) it makes a claim about the best human life, about what is required for human flourishing, and (2) it includes an account of how best to conduct one's life and oneself in keeping with that notion of human fulfillment, flourishing, and excellence.[165] Neither of these considerations is operative in standard character education as it has been conceptualized and implemented in American schools either historically or at present. In the American school setting, the term *virtues* is used to identify principles of right action (instead of human excellences, the classical Greek *arête*) that are to be adopted as morally obligatory. Their rightness is not connected to an Aristotelian account of what is biologically natural for the human species, nor is it about what uniquely fulfills our humanity. It never calls for a broad and general quest to learn about the human and the natural worlds, which Aristotle deemed essential for the discovery of the good human life. Instead, Lapsley and Narvaez say, we run up against a Kantian set of assumptions, for it is all about discrete actions conforming to principles of right action.

> In most accounts of character education one cultivates virtues mostly to better fulfill one's obligation and duty (the ethics of requirement) or to prevent the rising tide of youth disorder (character utilitarianism or the ethics of consequences). . . . [T]he point of virtues in most accounts of character education is to live up to the prescriptions derived from deontic considerations: to respect persons, fulfill one's duty to the self and to others, submit to natural law. When the goal of character education is to help children "know the good" this typically means coming to learn the "cross-cultural composite of moral imperatives and ideals." Rather than emphasize agent appraisal[,] the animating goal of many character educators is appraisal of actions, for, as Wynne and Hess . . . put it, "character is conduct" and the best test of a "school's moral efficiency" is "pupils' day-to-day conduct, displayed through deeds and words."[166]

Thus, against their stated intentions, character educators and the liberal supporters of a Kohlbergian approach have a great deal in common with each other and little in common with the Greek philosophical notion of ethics they claim to embrace:

> Character education, for all its appeal to virtues, seems to embrace the ethics of requirement just as surely as does moral stage theory, rather than an ethics of virtue. The most important moral facts for both paradigms are still facts about obligation, universal principles and duty. The most important object of evaluation for both paradigms is still action and conduct: it is still deciding the good thing to do rather than the sort of person to become. The fact [is] that character education is . . . thoroughly deontological and utilitarian with . . . little in common with virtue ethics.[167]

Lapsley and Narvaez propose that the difference between those who call themselves character educators and the champions of more liberal models is largely about the role of authority and hierarchy in moral education. Character educators embrace the transmission of ready-made moral principles on authority, whereas educators who are more liberal have a constructivist concern as well: they want to involve young people not only in the commitment to values but also in their formation. What I wish to call attention to here is what the different models of moral education in America share: a reliance on a notion of free will, decision making, and obligation to follow discrete moral principles as guides for action. That assumption is not universal across cultures (we just saw that it was absent for the Greeks). Instead, I argue in this book, it is highly specific to our own culture. And even the adversarial positions within our culture share a deeper and larger implicit religio-cultural framework that emerged from an ongoing cultural context and its particular Christian theological history. American moral education has inherited a past rooted in the teaching of the Christian catechism, and while it has gone in several disparate directions, it remains nevertheless beholden to theological notions of human nature and agency.

"The Inextricable Union of Person and Context"

What makes John Dewey's model of progressive education an alternative to both the character education championed by conservatives and the moral education promoted by Kohlbergians and other liberals is that it recommended contextual interventions into social structures rather than direct attempts to educate or manipulate the individual will. Dewey's model builds upon the social determination or shaping of action that Hartshorne and May exposed in the 1920s and which has been confirmed by a great deal of subsequent research. That social interventions and institutional incentives shape moral action undermines Kohlberg's moral cognitive developmentalist and the caring models just as much as it challenges traditional moral character education. Furthermore, the effectiveness in the classroom of Kohlberg's cognitive developmental model of moral reasoning and decision making has been shown to suffer from additional problems. For the research shows that, *pace* Kohlberg, "only weak associations between moral reasoning and moral behavior have been detected and these associations lack practical significance among school-aged populations."[168] Hence neither approach to the training of the individual will, whether by authoritative transmission of values or by the development of reasoning skills in decision making, shows evidence of effectiveness. Lapsley and Narvaez suggest that the conclusion we ought to draw from the research into moral development is that it is "at the intersection of person and context where one looks for a coherent behavioral signature."[169] "The *inextricable union of person and context*," they propose, "is the lesson both of developmental contextualism . . . and social cognitive approaches to personality."[170] As a result, "moral education can [i.e., must] never be simply about the character of children without also addressing the context of education, that is to say, the culture, climate, structure and function of classrooms and schools."[171] Lapsley and Narvaez end their essay on character education on a hopeful and instructive note, commenting that research shows there are effective ways to ameliorate just those "moral" problems that particularly conservatives identified in youth, namely, the use and abuse of alcohol, drunk driving, use of illicit drugs, early sexual intercourse, high rates of depression and suicide, violence, gambling—but *the effective interventions are systemic and structural rather than individual character- or will-based.*

> Schools characterized by communal organization, that is, by mutually supportive relationships among teachers, administrators, and students, a commitment to common goals and norms, and a sense of collaboration, tend to have students who report an attachment to school (an emotional bond to teachers or school and a sense of belonging), a belief in the legitimacy of rules and norms, and a high value placed on work. . . . [B]onding to school, was related, in turn, to lower levels of student misconduct and victimization.[172]

The Seattle Social Development Research Group, for example, launched a project in 1981 in eight local public elementary schools guided by a social development model according to which it was assumed that behavior is learned within social environments rather than by adopting and applying explicit principles or values. The presupposition was that "when socialization goes well a social bond of attachment and commitment is formed . . . [which] in turn orients the child to the norms and expectations of the group to which one is attached and to the values endorsed by the group."[173] The Seattle Social Development project "demonstrated long-term positive effects on numerous adolescent health-risk behaviors (e.g., violent delinquency, heavy drinking, sexual intercourse, having multiple sex partners, pregnancy and school misconduct) and on school bonding."[174] "But is this character education?" Lapsley and Narvaez ask. They remark that the answer "depends on whether character education is defined by treatment or by outcomes." The Seattle Social Development project has "generated [the] empirical outcomes that are claimed for character education" but has been guided by "a social development model" and not by a theoretical model "of virtue, morality, or character."[175] A similar project of the Developmental Studies Center in San Francisco has "documented the crucial role that children's sense of community plays in promoting a wide range of outcomes commonly associated with character education, including altruistic, cooperative and helping behavior, concern for others, prosocial conflict resolution, and trust in and respect for teachers."[176] What was important about the schools in the project was that they met children's "basic needs for belonging, autonomy, and competence." The sense of community in the schools that took part in the project was developed through "collaborating on common academic goals; providing and receiving help from others; discussion and reflection upon the experiences of self and others as it relates to prosocial values such as fairness, social responsibility and justice; practicing social competencies; and exercising autonomy by participating in decisions about classroom life and taking responsibility for it."[177] Thus a sense of community in the schools was "promoted through [changes in the] structures of the classroom and the school."[178]

Another important structural intervention studied was community service and service learning, the latter differing from the former in the extent to which it is linked to the academic curriculum. Service projects engage aspects of identity formation in adolescents and are instrumental in transforming social and moral civic identity, according to current research. These interventions, too, showed important positive outcomes in the behavioral issues identified as important by the moral character educators. So the positive outcomes in the various areas were brought about by "a developmental systems approach" to intervention in youth behavior. Lapsley and Narvaez conclude that the evidence points to using "a developmental systems approach" to moral education rather than the current "epistemological approach" of character education, which is "preoccup[ied] with core values," adding, "A developmental systems orientation is foundational to the positive youth development perspective that has emerged as a counter to a risks-and-deficits model of adolescent development."[179] Yet "not one of the youth developmental programs apparently viewed their competency-building and prevention work in terms of moral or character development."[180] The notion of what counts as ethics and moral training is clearly caught in a religio-cultural time warp that affects not only conservatives, who are more likely at present to acknowledge their religious roots, but also liberals, who tend at present to regard their outlook as secular. While Lapsley and Narvaez do not try to account for the conceptual bind that moral education seems to be in, they do recommend that a social systems approach that explicitly recognizes what it is doing is in fact *moral education* ought to be developed out of the successful systems approach already well established in addressing just those youth social problems that ironically the conservatives argued are evidence of the effects of the paucity in American schools of authoritarian moral character education.

> The conceptual framework for character education is adequately anticipated by a commitment to a developmental systems orientation. A developmental systems approach to [moral] character education draws attention to embedded and overlapping systems of influence that exist at multiple levels; to the fact that dispositional coherence is a joint product of personal and contextual factors that are in dynamic interaction across the lifetime.[181]

It turns out, in fact, that many schools have recently adopted a more systems-oriented approach to changing the culture of their school—and they have done so because it works. The Character Education Partnership now gives public recognition, through its National Schools of Character designation, to schools that use eclectic approaches as well as to those that implement the original character education model that was a revival of early twentieth-century moral education. When I met with Merle Schwartz, director of education and research at the Character Education Partnership in Washington, D.C., she talked a great deal about "school climate" as well as about moral behavior. Part of Schwartz's job is to go to schools all over the country as a consultant, helping to diagnose their problems and working with school representatives—administrators, teachers, students, staff, and parents—to devise situation-specific remedies that can help change both school climate and student behavior. I also visited several schools that had introduced mixed models, in which kids seemed engaged and happy rather than subdued and sullen, as they were in the Fillmore School. Nevertheless, the theoretical philosophical move that Narvaez and Lapsley recommend—redefining what ethics is about, what its domain is, and what moral development entails—is more a hope than a present reality.

Narvaez and Lapsley also propose adding to the structural intervention in social climate a new (or, really, the revival of an ancient Greek) approach to individual personal ethics. Rather than the pervasive model of instilling conformity to a set of objective virtues or principles, as standard character education envisions, or an open-ended discussion of the right ways to think about and determine right actions in various situations, as in values clarification and Kohlbergian rational decision-making models, Lapsley and Narvaez recommend the introduction into the moral education curriculum of an Aristotelian-type exploration of human flourishing—virtue in the real Aristotelian sense. Teachers, they say, should be asking the questions that occupied the Greek philosophers: What makes for a deeply satisfying human life? What does it mean for a given person or for people in general to flourish and thrive, and what does it take for that to happen? They suggest that the educational research on moral development now shows that the notion of thriving within a given context (the current buzzword is *developmental contextualism*) is the proper "basis for understanding the role of adaptive person-context relations in human development."[182] They conclude their essay on an Aristotelian note: "Perhaps a life course perspective on character will require additional constructs such as wisdom . . . , purpose . . . , personal goals . . . , spirituality and self-transcendence . . . , ecological citizenship . . . , and character strengths . . . to capture adequately the complexity of phase-relevant dispositional coherence and human flourishing."[183]

Some Final Comments on Character Education

We have seen through this brief account of moral education in schools that in recent decades in America both conservatives and liberals have focused on the individual rather than the community as the site, source, and focus of morals. For conservatives the focus is the training of the individual will, an emphasis seen from the earliest moral training in the adoption of the catechism to the various more recent forms of character education, while for liberals it is the more recent rational decision-making approach and its variants, which highlight individual free will. The only structural approach in moral education was that introduced in the early twentieth century by John Dewey and others influenced by him, and it is still evident in a certain tendency of Kohlberg's to introduce democratic structures that in turn underlie rational decision making. That approach persists today in some pockets of experiment: the democratic and caring school movements, and also the many service projects that connect students to the larger society and the world. The more liberal the model of moral education in a given school, the more likely it is to emphasize social service rather than the acquisition of individual character virtues. Nevertheless, the theoretical underpinnings of that aspect of character education are largely absent or unacknowledged, and such programs of social service continue to appear within a discourse of the training of individual moral character and will in the virtues. Historically, social structural interventions focusing on the group and the community have been seen only rarely, such as in the progressive movement in education in the mid-twentieth century.[184]

The initial thrust toward a social structural approach, at least to moral education, did not meet its demise because of the triumph of liberalism.[185] Rather, UCLA philosopher John McCumber has put together considerable evidence that the social philosophy of Dewey and others was eclipsed because of the McCarthy era's pointed attacks on philosophy departments and especially on those philosophers who taught and engaged in social philosophy.[186] To write about or teach social philosophy and to address issues of ethics in terms of social structure as formative of moral action became anathema in the political climate of the 1950s; what subsequently ensued was a period of forgetting, a forgetting that now allows social conservatives to attack liberals for decimating and fragmenting the social arena when, in fact, that occurred in part as a result of a right-wing attack on Communists, Socialists, and their sympathizers in universities.[187] Nevertheless, the fragility of social philosophy and moral education models based on it bespeaks the foundational nature of the free will perspective in America. Narvaez and Lapsley are trying in a small way to revive social philosophy as a basis for rethinking what ethics is really about and how virtue can be taught.

A Few Glimpses of Places Where Ethics Is Taught to Adults: Law, Business, Medicine, Local Government, and the Popular Press

While space and time do not permit an extended account of how professional schools in law, business, medicine, and government, for example, teach ethics to their students, we can take a peek into their practices of teaching ethics to gain some general impressions. I began my investigation of how ethics is taught in practical contexts with an Ethics Awareness Training workshop offered by Lockheed Martin's Utica, New York, office to students and faculty at Hamilton College, where I teach. Lockheed Martin is the largest defense contractor in the United States and is involved in such projects as replacing all the assets of the U.S. Coast Guard and designing and manufacturing launch platforms. The workshop was sponsored by the college's dean of multicultural affairs, and its purpose was to give students, and especially students of color, a sense of Lockheed Martin's commitment to nondiscrimination and multicultural fairness in the workplace. The Lockheed Martin trainers were the resource manager and a manager of the electronics section. They told us that we were attending exactly the same kind of workshop that every company employee is required to attend each year. Also, each manager in the company is required to offer one session of the ethics workshop every year. Lockheed Martin's ethics code focuses on six values: honesty, integrity, trust, respect, responsibility, and citizenship. "These values," the trainer suggested, "are the same as in your family."

After an introduction, we watched a video that began with the CEO of the company speaking about ethics. There followed a wide range of reenactments of actual case studies in eight ethical areas: changing practices, interpersonal relationships/harassment, retaliation, computer misuse, information security, competitive information, and international business courtesies. For each case study, after the problem was initially presented, we were given the opportunity to discuss the ethical issues involved and how we thought the situation should be resolved before we were shown how the situation was actually resolved. I found it notable that neither the company's lobbying of Congress nor its treatment of whistle-blowers, two topics about which I inquired, were part of ethics training or the domain of ethical concern at the company. Nor were issues of quality control or the substance of what the company was contracted to do. All the issues highlighted in the presentation involved only matters either of respectful personal behavior toward others in the workplace or of not using the company in any way for personal gain. The substance of the work, either its quality or its nature, was not to be subject to moral evaluation or criticism. Here we see the same kind of personalizing and individualizing of the ethical domain—analogous to the nuclear family— that we saw in standard character education and which is a product of the contemporary withdrawal of moral concern from the public arena to a narrow concern for only the private domain of personal relationships. And we also see here that character education in schools has become the model for the workplace as well: a code of personal virtues to which every employee is to commit him- or herself is the centerpiece of a program that then offers teaching examples of how to apply the principles so that individual employees can make the right decisions in similar situations.

Other field trips took me to an ethics and leadership camp for public officials that was held at the Markkula Center for Applied Ethics of Santa Clara University; to a meeting with an expert on legal ethics and the director of Stanford University's Center on Ethics; and to a discussion of medical ethics with a philosopher who is part of an ethical decision-making team at Upstate Medical Center in Syracuse, New York. I kept finding the same situation everywhere: that the domain of ethics was narrowed to issues of respectful personal behavior toward others and to avoiding conflicts of interest or personal gain, and that there were codes of professional ethics that offered the list of virtues we have seen in character education plus detailed rules about how to avoid conflicts of interest and not use the workplace for personal benefit. In every case structural issues—that is to say, issues that concerned the moral substance of the workplace's or profession's endeavors— were off the table. In law, for example, the *substance* of laws was not considered to be within the profession's ethical domain, and neither was the structure of delivery and access to legal services. In medicine, individual medical decisions on life and death and treatment issues were on the table, but the issue of how medical care is influenced by the pharmaceutical companies and the makers of medical equipment, as well as whether all should receive the same medical care regardless of race, income, and other inequalities, were off the table. The individualizing of ethics to matters of personal decisions and relationships fragmented the moral domain, thereby eclipsing the most egregious wrongs and short-circuiting critique. As a result, the moral critique of substantive projects or of structural incentives is limited to those who are most invested in the status quo (those high up in the hierarchy) or else to individual whistle-blowers who have to take on the whole system as a personal mission and who have and derive no protection from the various codes of ethics.

In the workshop at Santa Clara University I was persona non grata for raising the criticism that the ethics code and statement of values trumpeted by the Markkula Center for Applied Ethics did not pass what I call the "Holocaust test"—that is, even if all government employees obeyed the ethics codes that the center has been introducing into local governments, nevertheless those same employees would, without violating a single moral rule, be able to implement the kind of programs that the Nazis had the German civil service engage in, programs that contributed to the murder of millions of Jews and others. For example, government employees would be able to register and round up Jews, confiscate their property, and arrange for their transport to the gas chambers without violating "ethics" because they would not be doing it for personal gain nor would they be treating other employees or even individual members of the public discourteously. I began to see in this how it was possible for German bureaucrats to think they were carrying out their jobs ethically while contributing to mass murder: their moral objections to their substantive assignments in the bureaucratic machinery of mass murder were privatized, taken out of the ethical domain of the workplace. Individuals who tried to do otherwise could be accused of bringing allegedly "private" moral concerns into the workplace. The way that ethics is defined in both its operation and scope sets the stage for what can happen. It could happen here. And it is happening here in lesser ways right now. Look at health care, whose worst abuses are dramatically illustrated in Michael Moore's 2007 documentary film *Sicko*. Those abuses are clearly the result of institutional and corporate incentives and practices expressive of implicit social, cultural, and political values, not of aggregates of failed personal moral decision making.

The approach to ethics all around us not only trivializes our moral concerns but also lets us off the hook by nullifying our shared social and political responsibility for the moral content of our common ventures, projects and structures and policies that we didn't create but which we nevertheless maintain. As Randy Cohen, a writer of "The Ethicist," a column in the *New York Times Magazine*, put it:

> One way to understand right conduct is to imagine it on a continuum— etiquette, ethics, politics. . . . But I maintain that the difference between the two is artificial, if indeed there is a significant difference at all. . . .
>
> An ethics that eschewed . . . nominally political questions would not be ethics at all, but mere rule following. It would be the ethics of the slave dealer, advocating that one always be honest about a slave's health and always pay his bills promptly. But surely any ethics worth discussing must condemn the slave trade absolutely, not quibble about its business practices.[188]

I have spent this chapter introducing our assumptions about our ethical capacity and how we think we make our children into moral people. We found that our assumptions about our own moral capacity and about how our children come to be moral revolve largely around the notions of free will, personal decision making, and individual commitment. In the next chapter I turn to the Holocaust to investigate perpetrators and rescuers. I raise the question of whether choosing the good by an act of the free resolution of the will can explain the goodness of the rescuers, and whether the failure to abide by a free commitment and resolve to do the good can explain the perpetrators. The Holocaust will be the test case for a reevaluation of free will as the basis for moral agency.

2

Moral Lessons of the Holocaust About Good and Evil, Perpetrators and Rescuers

Good people can be induced, seduced, initiated into behaving in evil . . . ways . . . that challenge our sense of the stability and consistency of individual personality, character, and morality. . . . Thus any deed that any human being has ever done, however horrible, is possible for any of us to do.

—PHILIP G. ZIMBARDO

It may well be that one of the latent consequences of the pioneering . . . study [of rescuers of Jews in Nazi Europe] is its challenge to many of the assumptions of philosophers and social scientists about the character of human nature.

—RABBI HAROLD SCHULWEIS

In none of the cases [of the Polish rescue of Jews during the Nazi Holocaust] was it a conscious decision [by the rescuer] in which its possible implications were systematically considered.

—NECHAMA TEC

Some Insights from Teaching about the Holocaust

For a number of years I have been teaching a course on the Nazi Holocaust. I tell my students that it's not hard to understand the victims—anyone can imagine him- or herself as an innocent victim—but to really understand the perpetrators is another matter altogether. I've developed an exercise, a group drama project to enable students to get a small inkling of what the emotions of perpetrators might feel like. I have students form groups with a hierarchy from perpetrators to victims arbitrarily assigned, with some victims serving as agents of the perpetrators, as they did in the concentration camps. I then have the students enact over dinner in the dining hall a perpetrator-collaborator-victim situation: perpetrators control what victims can eat, how they are going to eat what they eat (with their hands or which utensils they may use), and when they eat; how and when they sit, stand, and talk; what they can say; when and if they can go to the bathroom, and other things of that kind. Causing any kind of physical pain or even the slightest harm is strictly prohibited. The exercise lasts for about an hour and is carried out entirely in public. Within very clearly defined limits, students get a small taste of upper-level perpetration, midlevel collaboration, and bottom-level complete victimization. Sometimes I have them reverse their places in the hierarchy and do it again. Later they meet to reflect together upon the experience, they write a diary of their emotional reactions, and we discuss in class what happened and how they felt. Students often talk about their own unexpected willingness to lord it over victims, even over "victims" who are classmates they've sat next to for weeks and some who have been in their weekly small discussion groups. How unexpectedly easy it is to engage in perpetration, they say. For many it comes far more easily than they had anticipated. Some find themselves, against all expectations, enjoying being on top, controlling the victims' behavior and humiliating the victims in small ways. Students playing the victims express surprise at how painful these restrictions and directives enforced by the perpetrators were to them and how under the thumb of the perpetrators they felt, how angry and belittled. Within an hour, the relationship of equal classmates getting together for a class group assignment in a liberal arts college in a bucolic setting begins to fade and the outlines of another reality begin to creep in, one of winners and losers: the entitled powerful, the weak and powerless, and the reviled victim-collaborators. Upon reflection, good kids begin to recognize how easy it could be to get carried away and become gratuitously cruel.

I introduced this brief sociodrama exercise into my Holocaust course some years ago as a result of a chance meeting. A father stopped by to pick up his daughter, who had been doing homework with my daughter, and we got to chatting as we waited for the kids to finish up. I told him about the course I was teaching on literature and films about victims, perpetrators, collaborators, resisters, and rescuers during the Nazi Holocaust and how hard it was to get students to imagine themselves not just as victims but even more as perpetrators or collaborators. They all romantically envisioned themselves as rescuers or resisters, hiding Jews in their attics or basements like the rescuers of Anne Frank and her family, or as fighting partisans like those of the Warsaw ghetto uprising. I knew that this father was a psychologist, but he told me that he was also an expert in psychodrama, and he offered to come to my class that year and try out an experiential exercise in what he called "sociodrama," a way to bring up, in a very small way, some of the feelings of what being in the Holocaust contexts we were reading about might be like, and not just the heroic ones. In subsequent years I took over the exercise and shortened it as well as absented myself from it so that my neutral role as teacher (and neither quasi-victim nor quasi-perpetrator nor quasi-collaborator) would remain untainted. It is a powerful experience for students even with all the further limitations and caveats that I have added. The sociodrama exercise with the students perhaps hints that there may be more continuity between normal life and the Holocaust than we would like to think. Is one of the important lessons of the Holocaust that ordinary individuals can all too easily slip into participation and collaboration in harming others; is it perhaps all too real in our own world? Of course I do not mean collaboration in mass murder like the Nazi genocide, but perhaps in subtler ways and much lesser forms collaboration is all around us.

My course always fills and has a long waiting list, a familiar phenomenon among those who teach courses in Holocaust history or literature. The fascination is both warranted and also at times suspect. I wonder about how thin the veneer of morality might turn out to be even in our own society. Does the Holocaust have something to tell us generally about moral psychology? About how ethics functions in an individual, in a group, between groups, and within hierarchies? Or are its lessons anomalous because of the extremity of the conditions in which they were played out, so any insights gained are relevant only to conditions so extreme as to be without import in normal life and ordinary society? We can think of other genocides: the systematic slaughter of 800,000 to 1 million Tutsis (and the rape of many more) by Hutus in Rwanda within a three-month period in 1994; the Khmer Rouge, led by Pol Pot, committing the genocide of approximately 1.7 million Cambodians between 1975 and 1979; the ethnic cleansing and rape in the Balkans in the 1990s; the systematic murder and rape of Mayans by the Guatemalan government for two years in the early 1980s; the Turkish genocide of the

Armenians in 1915–17 (a precedent the Nazis took to heart); and the Nazi genocide of the Gypsies and their earlier attempt at the genocide of the disabled, to name a few of the most familiar cases. Are these rare (or even not so rare) anomalies, or are psychological and social forces operating here that are closer to everyday life than we would like to think?

Perhaps the psychologist who suggested the sociodrama exercise to me and led the first round of it so many years ago had in the back of his mind the 1971 Stanford Prison Experiment, a well-known social psychology experiment devised and implemented by Philip Zimbardo, now professor emeritus of psychology at Stanford University. That experiment has been all over the news and public interest shows again of late because Zimbardo waited till 2007 to publish a full-length book, one geared to the general public, about the experiment and its aftermath. The Stanford Prison Experiment raised questions and provided some provisional answers about why and how people in perpetrator-victim situations turn brutal and even vicariously sadistic and also about the range of responses of both perpetrators and victims. Zimbardo devised the experiment because he wanted to determine whether the way to prevent vicarious evil is to prevent certain types of people—those who are sadistic, weak-willed, or of brutal character—from holding positions in which they would be tempted to commit acts that result in gratuitous suffering or even atrocities, or whether normal people, those without marked sadistic or brutalizing tendencies, become perpetrators simply because they are in power in situations that enable perpetration and victimization, such as prisons. In other words, how powerful is the *situation* to mold moral behavior versus how stable is *character* across all kinds of situations and how powerful is the *moral will* to control a person's own actions and to resist the temptation to cause gratuitous suffering? Zimbardo's experiment was devised to test the relative power of situational forces versus personality (moral character traits and free will) as causes of moral or immoral behavior. The question arose for Zimbardo because he had been researching the psychology of prisons, visiting many prisons, and team-teaching a course on the subject with a former prisoner. But studying actual prisons could not answer his questions about whether brutality was a consequence of the kind of people attracted to prison work because of their underlying sadistic personalities, which could be given freer rein in a prison setting, or whether instead the context was itself out of control and brought out the worst in everyone, prisoner and guard alike, and *produced* brutality rather than was the *result* of the brutal characters of prisoners and guards, their lack of moral compass or self-control. Zimbardo cleverly captures what he set out to test in the phrase "bad apples": were brutalizing guards "bad apples" or was the situation a "bad barrel" that corrupted "good apples"?

Although there were at that time extant psychological studies of prisons, Zimbardo decided to devise an experiment to test the relative strength of character versus situation as cause of brutalizing behavior—something the study of actual prisons couldn't tell him because the prisons' populations, both guards and prisoners, were already enmeshed in patterns of behavior whose causal threads could not be disentangled. So in devising the Stanford Prison Experiment, Zimbardo pointedly controlled for sadistic and brutal psychological tendencies by eliminating any of the applicants to his study (all college students) who displayed those characteristics on a battery of psychological tests; in fact, participants were chosen for their stable personalities. The twenty-four young men selected were divided arbitrarily and randomly into prisoners and guards who would enact a simulated prison that Zimbardo and his grad students would set up for two weeks in the Stanford Psychology Department building basement. Zimbardo's book *The Lucifer Effect* follows in extraordinary detail what transpired in the basement of the Stanford Psychology Department, practically hour by hour, between prisoners, guards, the warden, the head of the prison (a role into which Zimbardo cast himself with chilling consequences), and various visitors who played roles from prison chaplain to defense lawyer to member of a parole board. Even the parents and families of the young men who were playing the role of prisoner came one evening for a "prison visiting hour" to see their sons. After his blow-by-blow description, Zimbardo lays out the implications of the experiment for social psychological theory—that is, what we can conclude from the experiment about human nature, especially about human moral nature. He then briefly describes the many popular magazines and TV shows that have featured the Stanford Prison Experiment over the years as well as various replications and extensions of the experiment by psychologists across a number of different cultures, and also how his findings have been applied in the last thirty-five years to minimize brutalizing conditions in all kinds of contexts.

The upshot of the 1971 experiment was that situational forces easily turned normal college kids who were playing guards into sadistic perpetrators who humiliated the other college kids acting as prisoners. They progressively instituted degrading practices, including sexual humiliation, sleep deprivation, prolonged isolation, making some prisoners humiliate and psychologically abuse other prisoners at the guards' behest, and the like, as the experiment went on and the guards' control and desire for further control escalated. Despite the participants' awareness that they were all in a psychology experiment and were in real life similar college students, the simulated prison reality took hold: even though no physical violence toward the prisoners was allowed, the humiliation and sadistic control of prisoners by the guards reached such a height that the experiment had to be terminated within six days, instead of continuing for the planned two weeks. By that time four students (of nine) who enacted the role of prisoner had had emotional breakdowns and been released, the first within thirty-six hours of the beginning of the experiment. What my sociodrama exercise in my Holocaust course pointed toward, the Stanford Prison Experiment proves with a force that astonished even its devisers. Zimbardo and his grad students never anticipated the kind of deliberate psychological humiliation and extreme methods of control that the guards engaged in, nor how quickly it would develop and escalate. And they were incredulous that students playing prisoner had the kind of extreme emotional reactions (emotional breakdowns) that several did, and within such a short time. But that is exactly what happened. Good people had turned evil and with little encouragement and no prior training—only with some general rules that set some limits for what they could do to their prisoners. Knowing that they were acting in and enacting a pseudo-situation, that it was not a prison but actually the basement of an academic building on a college campus, that the situation was of short duration, that they were under no threat or compulsion, that little was at stake, and that those taking the roles of guards and prisoners were actually indistinguishable in terms of class, social group, and status, they nevertheless created a situation of tremendous psychological pressure and even torment for the prisoners. The participants were good kids from good homes; they had plenty of opportunities, and many had social and political ideals. Yet they quickly became perpetrators, in some cases merciless ones. Not one student acting as a guard walked out; not one openly criticized what they as guards were doing or objected openly to the brutality and humiliation and psychological torture inflicted by a fellow guard upon a prisoner. They all, to varying degrees, caused gratuitous suffering to victims.

I will come back to Zimbardo's experiment later to present his survey of the particular situational and social forces that he identifies as operative in the experiment. Suffice it to say at this point that the extent of the power of the situational forces to override any character traits or exercise of free moral will surprised even the social psychologists, Zimbardo and his graduate students, who were testing for exactly that effect. Most disturbing of all perhaps was Philip Zimbardo's own realization that he, too, in his role as head of the pseudo-prison, found himself increasingly concerned with order in the "prison" and at one point took extreme measures to avoid the possibility of a prisoner rebellion, including going down to city hall to try to persuade the city to let him use real jail cells in an unused jail to ensure that the "prison" would remain intact in the event of students storming it from the outside. (Zimbardo thought that one student who had broken down emotionally and been released might come back to spring his fellow prisoners from their pseudo-jail.) The city wouldn't agree for insurance reasons, and Zimbardo, still in his head-of-prison state of mind, instead dismantled the entire prison setup and removed the prisoners to the attic of the psychology building. Zimbardo found himself succumbing to a kind of paranoia—even he was losing perspective on what was really happening. The experiment ended when a psychology graduate student who had not been involved with the project dropped by and saw the state of the student-prisoners as they were being walked blindfolded to the bathroom. It was she who, from her outsider perspective, blew the whistle by confronting Zimbardo, telling him that the student-prisoners were in fact being tormented and that the entire experiment had to be immediately terminated on ethical grounds. All those involved, and not only those playing guards and prisoners, had gotten caught up in the pseudo-reality and lost perspective on what was really happening. By the next day, Zimbardo had regained enough perspective to disband the project, realizing that his own reactions were as distorted by the situation as anyone's else's—they were, in fact, a very personal demonstration of the theory he was testing about whether social processes could overwhelm individual (moral and other) character traits and judgment. Even being fully aware of the kinds of social psychological situational forces that could take hold did not prevent the social psychologist himself from succumbing to them.

The lessons about the power of situational forces to engage people in destructive and gratuitously cruel behavior exposed by the Stanford Prison Experiment have been widely noted—all the branches of the U.S. military have developed simulated situations based on the experiment to train American soldiers to resist the pressures exerted upon them by the enemy in prisoner-of-war camps if they should be captured. The U.S. Navy first applied the Stanford experiment for that purpose in the development after the Korean War of its Survival, Evasion, Resistance, and Escape (SERE) program, in which mock prisoner-of-war camps are simulated in order to train soldiers to resist coercive interrogation and abuse. The navy's training program also serves as a warning against the excesses that prisoner-of-war situations can bring out on our side as well as on the enemy side; the point is to make sure that any prisoner-of-war camp that the navy might set up will minimize any risk of our soldiers brutalizing enemy prisoners. The navy has taught this Stanford Prison Experiment lesson for more than thirty years. SERE has since been adopted by all the other branches of the military; however, recently a number of human rights critics and investigative reporters have claimed that since the start of the war in Iraq in 2003, rather than using the program to guard against abuses as the navy does, the army at its training facility in Fort Bragg, North Carolina, has been using the same SERE program in reverse, turning it against its original intentions and using it to teach soldiers to harness the power of brutalizing situational forces in enemy interrogations. Rather than serving as a warning of the ease with which the prison situation can devolve into torture and hence as a stopgap to restrain excesses, the army at Fort Bragg has apparently turned the SERE program into a how-to exercise (rather than a how-not-to exercise) by using techniques derived from the Stanford experiment to train soldiers in techniques of psychological and physical "soft" torture. Moreover, these techniques have been taken from Fort Bragg and applied at the Guantánamo Bay prison in Cuba and from there transferred to the Abu Ghraib prison in Iraq.

During the military trials of the soldier-guards who were caught abusing prisoners at the Abu Ghraib prison in Iraq, Philip Zimbardo was brought in as consultant by the lawyer of one of the accused guards who had been directly in charge and who was brought to trial by the military once the incriminating pictures were leaked (by a whistle-blowing soldier) of naked prisoners wearing dog collars being led around by guards and of other sexually humiliating poses and photos of torture. Zimbardo includes a lengthy chapter in the book about his investigation of the torture at Abu Ghraib and of who was responsible for the torture of Iraqi prisoners there. Was it a few "bad apples," as the American military and U.S. government claimed repeatedly, or was it a "bad barrel" in which normal American kids found themselves and succumbed? Or, more specifically, what proportion of blame ought to be assigned to the "bad apples" versus the "bad barrel," and how big ought the "barrel" to be and who ought to be included in it? The Abu Ghraib torture scandal and the debate over the use of "soft" torture techniques advocated by the Bush administration, especially by Secretary of Defense Donald Rumsfeld, Vice President Dick Cheney, and Attorney General

Alberto Gonzales, who called the Geneva Conventions against torture of prisoners-of-war as an interrogation technique "obsolete," suggests that we Americans ought not automatically exempt ourselves from the possibility of succumbing to being "evil" and put infinite distance between ourselves and the evil of, for example, even the Nazis—or the Iraqis, for that matter. Zimbardo quotes law professor Jordan Paust, who was a former captain in the U.S. Army Judge Advocate General's Corps, as saying that "not since the Nazi era have so many lawyers been so clearly involved in international crimes concerning the treatment and interrogation of persons detained during war," and first and foremost on that list of lawyers is then attorney general Alberto Gonzales.[1]

In Abu Ghraib we can see the exploitation for nefarious purposes of research into the psychology of evil that was sparked by the study of the Holocaust. Some historians and psychologists, and even some victims (the Italian Jewish writer and chemist Primo Levi, an Auschwitz survivor, comes to mind especially), have explored the Holocaust as a great laboratory of human nature, of the conditions under which ordinary people commit evil. After the war significant social psychology agendas and experiments arose in an attempt to explain the bureaucratized and institutionalized mass murder that the Nazis carried out; among these were the obedience experiments of Stanley Milgram, Philip Zimbardo's high school pal and later colleague, and the investigations by Theodor Adorno and others of the psychology of authoritarian personalities and prejudice.[2] But there is much to learn from detailed examinations of the Nazi Holocaust itself. How do people come to do evil, and was it indeed ordinary people who turned evil, or made individual evil decisions, or were the monstrous acts committed by *monstrous people* unlike ourselves in every way or in significant ways? Without widespread complicity, how could the Nazis have enlisted all branches of government and industry and the society at large in the killing process?

Two Cases of Perpetration During the Holocaust: The Nazi Doctors and the Ordinary Men of Reserve Police Battalion 101

We are all potential murderers, Zimbardo concluded from his research. Two particularly telling cases revealing how normal people turned evil during the Holocaust come from an in-depth study by the psychoanalyst Robert Jay Lifton of the Nazi doctors who oversaw and put into effect the program of systematic murder in Auschwitz and from an equally sobering study by historian Christopher Browning of middle-aged army reservists in the Hanover, Germany, area who were called up to active duty in order to carry out mass roundups and murder of Jews behind the lines as the German army moved eastward. In both cases, people who before the war had been engaged in work appropriate to the normal activities and ends of their organizations and professions became murderers. Lifton and Browning investigate in detail how this could happen, showing that there was little or no coercion or retaliation that could be used to explain or excuse these perpetrators.

Lifton begins his study of Nazi physicians with the observation that historically there have been other regimes that have used physicians for evil rather than for good: he mentions the Soviet use of psychiatrists to certify and then imprison political dissidents as mentally ill, doctors brought in as torturers in Chile, medical experiments and vivisections performed by Japanese physicians on prisoners during World War II, and the CIA's use of American physicians and psychologists in experiments with drugs and mind manipulation. Nevertheless, to date only the Nazis have made physicians the "high priests," so to speak, of what Lifton calls their "biomedical vision," the racial ideology at the center of their genocidal project. National and racial "healing" was to be undertaken through "medicalized killing"; Lifton coins the phrase "killing in the name of healing" for the Nazi project, which placed physicians in a central role in genocide. While the phrase "Auschwitz doctor" may bring to mind most easily the demonic experiments of Mengele, the physicians at the heart of the genocidal mass murder taking place at Auschwitz and other killing centers are more Lifton's concern and are remarkable for what Lifton calls their "ordinariness." What Lifton means by this, in part, is that their engagement in mass murder, which is at the extreme end of human behavior, does not seem generally to be motivated by the demonic emotions that one would usually associate with that kind of behavior, such as frenzy, rage, or sadism. A friend to whom Lifton confided this responded perceptively, "But it is demonic that they were not demonic." Indeed. The normality of the doctors both suggests that evil became the norm and also hints that evil was not only normalized but perhaps even is or can be normal and hence be all around us yet beneath our radar. That is also what Lifton's introduction suggests in its title, "This World Is Not This World," taken from the words of a Jewish survivor who spent three years at Auschwitz and whom Lifton interviewed decades later in his home in Haifa, Israel, where he was a dentist. "This world is not this world," the dentist remarked, sighing deeply as he looked about the comfortable room with a beautiful view of this seaside city on the slopes of Mount Carmel. Lifton takes his words to mean that underneath the normal lurk "darkness and menace." The normal or ordinary can disguise a dark underbelly—perhaps one of the most important lessons of the Holocaust.

It is not surprising that the Nazi project of mass killing began before the war, in 1939, with the rounding up and murder of the disabled, especially children, a program the Nazis called "euthanasia," thus implying that the death was to the benefit of the patient by lessening suffering—which of course it was not. And even before that there had been a program of forced sterilization instituted as soon as Hitler took power in 1933. The Nazis did not, however, invent the biological rationalizations for sterilization and murder of the disabled and of those deemed racially inferior according to pseudo-biological ideologies. In the 1920s the United States was the worst offender in forced sterilization of the disabled and criminals and in the legal prohibition of interracial marriage, all done ostensibly in the name of biological imperatives. In Germany, *The Permission to Destroy Life Unworthy of Life*, written by two highly regarded academics, the jurist Karl Binding and Alfred Hoche, a professor of psychiatry, was published in 1920 and called for the killing of the disabled, the deformed, the mentally ill, and the incurable as a therapeutic goal, as "purely a healing treatment" and "healing work."[3] Lifton points to an earlier embrace of direct medical killing by Adolf Jost in his 1895 book *The Right to Die*, a right that Jost saw not as the individual's but as the nation's over the individual. It was in this ripe atmosphere and with this history that Nazism took root and took over. To implement the program of sterilization of the disabled and other biological "unworthies," the Nazis put into effect a system of "hereditary health courts," manned by two physicians and a district judge, to determine who should be defined as "hereditary sick," a category that included people with various hereditary and non-hereditary diseases and mental conditions, and sterilized on the grounds of the biological "purification" of German "blood." Physicians were required by law to report to these courts any patients who fit any of these categories, and so from the beginning of the Nazi regime doctors were enlisted to serve a kind of policing function for the bio-state. It is estimated that somewhere between 200,000 and 350,000 people were sterilized as a result. During the height of the implementation of the sterilization policy, the 1935 Nuremberg Laws prohibiting and dissolving marriages between "Germans" and Jews were also instituted. I put Germans in quotes here because Jews in Germany were

German citizens and hence the introduction of such a distinction and opposition is in itself a mark of complicity with the Nazi racial program. Thirteen percent of physicians in Germany were Jewish, and in a number of major cities the figure rose to 50 percent, some of whom were among the most distinguished and prominent doctors in Germany, some of international repute. The Nazi medical policy of racial purification included at this early stage the intimidation of Jewish doctors, a propaganda campaign against them as "anti-healers," and the gradual institution of limitations on their practice of medicine (by 1939 an amendment had been added to the Nuremberg Laws revoking the medical licenses of all Jewish doctors).[4] Jewish doctors had to be pushed out of the profession if medicine was to be turned away from its professional integrity as the healing of individuals and co-opted, politicized, and Nazified toward the purpose of the "healing" of the *Volk* as a single racial biological organism.

Lifton says that the majority of Nazi doctors whom he interviewed approved of the forced sterilization laws at the time they were passed, and hence their early enlistment in the Nazi state was easily won. (There is a slippery slope here, to which I shall return when we discuss the social processes at work in perpetration.) Lifton quotes deputy party leader Rudolf Hess proclaiming at a mass meeting as early as 1934 that "National Socialism is nothing but applied biology."[5] Lifton suggests that the Nazi genocide is the only mass murder that began with a program of sterilization as a first stage; from this we can infer the centrality of the biomedical rationalization for murder. That the crimes of the state were to be camouflaged by the use of doctors as its executioners was there from the beginning. The "euthanasia" project, despite the secrecy and deception with which it was carried out, came to public attention and, unlike the forced sterilizations, was ordered stopped due to an outcry both from the public and from several prominent Protestant and Catholic clergy who would not go along with the Nazi redefinition or rationalization (although the killing of disabled children was not actually discontinued, and, of course, there was no such outcry from the public when it came to the murder of Jews). Hence the centrality of the biological rationalization or ideological lie in effecting the corruption of doctors to serve a pivotal role in the system of murder. It was not physicians who protested against the killing of the disabled but religious leaders and the public at large. The Nazis' placing doctors in the position of killers served as an enactment of the subterfuge that were engaging in the biological improvement of the species rather than straight-out murder. Lifton suggests that what he calls the "medicalization of killing" in the Nazi biological and racial ideology of "killing [the biologically inferior] in the name of healing [the race or the species]," in its choice of language brings to mind the imagery of engaging killing in a legitimate healing process, thereby "destr[oying] the boundary between healing and killing," and thereby redefining killing as "a therapeutic imperative." Lifton describes the process of the Nazi takeover and reorganization of medicine, and the introduction of the slippage of the meaning of medicine from the healing of individuals to the focus on the alleged health and healing of the German *Volk* as a mystical entity, a Romantic notion of the merger of all in an imagined racial body, the *Volk*, to whose national "cure" all were to be dedicated. Jews were redefined as the "disease" of this national body, or the cause of the "disease." He quotes one Nazi physician as saying that it was necessary to develop "a totality of the physicians' community, with physicians having total dedication to the *Volk*," in what this doctor termed "biological socialism."[6] Of all the professions, doctors in Nazi Germany had one of the highest percentages of party membership and of representation in the SS and SA. But all professions and institutions, not only medicine, were redefined, politicized, in terms of the Nazi totalitarian vision and interpretation of its interests.

Such biological-ideological linguistic formulas and their institutionalization marked the difference between the historical violence of anti-Semitic pogroms such as Kristallnacht, with which the Nazi murder of the Jews began in 1939, and the subsequent Nazi genocidal system that enlisted all the organizational structures of German society itself in the murder project. The desired routinization and psychological numbing were also part of the reason that the organized roundup and mass shooting of Jews into mass graves as the German army moved east (sometimes by units such as Reserve Police Battalion 101, but more often by the Einsatzgruppen, special forces of the SS), in which all told 1.5 million Jews died, eventually was replaced with the even more impersonal and routinized death by gas and disposal of the bodies in the crematories of the six death centers set up by the Nazis in eastern Europe. That routinization and bureaucratization, the dividing up of the components of the sequence of murder into discrete parts performed by different organizational sectors, induced an emotional tenor of (false) calm, a false normality and legitimacy and diffusion of responsibility. The characterization of the murderers as "ordinary" and of the carrying out of genocidal murder as "normalized" function as part of the lie, for via the lulling of the anticipated feelings that murder is assumed to express, they enact the lie that what was going on was not murder but something else. Even today it is possible to catch ourselves inadvertently seduced by the Nazi language and (what must be called, I think) their system of rationalization, and therefore we begin to slip into accommodation and complicity. Nevertheless, when we think of doctors succumbing to linguistic sleights of hand that equate heal-ing with killing via a metaphorical use of language that defines Jews as an evolutionary parasite or a "cancer" on the German nation's "body" that needs to be cut out or as vermin that need to be gassed via a pesticide (that's what the Zyklon B gas used in the chambers was), we wonder how this could have happened.

One is almost incredulous when Robert Jay Lifton quotes a Nazi doctor as saying, "I would remove a gangrenous appendix from a diseased body. The Jew is the gangrenous appendix in the body of mankind." Now that we see the linguistic sleight of hand involved in the question, we wonder why any-one, especially doctors who were educated people, would believe this rather absurd lie. More broadly, how do people come to hold beliefs, and under what circumstances do they adopt or reject a belief? And who goes along and who refuses to do so? We will return to these questions in the discussion of those who risked their lives to rescue Jews during the Holocaust. But what we can already point to is the role of language, which is to say the in-terpretation of situations and of people's actions in those situations, as cru-cially important in both perpetration and rescue.

The evidence seems to suggest that *it is the uncritical adoption of a particular interpretive frame for a given situation that is at the center of perpetration when it comes to normal people.* Acceptance of the implicit social meaning and rules of a situation involves the passive acceptance of the authority of those defin-ing and redefining social (and institutional) reality through language and enactment. Normal people are potential followers, rather than architects, of evil. Lifton returns often to the normality, the ordinariness, Hannah Ar-endt's "banality," of the perpetrators but qualifies that judgment in two ways. First, Lifton proposes that while the men who committed the atrocities may have been banal, their evil was certainly of extraordinary proportions. Second, that kind of acting out of evil, no matter what language was used to disguise it, nevertheless changed those who committed it. No longer were they ordinary, nor did they have simply ordinary motivations. As a result, there is also, he suggests, a limit to our capacity to understand them completely, which is to say to reconstruct and relive in imagination the perpetrators' inner lives. Nevertheless, we can come to a much greater understanding.

Robert Jay Lifton calls the Nazi state a "biocracy," a sort of theocracy for which the Nazis claimed divine sanction and which put in the place of a theology a biological ideology of a sacred racial order, namely, themselves. The use of doctors was an attempt to legitimate Nazi genocidal ideology and murder by giving it the authority and respectability of science, but the Nazis did not engage in the legitimate discoveries of science in any honest way. The Nazi strategy was to enlist and corrupt physicians and other experts—Lifton points to physical anthropologists, geneticists, and racial theorists of all kinds—in an attempt to give their racism scientific legitimacy and authority. It was what we might call part of the cover-up, and what Lifton calls "a pretense of medical legitimacy." The corruption of doctors and the medical establishment as well as the corruption of other professions—perhaps law was second only to medicine in the Nazis' enlisting of it to legitimate discrimination and finally genocide—was a strategy to gain the veneer of legitimate authority for their policies and actions. That strategy used first and foremost the distortion of language to deny the obvious meaning, especially the *moral* meaning, of the actions undertaken.

"The Nazi doctor knew that he selected," Lifton writes, "but did not interpret selections as murder. . . . Disavowal was the life blood of the Auschwitz self."[7] So taking a role in the social drama, using its language and enacting its meanings, gave away the store. In the case of the institutions and professions corrupted by the Nazis, taking part in them could never be innocent but entailed unavoidable complicity and falsification; no person who took part in them could remain in any way an innocent bystander, untainted.

Perhaps the single best portrayal of the way that familiar language, daily encounters, and common institutions were corrupted by the Nazi regime can be found in Bertholt Brecht's play *The Private Life of the Master Race: A Documentary Play* (also called *Terror and Misery of the Third Reich*). The play is a series of vignettes of the first five years of the Nazi regime, documenting how each institutional and personal context was politicized so that the winners would be German Christians and the losers would be Jews. Situations were redefined to cast Jews as the villains and German Gentiles as the innocent victims of Jewish perfidy. A stunning example is captured by Brecht in a scene called "Augsburg 1935: In Search of Justice." In it a Jewish small businessman named Arndt, owner of a jewelry store, is trying to collect money from an insurance company for the destruction of his business by three Nazi storm troopers. In a scene in the judge's chambers, the police inspector and the judge, who are in collusion, conclude that the case must be reconfigured as the provocation of the storm troopers by the Jewish businessman Arndt; hence the Nazi thugs are the victim and it is they who are due restitution from the Jew. Later the prosecutor comes in and says to the judge, "But your national instinct must tell you, my dear Goll, whom you *should* do right by." He goes on, "I must stress one thing: I'm advised—and my advice comes from the highest circles in the S.S.—that by now somewhat more backbone is expected from German judges." Then the prosecutor continues, "But our Minister of Justice made an excellent remark which might give you something to hold on to: 'Whatever's useful to the German Folk is just.'" The judge concludes, "I see it as merely an obvious case of Jewish provocation and nothing else." Later, speaking of the case to another judge, Judge Goll says, "My God, I'm willing to do anything, please understand me. . . . I decide this and I decide that as they require but at least I must know *what* they require. When you don't know that, there is no justice anymore. . . . I must be told which decision is in the interest of the higher authorities. . . . After all I have a family, Fey. . . . I'm willing to do anything." The Jewish victim ends up as the defendant as the court decides to make him pay for the destruction of which he was the victim. In the Brecht play we see how in the initial phase of Nazification the motives of fear and greed were completely on the surface and the lies self-conscious and deliberate. Living and telling lies over time, however, induced self-deception and rationalization, a knowing that hid behind a pretense of not knowing—Lifton's life of "disavowal."

Brecht's play brings out the moment of the corruption of institutions, the first lies. By the time of the Nazi doctors' committing and overseeing mass murder in the death camps, those lies were embedded in ongoing institutions and daily social relations. Nevertheless, that they were lies could still become apparent from any wider perspective, that of history, cultural meanings, the professional ethics of medicine and law that predated the Nazification of universities, religious teachings, and so on. Even if one pretended that killing was healing, and called it so, killing could not completely match the actual experience of healing, nor could the corruption of justice do anything but masquerade as justice. Experiences that doctors and judges would have had both in prior life and in other parts of their lives exposed the lie. As a consequence, there could be no naive use of the Nazi language of rationalization and subterfuge. Its very use made a person complicit, a sellout, part of the cover-up. Yet the ability (and courage) to discern and openly confront the lie as a lie took a highly developed and well-tuned capacity for self-reflection along with a broader perspective; alternatively, it took belonging to a different social group or community with which one had such a profound identification and for which one had such deep respect that *its* interpretation of the situation trumped the Nazi one. As Spinoza would have known, the people who could resist social forces and have independence of mind were either those who were rigorously trained in self-reflection and who also identified with a much wider and more pluralistic world and a longer-term perspective (as his *Ethics* aimed to provide), or those who came from and profoundly identified with a different community, one that had a different (and most likely critical) interpretation of the situation. I have just described (and anticipated) who the rescuers were: they were either social outliers, people who in some sense were and felt apart from the Nazified situation (like Zimbardo's graduate student who happened upon the scene of the disoriented, blindfolded prisoners and blew the whistle), or members of other groups in the society that opposed or were critical and contemptuous of the Nazis (for example, the underground partisan resistance or local religious minority groups).

I suspect that one of the purposes of the corruption of language was to make everyone complicitous, implicated in the guilt of the regime even if only to a small degree, so as to weaken everyone's capacity for resistance. Guilty behavior contributed to creating a fanatical identification with the group. It led to an embrace of the Nazi distortion and deception about the meaning of the situation, for going along with a situation produces belief in it ex post facto, especially when there is little extrinsic reward, such as payment, for doing so. Perhaps this helps to explain the lack of emotion appropriate to the reality of the enormity of the crimes that Robert Jay Lifton found in the physician-perpetrators to a man. In the Brecht play, implicit in the act of giving in to the use of the language was fearful submission to a veiled threat. The language served as a kind of litmus test of loyalty and an aggressive invitation to comply. Gaining even the prisoners' complicity in the concentration camps by engaging them in tasks that contributed to the system of murder was part of a deliberate strategy to render them passive and submissive. It was a strategy aimed at reducing their self-respect and breaking their will by putting prisoners (let alone guards and everyone in the system) in situations in which moral compromise was unavoidable, inducing guilt. Moral compromise, inducing guilt, was the weapon of choice, the means of mass social control, and it infected all aspects of Nazi society and institutions. Hence the lie could never be innocent, for innocence would leave people too autonomous and ready to rebel even at the cost of their lives. The inducement of guilt makes people not just passive in the face of authority but also in need of the very rationalizations that the Nazi ideology and distortions of language provided. Here we are getting a glimpse of the negative side of social processes. The social motivations that contribute so much to success in life can, in contexts of perpetration, make us collaborators and even murderers. Perhaps the point is that to be human is to be a social animal, as Aristotle so presciently put it, but the glorious human capacity of social intelligence comes not only with all its beauty of relationship and attachment and exquisite attunement to social cues but also with its concomitant of collusion and collaboration. That is certainly part of the story of many of the Nazi doctors at Auschwitz whom Robert Jay Lifton studied.

Ordinary Men as Mass Murderers

The ordinary men of Reserve Police Battalion 101 whom Christopher Browning reports upon were in most respects at the other end of the social spectrum from the Nazi doctors. They were not in any sense "high priests" of the killing process, neither offering an ideological cover-up nor legitimating murder by their respectability and social authority. They were not movers and shakers but local people enlisted and caught up in unexpected and vast evil to which, however, they voluntarily contributed. I think they are analogous in some respects (although not in the magnitude of the mass murders they committed or in the organized and clearly instituted system of murder they contributed to) to the low-level American soldiers caught up in torture at Abu Ghraib prison, for both groups were swept up in a system of evil beyond their understanding and anticipation. Nevertheless, humiliation and torture, even leading at times to death, are not systematic mass murder of tens of thousands, so the cases are different in the magnitude of evil and in its organization. Moreover, there is a stark moral clarity about the obviousness of the evil of the murder of children, old people, families, even whole villages, while the situation of torture in Abu Ghraib prison was not quite so crystal clear—but yet clear enough. It was because of that clarity that

Browning chose to study this particular battalion, as it starkly reveals how far normal people will go in complying with, or being guided by, social situational forces even when they are not coerced and even when the social system engaged in involves murder. Reserve Police Battalion 101 represents a warning about the limitlessness of the possibility of engaging and enlisting normal people in evil through framing (interpreting) the situation in certain ways and defining the roles that those involved will take. By the men's "voluntary" willingness to commit mass murder I simply mean that they were not externally coerced; in fact, perhaps one of the most shocking things about their engagement in murder was the repetition by the leadership of the battalion just before mass murder operations were to begin that participation was optional and no repercussions would ensue from refraining. Nevertheless, murder they did.

Browning begins by pointing out that most of the mass murder of Jews in the Holocaust took place within an eleven-month period from March 1942 to February 1943. Yet 20 to 25 percent of the 6 million Jews killed in the Holocaust had already perished by March 1942 (including those remnants of my father's family who had not settled in either the United States or Israel but were still in Lithuania when the Nazis and their Lithuanian collaborators killed them in late 1941), and within the next year almost all of Polish Jewry would be murdered. So the mass murder of Jews was, as Browning put it, a kind of blitzkrieg operation: not incremental or gradual but a massive mobilization using vast numbers of shock troops. Of the 3.5 million Jews living in Poland before the war, only fifty thousand survived. Poland was the central locus of the mass murder of Jews; it was where five of the six killing centers were built to which Jews from all over were brought by train, as well as the locus of a great deal of the mass shootings before the killing centers were fully operational. There was also the rounding up and imprisonment (and starvation) of Jews in vast ghettos prior to the killing, especially in Polish cities. The manpower needed to do all this was staggering, particularly during a time of war.

The rank and file of the reserve police battalion were middle-aged and either working-class or lower-middle-class family men, with limited geographic mobility and little education except for vocational training that had ended at age fourteen or fifteen. At an average age of thirty-nine (more than half were between thirty-seven and forty-two), they were too old to be in the German army and so were drafted instead into the order police as raw recruits with no experience in occupied territory. All had grown up before the Nazi era, with the morals and norms of that prior era. Most came from Hamburg, one of the least Nazified areas, and from a social class that, according to Browning, had been anti-Nazi in political culture. They hardly seemed a likely group to recruit to engage in mass killing in the name of Nazi anti-Semitic ideology. Like Zimbardo's college students, who showed no evidence of sadism or other tendencies toward brutality or instability, the recruits of Reserve Police Battalion 101 gave all the indications of being unlikely to commit acts of gratuitous cruelty, torture, or murder. The battalion members are perfect examples to highlight the power of situational forces in motivating evil versus character traits and the freedom of the will to refrain from evil. Of course, that that was the case, unlike in Zimbardo's experiment, was completely inadvertent.

Reserve Police Battalion 101 was first charged with rounding up the Jews of the Hanover region and deporting them to the east. The battalion had been in occupied Poland for less than three weeks when they were presented with their first "action" to engage in the mass murder of Jews. That was at Józefów in July 1942. The head of the battalion, Major Wilhelm Trapp, a career policeman, addressed the unit about rounding up the eighteen hundred Jews of this small village, separating out the able men, who would be sent to a work camp, and shooting the others—women, children, and the elderly—and dumping their bodies into a mass grave. At the end of his address, Trapp remarked that any members of the battalion, especially the older members, who were not "up to" taking part in the assignment could remove themselves, giving potential decliners a ready, face-saving excuse. Trapp emphasized how "unpleasant" the task was for him and for the group but added that the assignment came from the highest authorities. In order to encourage participation in what would be *for the battalion* an "unpleasant" task, he included in his pep talk some anti-Semitic propaganda about Jews in the village being allied with anti-Nazi partisans and Jews in general being the cause of German misfortunes. This first instance of mass murder became prototypical of what was ahead.

The order police were at the disposal of the SS and the police leader in the Lublin District, Odilo Globocnik. Globocnik was put in charge of the extermination of the Jews and the disposal of their property in the entire General Government area of Poland, which had about 2 million Jews, of whom nearly 300,000 resided in the Lublin district. There were SS units of security police, Gestapo, and Kripo, but the three order police battalions in the Lublin area provided the greatest available police manpower in the area, with about 1,500 men. Gassing began at the Chelmno death camp in early December 1941, at Birkenau in February 1942, and at Belzec in March. Sobibor began its killing operations in May, and Treblinka in July. Between mid-March and mid-April 90 percent of the 40,000 Jews of the Lublin ghetto were murdered, either by being shot on the spot or by transport to Belzec. By mid-June 100,000 Jews of the Lublin district had been murdered, along with 65,000 from Krakow and Galicia, the neighboring districts. While the murder of the Lublin-area Jews was going on, Jews from Germany, Austria, and other areas under Nazi control were being transported to the Lublin district, some directly to the death camps but others to the ghettos for temporary housing before extermination. Reserve Police Battalion 101 arrived in the Lublin area on June 20, 1942. When they arrived, there is no indication that the members of the battalion knew what they were going to be used for. Globocnik had called a temporary halt to gassing in the death camps in mid-June due to temporary logistical problems, and as a result he decided to institute mass murder by firing squad as a substitute just as Battalion 101 was arriving in the area. They were immediately assigned that function, beginning with the massacre at Józefów.

When Major Trapp gave his speech to the men of Battalion 101 just before they set out for their first mass murder of Jews at Józefów, he told them for the first time that they would be rounding up Jews, and then murdering by firing squad the old, women, and children, and sending the men for slave labor. Upon hearing what they had to do and that they could refrain if they "were not up to it," about thirteen men (of the roughly five hundred) immediately took Trapp up on his offer not to participate in murder but to be reassigned, without repercussions, to some other aspect of the job. A company was ordered to surround the village; another was ordered to round up the Jews and take them to the central marketplace, but killing on the spot babies and anyone who could not walk there. In the marketplace the young male Jews for slave labor were taken out of the group and the rest were loaded on battalion trucks to be taken to the forest to be murdered by firing squad. Trapp himself did not go to the forest but stayed in the village talking to its local priest and mayor. Some men complained later that Trapp never went to the forest to witness the killings. Trapp was said by one of the policemen to have announced on one or more occasions something to the effect of, "Such jobs don't suit me. But orders are orders." He was also observed by several on various occasions weeping, on one occasion saying that "everything was terrible." His driver reported that he had confided to him, "If this Jewish business is ever avenged on earth, then have mercy on us Germans."[8] Nevertheless, Trapp did nothing to stop the killing, nor did he refrain from arranging all the details and doing all the organizing and implementing of them so that the mass murders were carried out. When the roundup was completed, it was the battalion doctor and the company's first sergeant, according to Browning, who instructed the men in the murderous task ahead in the forest and how it was to be carried out.[9] Another policeman in the battalion later gave the following testimony about what precisely the physician had instructed them to do:

> I believe that at this point all officers of the battalion were present especially our battalion physician. . . . He now had to explain to us precisely how we had to shoot in order to induce the immediate death of the victim. I remember exactly that for this demonstration he drew or outlined the contour of a human body, at least from the shoulders upward, and then indicated precisely the point on which the fixed bayonet was to be placed as an aiming guide.[10]

The policemen were instructed by the doctor to deliver neck shots to their victims. This policeman remembered the physician well because they had become friends as fellow musicians—he was a violinist and the doctor an accordionist—and they would entertain together on social evenings.

The shooting of Jews in the forest took place in the following way. The Jewish victims, in groups of about thirty-five to forty at a time, were paired face-to-face with policemen. Together executioner and victim walked down into the forest to the place that the other SS captain, Julius Wohlhauf, had spent the day looking for as the best spot for the mass murder. The Jewish victims were then ordered to lie facedown in a row. The policemen were placed directly behind the heads of the victims, and they all fired in unison at their Jewish victims. Then a second group of policemen and victims walked together to a slightly different location so that each group of victims would not see the bodies of the previous group. The first group was then paired with their next round of victims, and they shot again. The two contingents alternated as Jews were brought to be killed in groups of about forty at a time. This rotation went on all afternoon until there was an alcohol break for the shooters. Then the systematic shooting resumed and went on without a break till nightfall.

When the doctor gave his instructions in the marketplace about exactly how to kill the Jewish victims, just as the policemen in the firing squad were to be sent off to the forest, several members of the battalion who had not previously opted out of killing decided at that point to withdraw themselves. One said that the task of murdering Jews in this way was "repugnant" to him and he asked for reassignment. He was immediately reassigned to guard duty without reproach or repercussions. Another group, after it had been shooting for a while, asked to be released from shooting, and they were immediately reassigned to the trucks, again without reproach or repercussion. Others were released at midday when they asked to be. Others, although not asking to be released or reassigned, passively resisted simply by shooting past their victims; a couple hid in the garden of a Catholic priest or remained in the marketplace to evade firing squad duty, while others procrastinated with their searches of Jewish homes to avoid being assigned to the firing squads. Everyone was aware that opting out was possible and that no reprimand, shaming, or repercussions would ensue. There were no negative consequences for those who managed to avoid shooting or stopped after a while.[11] Browning estimates that of those assigned to the shooting squads, about 10 to 20 percent either withdrew openly or evaded shooting in one way or another. So he concludes that at least 80 percent of those assigned to shooting did so and continued to do so until 1,500 Jews from Józefów had been murdered.

The men of the battalion had every opportunity to refrain from killing: they had backgrounds that did not dispose them to be enthusiastic Nazis; they had unenthusiastic leadership; they had repeated opportunities to withdraw without retribution; they had multiple occasions within a killing operation to absent themselves; they were not convinced anti-Semites. Nevertheless, the unlikely men of Reserve Police Battalion 101 were directly responsible for the mass murder of 83,000 Polish Jews, over time shooting 38,000 and rounding up 45,000 and putting them on trains to the nearby death camps.

The chilling implications of Browning's research suggest the extraordinary ease with which the situation brings forth individual behavior. Reserve Battalion 101 brings to mind the Stanford Prison Experiment and Zimbardo's description of the Abu Ghraib prison. It is not that individuals are not responsible; they surely are and should be held accountable. In the second half of the book I will propose how and why individual responsibility is both possible and necessary. Clearly, those who shape situations and their meaning and institute implicit rules and standards bear a great deal of responsibility, too. The degree to which people lack independence of mind and moral viewpoint stands out here as perhaps the most salient and glaring moral problem that needs to be anticipated and for which mechanisms must be put in place to address. If this is not the failure of moral character and the ethical resolve of the free will, I don't know what is. Human beings are not who or what we think they are.

Rescue: National, Communal, and Personal

Now for the good news. The largest-scale operations to rescue the Jews of Europe were also driven by social and situational forces, forces that enabled people to climb to heights of nobility and self-sacrifice despite terrible threats of torture and death not only to themselves but also to their loved ones. Here we find unexpected goodness in unexpected places. Not only did neighbor save neighbor and nations save their own, but at times the deepest empathy could be seen across the starkest human differences. There were whole nations whose political, religious, and social leaders came together to rescue their Jewish populations from the Nazis; there were small communities and organizations, religious, political, and philanthropic, that bucked the larger society and set up networks of Jewish rescue; there were great heroes who took it upon themselves to lead personal efforts to save as many Jews as they could; and finally there were individuals who almost invariably were asked by a Jewish friend, neighbor, or acquaintance to help, often in initially small ways, but who over time came to do acts of extraordinary kindness, risking their lives at every turn not only for friends but also for Jewish strangers and their families and other victims. Yad Vashem, the Holocaust memorial museum in Jerusalem, gives formal recognition to these righteous Gentiles. Who rescued and why? Were rescuers made of different stuff? Are there common features of rescuers that distinguish them from perpetrators? Can we predict who in another Holocaust would fall into each group? Were all rescuers brought up well and taught to be good, while perpetrators lacked moral training and good parenting and schooling? A more subtle version of this line of thinking raises the question of whether those who went along and committed atrocities in the Holocaust, such as the men of Reserve Police Battalion 101, were more inclined than other people to have authoritarian personalities, submissive to their betters in the hierarchy and ruthless with those beneath them, and hence more likely to submit to authority whether right or wrong. Were the rescuers, by contrast, noble loners who defied the herd, stereotypic Romantic heroes all?

There were three nations that saved their Jewish populations, and they did it as whole societies, by collective action from leaders down to the person on the street. Denmark is the most well-known case, while that of Italy is somewhat less well known, and the case of Bulgaria is surprising and largely forgotten. Of these three countries, two were allies of the Nazis, and the third, Denmark, was in many ways a special case and different from other European countries conquered by the Nazis. Nechama Tec, in her excellent study *When Light Pierced the Darkness: Christian Rescue of Jews in Nazi-Occupied Poland*, points out that across Europe rescue was directly proportional to the absence of direct Nazi control; within indirect control, the leeway a country had to maneuver was crucial in whether or not it saved its Jews. Where the Nazis had control of the governmental machinery, Tec says, they would do whatever it took to exterminate the Jewish population and would brook no interference. So two of the countries that saved the Jews were Axis allies and hence nations whose internal affairs were not directly controlled by Germany. Even so, Italy, at Nazi insistence, introduced its own Nuremberg-like laws in November 1938 disenfranchising Italian Jews and removing them from all the professions and the workplace. Nevertheless, even with the isolation of Jews from Italian society and the gradual discrimination set in place, Italians not only saved Jews within Italy but made an all-out effort to save Jews who came under their jurisdiction in the countries they came to occupy, including parts of Yugoslavia and areas of Greece and southern France.

Denmark was a special case: the Danes refused to hand over Danish Jews, who numbered only eight thousand, to the Nazis, and the Danish underground, with the knowledge and help of the people and the king of Denmark, arranged the rescue of the Jews by boat to neutral Sweden. Denmark had surrendered to the Nazis on the very day it was invaded, April 9, 1940, and had promised loyal cooperation. Perhaps because of the declaration of cooperation and also because of the Nazi racial ideology that stipulated that Scandinavians were of the highest Aryan racial rank, Denmark retained a great deal of autonomy, and the Nazis left it virtually alone as a model protectorate of the Reich. Danish officials insisted to the Nazis when they brought up the matter that Denmark had no "Jewish problem." Meanwhile, Danish courts continued to give stiff penalties to those who participated in outbreaks of anti-Semitic violence and even to those who published or spread anti-Semitic propaganda. Eventually the rising resistance of the Danes and the reprisals of the Nazis led to the end of the "model protectorate" when the Nazis took one hundred prominent Danes hostage, including a number of Danish Jews, among them the country's chief rabbi. In protest the Danish government resigned, and the Nazis instituted direct control of the Danish government on August 29, 1943. At that point Hitler decided to enforce the Final Solution in Denmark, and the Germans began immediate plans to round up the Jews of Denmark to send them to death camps. An unlikely friend, a German diplomat in Denmark, Georg Ferdinand Duckwitz, secretly made sure that Sweden would accept and grant asylum to the Danish Jews, then leaked the news of the German threat to the Danes. At that point Danish civil servants began to make individual efforts to warn the Jews and also arrangements to hide them. The operation was organized by the Danish underground and paid for in large part by wealthy Danes. Jews were placed in hiding and then transported clandestinely, mostly by small fishing boats but some by larger vessels, over the Oresund Strait from Zealand to Sweden, with the cooperation of the Danish harbor and civil police. Some Jewish refugees were smuggled on the regular ferries between Denmark and Sweden in freight cars that had been stolen by the underground after being inspected and sealed by the Germans. The Danes managed to rescue even the 472 Jews who had already been rounded up by the Germans and put in a detention camp before being sent to the Theresienstadt concentration camp. More than 99 percent of Danish Jews survived the Holocaust, the largest percentage of any country in occupied Europe.

Like the Danes, the Italians acted with a kind of universal tacit understanding among important military and diplomatic leaders and also among the lesser leaders and the rank and file within the bureaucracies and the military, who together took common action to save Jews. In the 1987 documentary film *The Righteous Enemy*, Joseph Rochlitz tells the story of his father's rescue by the Italian military by piecing together the larger story of the Italian rescue of between 25,000 and 30,000 Jews under their jurisdiction in occupied southern Europe: the Adriatic coast of Yugoslavia, southern Greece, and parts of southern France. Rochlitz managed to interview a number of the Italian bureaucratic officials, diplomats, and military leaders who participated in the rescue, as well as a Yad Vashem historian, the French Nazi hunter Serge Klarsfeld, and several survivors, including his father. Between these interviews he interposes extraordinary footage of the Italian camps where Jews were protected and pictures of the important figures in the story during the war. Rochlitz's father, Imre, an Austrian Jew, was a child of thirteen when he fled with his family to Zagreb, Yugoslavia, after the Anschluss, the 1938 annexation of Austria by Germany. When Yugoslavia fell to the Nazis in April 1941, the Adriatic coast of Croatia was put under Italian military control, where it remained until 1943. In Croatia the Nazis installed the fascist Ustashe, who began a campaign of murder that rivaled the Nazis' own. The Ustashe murdered some 750,000 people in their death camps, mostly Serbs but also Gypsies and Jews. The Ustashe set up death camps along the river Sava, and it was to one of those camps that Imre Rochlitz was sent just before his seventeenth birthday. His job was to go with other prisoners and guards each morning to pick up the dead, about four hundred corpses a day, and dig mass graves. Through the intervention of an uncle who was a decorated World War I veteran, Imre was released after three weeks in the Jasenovac camp.

The interior of Yugoslavia was under the Ustashe but the coast was under Italian control and occupation. Thousands of Jews tried to get into the Italian zone, and by mid-1942 between 3,000 and 4,000 Jews had managed to go by train to Split in Dalmatia and to safety under the Italians. The Italians were under increasing pressure to expel Jews and hand them over to the Ustashe to then be turned over to the Germans, but the Italian diplomatic officials and the Italian military repeatedly refused to comply. Once Auschwitz was functional the Germans put increased pressure on the Italians to give up the Jews who had taken refuge in Dalmatia, but the local Italian commander, Negri, responded to the Nazi demand by saying that it would be "contrary to the honor of the Italian army" to take measures against the Jews. When the Germans, in frustration, took the matter up with Mussolini himself in August 1942, Mussolini agreed to the deportations.

An Italian count, Luca Pietromarchi, was the Foreign Ministry official in charge of the Italian occupation of Croatia. He was of the papal nobility, and his wife was of Jewish descent. He had orders from Mussolini to participate in the Nazi Final Solution, yet he knew that the Italian army had already taken it upon itself to protect the Jews. Nevertheless, because he could not openly defy Mussolini, he developed with others in the Foreign Ministry and in the military an elaborate system of bureaucratic delay and other means of subterfuge to protect the Jews. Pietromarchi kept a detailed personal diary of how he subverted the orders to hand over the Jews, a diary that was discovered after the war and is a major source of knowledge about how the Italians saved the Jews under their jurisdiction in Croatia.

When asked in 1987 for the film why they saved the Jews, diplomat Roberto Ducci replied, "We happened to be human people and Christian people, so we protected because we did not share at all the idea of eliminating all the Jewish race. And therefore we did all, whatever was possible not only in Croatia but in other zones that were under the Italian occupation, were occupied by Italian forces in order to refuse the request by the Germans." Joseph Rochlitz persisted and asked why the Italians defied the Germans, who were their allies, after all. Ducci then replied, "One can be an ally but not become a criminal." That says it all. The Italians, that is, significant leaders in both the military and the bureaucracy, had complete clarity in their interpretation of the situation, never denying that mass murder was the German goal and policy, and they acted without any hesitation in engineering creative ways to save the Jews. As time went on, they took increasingly daring actions and also extended their saving strategies into more and more areas, trying to save Jews even in areas contiguous to those they controlled—for example, in Salonika, where the majority of Greek Jews lived. Some of the most dramatic rescues took place within southern France after the Germans removed from the control of the Vichy government parts of the Riviera and turned them over to the Italians in November 1942 in order to prevent an Allied invasion. When the Italians took over control, Marshal Philippe Pétain had already handed over 42,000 Jews (mostly not French) to the Nazis. Things changed dramatically with the arrival of the Italians. When the Vichy police tried to continue to round up Jews to deliver to the Germans, the Italian military stood in their way. The local Italian military commanders, for example, blocked the train tracks in Grenoble, so that no Jews could be transported, and forced the French to release the Jews who were already being held. The head of the Nazi SS in France was furious, as was the Vichy government, which was incensed at the Italians asserting their authority within their own country. The Vichy government tried to gain favor with the Nazis by exposing the Italian protection of the Jews to them and promising that if they regained control they would hand over many more Jews. As the Germans kept stepping up the pressure, the Italians created a Racial Police force under Guido Lospinoso, inspector general. But Lospinoso interpreted his mandate as saving the Jews of southern France rather than delivering them to the Germans, and he had as his adviser on Jewish affairs a prominent Italian Jew, Angelo Donati. In March 1943 Lospinoso developed a plan to transfer the Jews of the Italian zone of southern France to a region inland with the ostensible reason that there they could not pose a security risk in the event of an Allied invasion but where they would be protected. The Italians requisitioned private residences and numerous hotels to house the Jews. Serge Klarsfeld, who was saved in Nice as a child and later became a major Nazi hunter (capturing among others Klaus Barbie, head of the Gestapo at Lyon), remarks that the Jews in the Italian zone in France felt they were in the Promised Land. Fifteen thousand Jews were saved by the Italians in Nice alone.

Within Italy proper 85 percent of the 40,000 to 50,000 Italian Jews were saved. Italian Jews had been deeply integrated into Italian society since the nineteenth century and there had been a Jewish community in Rome since before Christianity. Approximately 8,000 Italian Jews were rounded up by the Nazis and their Italian Fascist collaborators in areas of Italy they directly controlled after Italy surrendered to the Allies on September 8, 1943. Rochlitz admits that from 1943 to 1945 the Italian Fascist persecution of the Jews within Italy was worse than that of Vichy France. Nevertheless, that does not erase the Italian rescue of between 25,000 and 30,000 Jews within occupied Europe and the rescue of most of the Italian Jews; despite the Vatican's tacit support of the Nazis and of the Final Solution, hundreds of Jews were hidden within the Vatican itself, and thousands more in convents and monasteries throughout Rome; the cardinal of Florence publicly asked for Jews to be sheltered and found convents willing to take Jewish refugees.

Bulgaria was, like Italy, an ally of the Nazis, and Bulgaria saved its Jews even though the character of the Bulgarian Jewish community and its place in the society were quite different. Bulgaria had joined forces with Germany quite early and had instituted anti-Jewish policies. Like Denmark and Italy, Bulgaria had been rewarded by the Nazis for its cooperation with a great degree of autonomy in governing its own country. In 1941 the Bulgarian government had introduced a package of anti-Jewish laws to identify Jews, expropriate their property, and concentrate them. But the Bulgarian public objected, and as deportations grew nearer, protests erupted from a number of different powerful groups: the Bulgarian Orthodox Church, the intelligentsia, the Communists, the underground, and some sectors of the government itself. The Jews of Bulgaria, while having had full economic rights and citizenship before the Holocaust, had been a culturally isolated minority (a mere 1 percent of the population) and were not assimilated and integrated into the society as the Jews of Italy and of Denmark had been. The community had played no significant role in Bulgarian political or cultural life. Nevertheless, the Orthodox Church hierarchy adamantly and forcefully protested anti-Semitic persecution and defied the government's anti-Jewish policies. Orthodox priests conducted mass (mock) conversions of Jews and issued false identity papers with much earlier dates of conversion; some performed marriages of convenience between Christians and Jews to help shelter Jews from persecution; and finally in September 1942 the metropolitan (a rank above archbishop in the Orthodox Church) Stefan of Sofia delivered a powerful sermon pronouncing that Jews had been punished enough for the death of Christ by their wandering and that their fate should be in God's hands, not human hands. When the government decided to cede to the Nazis only "foreign" Jews living in Bulgaria in the newly annexed areas of Macedonia and Thrace, the head of the Orthodox Church intervened on behalf of the Jews, as did the vice president of the Bulgarian parliament. And when the protests of these prominent figures did not stop the deportations, an order from the Bulgarian king prevented the deportations of Bulgarian Jews from Bulgaria proper.

Nechama Tec was surely correct in citing the level of control that the Nazis exerted in a country as the most important factor in whether or not Jews were saved. It is a common misconception (even in the Netherlands itself), perhaps because of the popularity of *The Diary of Anne Frank*, that Dutch Jews did not perish in large numbers in the Holocaust because they were saved by their fellow Dutch. Yet 75 percent of Jews in the Netherlands perished. Of approximately 140,000 Jews living in the Netherlands before the Holocaust (1.5 percent of the population, including some 25,000 German Jewish refugees), only 35,000 survived. After the German victory in 1940 over the Dutch, a top Nazi official, Artur Von Seyss-Inquart, was placed in power in the Netherlands, and he immediately set about ruthlessly carrying out anti-Semitic policies. An early rebellion against the Nazi government, in which the Dutch called a general strike in 1941 to protest the Nazis' anti-Jewish measures, was put down ruthlessly with reprisals against both Christians and Jews, many of whom were deported to concentration camps. A climate of resignation ensued and there was little Dutch opposition of any magnitude until 1943. The Nazis began to deport Dutch Jews to death camps in 1942 and made Camp Westerbork, originally set up to shelter Jews fleeing from the Nazis, into a transit camp for the death camps in Poland. The collaboration of the Dutch with the Nazis was high—it was the Amsterdam city government, the Dutch municipal police, and the local railway workers who engineered the roundup and deportation of the Jews. Holland also had a strong Dutch Nazi Party and a local branch of the SS. After the war, trials in the Netherlands convicted 50,000 Dutch citizens of collaboration with the Nazis.[12] Tec concludes, "Whatever the reasons, the Dutch did not succeed in saving many Jews"; the evidence suggests that most Dutch Jews perished in the Holocaust because saving actions were limited.[13] And while the German implementers were extremely ruthless, the Dutch on the whole, both the general public and government officials, were at best indifferent and at worst collaborative bystanders.[14]

Even in Poland, a nation with a long history of virulent anti-Semitism, a large and poorly integrated Jewish community, and a central role in the Final Solution, there were still Poles who endangered themselves to hide, help, and save Jews. Of 3.5 million Polish Jews, only about 150,000 survived the war; 40,000 to 60,000 of these passed as Christians within Poland. Their survival took tremendous effort on the part of some Poles to hide and house them, to feed them when food was scarce, to find medicine and medical care when they became ill, to help them rid themselves of identifiable Jewish dress and other markers that would give them away, to make sure that neither their children nor their neighbors betrayed the Jews and the Christian rescuers, and more. The danger was extreme for the Polish rescuers, whom the Nazis saw as members of the "Slavic race," which they considered above the Jews but still inferior and to be made into a slave caste to themselves. Indeed, the Nazis waged a war against the Polish elites, murdering large numbers of the professional, clerical, intellectual, and political classes. Three million non-Jewish Poles perished as well as 3 million Polish Jews.

Who were these rescuers and why did they help? Rescue was a collective affair, depending on situational factors and social belonging as much as perpetration did. The Oliners' ten-year study of rescuers of Jews, in which they interviewed more than seven hundred rescuers and nonrescuers who had lived in Poland, France, Germany, the Netherlands, and Italy during Nazi occupation, emphasizes the many groups and networks that were dedicated to aiding Jewish rescue. In Poland, Zegota, the Council of Aid to Jews, was formed by Catholic intellectuals and members of moderate and leftist Polish political parties; in Holland, it was the National Organization for the Assistance to Divers. In France, various religious groups helped Jews, including the Protestant Comité d'Inter-Mouvements Auprès des Evacués and the Hugenot congregation of Pastor André Trocmé, both in the Haute-Loire region of France; Father Marie-Benoît's network in Marseilles; and L'Amité Chrétienne. In Germany there were the Caritas Catholics; in Italy there was the Assisi Underground. Many secular resistance movements, such as Franc-Tireur, France Combattante, Libération, and Liberté, also participated in Jewish rescue.[15] And there were a number of Jewish assistance, rescue, and resistance organizations as well as significant Jewish membership in the aid and resistance organizations mentioned, but they are not our current concern. In the partisan and resistance movements, Jews and Christians were on an equal basis, with Jews often taking the initiative and leadership in rescue and resistance, whereas in rescue alone, Jews were victims and the Christians powerful givers of aid.[16] The Oliners concluded that "although [the rescuers] often acted in secret, they did not act in a social vacuum. . . . Rescuers lived in a world apart from others, but they were simultaneously embedded in relationships with others. Their activities were secret, but they were not socially isolated."[17]

Nechama Tec found in her study of Polish rescuers that what they had in common was one form or another of outsider identity, a less than full sense of belonging. "Being on a periphery of a community," she writes, "means being less affected by the existing social controls. With individuality, then, come fewer social constraints and more freedom. . . . [F]reedom from social constraints and a high level of independence offer the opportunity to act in accordance with personal values."[18]

The rescuers interpreted the situation in ways that defied and countered the Nazi interpretation and that of its anti-Semitic sympathizers. Being a perpetrator, a collaborator, or a rescuer depended largely on which group one was in or primarily identified with as well as on other factors such as a higher level of optimism that one would not be caught, a kind of derring-do, as well as on opportunity and location. For none of these rescuers did rescue seem to be a matter of thought-out, conscious choice. Instead they characterize rescue as arising from social interpretation, from "characteristic ways of attending to routine events," characteristic ways of understanding them that led to action. And in fact most rescuers said that "they had no choice" and that it was the "ordinary" thing to do.[19] Rescue was the obvious action that resulted from viewing the situation in the way they did, and in the way that those like themselves with whom they identified and with whom they acted in concert did, too.

In the case of the Italians, Susan Zucotti has suggested in her study *The Italians and the Holocaust: Persecution, Rescue, and Survival* that an important contributing factor was the Italian penchant to be skeptical and even contemptuous of authority, as well as a general warmth and humaneness.[20] So Italians, too, rescued according to situational factors, the group context, and an alternative (anti-Nazi) interpretation of the situation. The availability and social authority of a standpoint other than the Nazi one, or an anti-Semitic one that agreed with the Nazi viewpoint, seem to have been crucial conditions for rescue. Essential to the possibility of rescue was the presence of minority groups who could be independent of the groupthink or totalism of the dominant society and of the seduction of the Nazi influence despite the latter's violent attempts at stamping out anything but the official interpretation through language and enactment of that language in institutions, roles, and actions. *Pluralism was the only antidote*; hence in Poland and places like it that were dominated by the murderous ideology and plans of the Nazis and their collaborators, what saving actions there were came from the barely functional social outsiders the Nazis and their Polish followers had not managed to terrorize into silence and inaction. If some few individuals retained independence of mind, it was nevertheless the independence of minority social groups from the dominant one, and of the differences in their history and group experience from those of the dominant group, too, that primarily provided possible alternative interpretations of events from which rescue could arise. One clear indication of the validity of this theory of rescue during the Holocaust is the nearly ubiquitous matter-of-factness of the rescuers' explanations of their behavior. Tec mentions both this and the unplanned, spontaneous nature of the rescue as salient characteristics that fit across the board.[21] Rescuers would reply to questions about why they rescued with the answer that it was the obvious, natural thing to do, the only thing to do. As the Italian diplomat Guelfo Zamboni, who saved hundreds of Jews in Thessaloniki, put it, "But faced with pleas for humanity what would you have done? . . . How else was I to save them?" His words are a perfect example of the felt obviousness of rescue under the circumstances, circumstances that the Nazis tried to enforce as an equal obviousness to murder. The alternative interpretations offered by the social and situational contexts of the rescuers entailed the actions they took. Just as perpetration flowed from Nazi language and its institutionalization, so did rescue from the alternative perspectives and social identities. Of course, perpetration and rescue are neither equivalent nor parallel: the experience of lying and murder is not the same as the experience of rescue. The emotions are not the same, and rescue produces heroes while perpetration produces villains, even if the individuals involved in them began once upon a time in similar innocence.

So we can see that one of the important lessons of the Holocaust is that social systems can be good or evil and rope people into good and evil. They are not mere aggregates of good people or bad people. In a social system everyone bears responsibility, but those who created and maintain the system bear the most responsibility. This is Zimbardo's "bad barrel." Zimbardo identifies three levels: the systemic, the situation, and the individual. Good and evil are characteristics of all these levels, not just of individual people or their decisions. In fact, the systemic level is the most important for ethics, for it is there at the structural level that ethics is mostly determined. So we need to bring our moral valuations to that level first and foremost. Zimbardo likens the shift in viewpoint in psychology to a shift in medicine from an individual health model to a public health model—for example, how systems of clean water and modern sewage reduce disease or the distribution of health resources fairly and equally across the society produces the most widespread improvements in health. The systemic view is what I am advocating, too. But that systemic viewpoint cannot be like the Nazi shift from healing individuals to healing a metaphorical corporate body that is the nation. Instead, the proof of the moral quality of the system has to be in the individuals it produces. The criterion is the benefit to individuals, on one hand, and the moral quality of individuals, on the other. A good system will benefit all the individuals and also produce individuals who do good rather than evil. And that's exactly what we found in Denmark and with the Italians, especially in the areas they occupied during the Holocaust. Both systems were good for the many and the weak, and both produced good, humane individuals who did more good than they ever would or could have done on their own. Of course we must realize that the capacity to do good is sometimes limited by the gun at one's head; such is the case for those extraordinary Poles who were contending with an evil social system.

It is not that individuals do not bear responsibility for their actions. They do. Yet responsibility must also be more widely shared by going up the hierarchy of the situation and the system. The producers of the social system—the architects of its implementation, its overseers—bear more responsibility than its lowest-level functionaries who carry out its effects through their actions. These last in many respects have the least freedom of action in the entire setup.

If we look back at character education in schools, the model was flawed in many ways. The basic bearer of moral valuation is the school as a social system. And everyone in that system is part of the system's own moral character as a system of relations and purposes. So the whole idea of inculcating into kids a kind of polite overlay that distracted from and masked the moral nature of the school itself, whether good or bad, was flawed. A school is not an aggregate of individuals being nice and making morally good personal choices; its moral character is fundamentally in the social world it constructs and which shapes the roles people take, how those fit together, what kinds of relations are produced by the incentives the system puts in place, how those incentives are played out (competition versus cooperation, for example) and the moral value of the purposes they enact, and the ways power and hierarchy are established and distributed.

But what are these mysterious social and situational forces that drive so much of our behavior, making of us heroes and villains, often against our best intentions and with so little awareness of their hold upon us? The next chapter addresses just these questions by offering a survey of what social psychology has discovered about the situational and social shaping of action. I'll end here with a description of the small remote rural village of Le Chambon-sur-Lignon in southern France, which became a refuge for Jews after the fall of France to the Nazis. The story is important because it brings together and illustrates in sharp outline the several factors that came together to produce an instance of extraordinary goodness in a time of peril. This community of approximately five thousand opened their homes, hiding Jews in their houses, farms, schools, and public institutions for months and even years and then finally getting them over the Alps to Switzerland. They saved around five thousand Jews.

Le Chambon is about 350 miles from Paris in what was then an isolated area of south central France, a poor farming area. By 1940 France was home to approximately 350,000 Jews, half of whom had fled to France as refugees as the Germans advanced. France had traditionally been a place of asylum, and Le Chambon had a history of granting safety to those in need, sheltering refugees from the Spanish Civil War in the late 1930s. Within a few months of Marshal Philippe Pétain becoming head of the Vichy government after the surrender of France to the Nazis in June 1940, a set of anti-Semitic laws was signed into law removing Jews from French public life. An official anti-Semitic propaganda campaign was also put in place. In 1941 the

French police under Vichy orders began the roundup of "foreign" Jews, who were imprisoned in internment camps before being handed over to the Nazis for extermination.

The village of Le Chambon was home to a minority community of French Huguenots, the descendants of Calvinists who had converted to Protestantism in the sixteenth century and had a long history of persecution and discrimination within Catholic France. In 1934 the village hired a new pastor, André Trocmé. Trocmé was a conscientious objector and a pacifist whose views were unpopular not only in general in France but even within the Protestant community. He and his wife, Magda, an Italian, were worldly, highly educated people who found themselves relegated to this rural spot because of the marginality of their views. Trocmé founded a private boarding school, the Collège-Lycée Cévenol International in 1938, bringing in fellow pacifist minister Edouard Theis as director. The Trocmés had been wildly unpopular for opposing the French fight against the Nazis, and once France had fallen they were equally unpopular for their outspoken opposition to Vichy's collaboration with the Germans. In Le Chambon they insisted that Pétain's was a "dishonorable armistice" and viewed the Vichy regime and Pétain at its head with utter contempt. The day after France signed the armistice, the two pastors, Trocmé and Theis, delivered a Sunday sermon to rouse the community to action. A copy of that sermon survives, and in a 1989 film, *Weapons of the Spirit*, a section of it is read aloud in English translation by the Trocmés' daughter, by then middle-aged: "The duty of Christians is to resist the violence that will be brought to bear on their consciences through the weapons of the spirit. We will resist whenever our adversaries will demand of us obedience contrary to the orders of the Gospel, and do so without fear but also without pride and without hate."

Magda Trocmé, in her interview in the film, offers this explanation for their opposition to Vichy and for the actions to save Jews that she and her husband spearheaded:

> If we didn't obey Pétain or believe what the Germans were saying, it's not because we were so clever. We just had different pasts. I was Italian. I'd known the rise of fascism. My husband's mother was German. We had many relatives there. . . . We'd seen Nazism develop. So we reacted not just as Christians but also because of our backgrounds. In a way we were prepared and the village was prepared by its Huguenot history.

The Huguenot history was important in several respects. First, it was a history that emphasized the Hebrew Bible (the Old Testament), and the Calvinists read themselves into the history of the Jews and identified, I think, with the persecuted Jews. One of the simple rural rescuers speaks in the film of how, when speaking to other rescuers, they referred to the Jews as "Old Testaments" to hide from public disclosure what they were doing. One or another would say, for example, "I have room for three Old Testaments." Second, as a minority community with a long history of discrimination and marginality, the Huguenots were skeptical of normative French authority and viewpoints. Third, they had their own alternative sources of moral authority and of authoritative opinion and interpretation of the situation and of the world. Finally, they had a history of granting asylum and a system of rescue that was already in place. And of course they were blessed with outstanding, inspirational, and courageous moral leadership—that was incredible luck. All these factors came together to produce the conditions for rescue. The interviews that Pierre Sauvage had with the rescuers forty-five years later also reflect the matter-of-factness with which these extraordinary rescuers describe what they did. This, too, is crucial, because it indicates that this community had a different but to them obvious and natural way of interpreting events, which automatically led to the actions they took. One woman who ran a boardinghouse for children at the time comments in the documentary, "It happened so naturally. We can't understand the [present] fuss. It happened quite simply. I didn't have that many. Usually they were passing through. They'd often arrive at night . . . they slept on the floor. We'd manage somehow. We gave up our beds when nothing else was left. . . . I helped simply because they needed to be helped." Another rescuer comments, "The Bible says to feed the hungry and visit the sick. It was a natural thing to do." And another rescuer, in response to Sauvage's question of why they rescued, shrugs and says, "I don't know. We were used to it." The case of Le Chambon clearly indicates that if a society is to be humane and decent, it is absolutely vital that it nurture minorities and support the broadest pluralism. Because of their history, the people of Le Chambon could see what others could not and act differently when others went along. They had clarity of interpretation, of moral vision, when most others were taken in.

Le Chambon brings to the fore a crucial insight about why people do good and commit evil: people interpret the world through others' eyes. Interpretation of the world is a social phenomenon, and not for the most part a private one. (As Spinoza said, independence of mind is a private virtue—and a rare one, too, he believed.) So the crucial questions become: Whose eyes? Which authority structure? Which authoritative interpretation of the world? We are inevitably informed by our contexts, and generally act as they induce us to.

Looking at ethics through the lens of free will decision making assumes that situations are objective and that it is the choice of action that is good or bad. But the evidence from the Holocaust—and from social psychology—suggests otherwise. It suggests that situations are subject to different interpretations and that interpretations are largely driven by community and social authority.

3

The Overwhelming Power
of the Group and the Situation

What social psychology has given to an understanding of human nature is the discovery that forces larger than ourselves determine our mental life and our actions—chief among these forces is the power of the social situation.

—MAHZARIN BANAJI

Evil that arises out of ordinary thinking and is committed by ordinary people is the norm, not the exception. . . . Great evil arises out of ordinary psychological processes that evolve, usually with a progression along the continuum of destruction.

—ERVIN STAUB

Values of independence are phony, really. There is no such thing. There can be no pride in living without oxygen. We're not made that way. It is nonsense to try and give up symbiosis and become an independent self.

—HEINZ KOHUT

The Social Psychology of Situations

Social psychologists offer ample reasons to believe that the moral lessons we gathered from the studies of the Holocaust are not unique or aberrations but instead what goes on all the time. Social and situational forces rather than free will decision making and stable personal character largely determine what people do, what actions people take, whether they harm or help. In the

Holocaust social and situational forces were more extreme and blatant cases of business as usual, so they were more visible to the eye. Some early social psychology experiments came about in an effort to understand what had happened in the Holocaust, especially to understand why seemingly normal people became perpetrators of mass murder or engaged in tasks contributory to mass murder. Stanley Milgram and Philip Zimbardo are the two social psychologists, independently of each other but each over a lifetime, whose experimental work has focused on the social and situational forces and conditions that lead people to engage in harming others, in inflicting pain, doing evil. It turns out that Milgram and Zimbardo were high school class-mates at James Monroe High School in the Bronx, both graduating in 1950, and even back then both were fascinated by the power of social situations to shape people's behavior. Milgram, being Jewish, was riveted by the Holocaust and how it could have happened, how Germans could have so readily and obediently murdered Jews. In school, Zimbardo says, he and Milgram would hang out and talk about the successes and failures of their friends and ac-quaintances. "Not coming from well-to-do homes," he says, "we gravitated toward situational explanations and away from dispositional ones. . . . [We saw that t]he rich and powerful want to take personal credit for their success and to blame the faults of the poor on their defects. But we knew better; it was usually the situation that mattered." Zimbardo comments that it was only after Milgram's death that he discovered they had both been fascinated and even influenced by the television show *Candid Camera*, which began air-ing in 1948. Its creator, Allen Funt, would devise scenarios that would chal-lenge the usual perceptions or ways of acting of unsuspecting passersby, whom he would be filming with a hidden camera. Then we TV viewers, just like psychology experimenters observing from behind one-way glass, would watch surprising human behavior from behind our hidden "screens." Zim-bardo remarks that Funt was "one of the most creative, intuitive social psy-chologists on the planet" and exposed "truths about compliance, conformity, the power of signs and symbols, and various forms of mindless obedience." When Zimbardo interviewed Funt for an article in *Psychology Today* he dis-covered that Funt had been a student of Kurt Lewin, one of the great found-ers of experimental social psychology, at Cornell and had been as influenced by Lewin in his own creative way as had Milgram and Zimbardo in their disciplinary way.[1]

In his famous experiments Stanley Milgram devised ways to test people's normal obedience to authority and willingness to harm others on that authority. Milgram, a psychologist at Yale, set out to test the obedience of normal adults and whether they would be willing to inflict pain on others when instructed to do so by an appropriate authority for an apparently appropriate purpose. These have become known as tests of blind obedience to authority. Milgram comments about his own motivation that "the impact of the Holocaust on my own psyche energized my interest in obedience and shaped the particular form in which it was examined."[2]

Working at Yale University, Milgram in 1961 advertised in the local papers for subjects for his initial experiment, which was characterized as a scientific experiment to study memory and learning. Those accepted would be asked to volunteer for an hour at their own convenience and would be paid for their time. The subjects chosen came to the university laboratory and were met by a researcher in a lab coat who told them that they would be part of an experiment to determine ways to improve learning and memory through the use of punishment. The subject was told that he would be the "teacher," and there would be a "learner" on the other side of a screen or wall to whom he would say a series of words that the learner was to remember and then recite. If the learner produced the correct words, the teacher was to say "good" or "that's right," but if not, the teacher was to press a switch on a device that delivered an electric shock to the learner. Each wrong answer meant that the next shock would be 15 volts higher than the previous one, and there were thirty levels of shock, each marked with a phrase such as "strong shock," "very strong shock," and so on, up until "danger, severe shock" at the twenty-fifth level and "XXX" at the highest levels (435 and 450 volts). The subjects were told that the shocks might be painful to the learner but would not cause any real harm. The researcher in the lab coat then had the two participants draw straws to see who would be the learner and who the teacher. This, of course, was rigged: the learner was a confederate of the researcher and would merely pretend to be shocked, since the shocking device did not actually give shocks. At a certain level of shock, the learner was instructed to complain of the pain; at a higher level, the learner was to admit to a heart condition. As the shocks got higher the learner was to shout out in pain and at a very high level say that he couldn't stand the pain, his heart was hurting, and he wanted to get out of there. After 300 volts the learner was no longer to complain but maintain silence. During the experiment the researcher in the lab coat calmly but insistently explained to the experimental subject that the experiment must go on and that he was to follow the rules. In the initial version of the experiment the lab-coat-wearing re-searcher remained in the room with the volunteer subject.

Milgram wondered how many of his volunteer subjects would actually go along with it and how far up the scale of pain and harm they might venture. In anticipation of carrying it out, Milgram presented the plan to a group of forty psychiatrists and asked them to predict how far they thought these average American subjects would go in punishing a learner. Most of them predicted that less than 1 percent of the subjects would go all the way to the end of the scale and deliver the highest voltage. Only sadists, they predicted, would go all the way, and most people would quit at about level ten. This prediction turned out to be wildly off base. In fact, about two-thirds of the subjects (65 percent) went all the way up to delivering the maximum shock of 450 volts—beyond the point at which the learner had stopped responding and crying out in pain and had become ominously silent. The most obvious assumption for the subject to make at that point was that the learner was now unconscious. If a subject expressed concern about being responsible for hurting the learner, the researcher reassured the subject that the researcher was taking responsibility; if the volunteer subject expressed concern for the learner and the desire not to continue, the researcher told him to continue, and—astonishingly to everyone concerned, and to us today—a majority continued. Milgram repeated this experiment with nineteen different small variations over the course of a year. For example, he added women; he varied the distance of the researcher from the experimental subject; he introduced confederate peers to the subject who rebelled; he even took the whole ex-periment out of Yale University and set it up in a run-down office in Bridge-port, Connecticut, to see if subjects would still comply with scientific experimenters in a private experiment without the prestige of Yale behind it.

By introducing small variables, Milgram found that he could get 90 per-cent of the subjects to comply if the actual pressing of the switch was done not by the subject but by another person (a confederate of the research team). But if he introduced confederates who refused to go along, 90 percent of the experimental subjects would follow their lead and refuse. If the learner was more remote, compliance was increased. All in all, Milgram tried the experiment on more than one thousand Americans of all kinds. The results defied expectations—of the psychiatrists and certainly of people themselves—but were very powerful and reproducible. Since then, eight replications of the study have been done in the United States, and nine in European,

African, and Asian countries. There was comparable obedience in all the different countries and over different time periods.[3]

The Milgram experiment, like the Stanford Prison Experiment, came under scrutiny for ethical concerns about human subjects research, since the subjects were put in a situation of significant (although brief) stress. Nevertheless, a slightly modified version of the Milgram experiment was repeated in 2006 by psychologists at Santa Clara University for ABC television's newsmagazine program *Primetime*. In this version, the maximum shock was set at 150 (rather than 450) volts, a level that had been a turning point in the original experiments: almost all of the experimental subjects who administered 150 volts went all the way. (Of course, in the 2006 experiment these were pretend shocks, as they were in the Milgram experiment.) The initial reason for conducting the Milgram experiment yet again was the suspicion that times had changed and people were no longer obedient in the way that they had been forty-five years earlier. Perhaps that kind of obedience had been an artifact of a particular era. But that suspicion turned out to be totally wrong, and the results disturbing: the subjects in 2006 were just as obedient in the same percentages and under the same conditions as those nearly half a century earlier. Astonishingly, just as Milgram had discovered, no intimidation or fear, no political conditions of war or threat were needed to induce normal people to harm others and to keep going even amid victims' cries of pain and pleas for release. Nor was isolation or brainwashing necessary. It didn't even take Nazi-like conditions to produce Nazi-like behavior in normal people—and within an hour. The results this time around, forty-five years later, were as chilling as Milgram's.

The *Primetime* film footage captured a number of the volunteer subjects as they went through the experiment and then reflected on what they did and said during the experiment in interviews with *Primetime* anchor Chris Cuomo. Interspersed with both the footage of the experiment and the interviews with the subjects are Cuomo's interviews with social psychologist Jerry Burger. The setup followed the original Milgram design: there was a room with a lab-coat-wearing researcher sitting at a desk behind the subject, who was sitting with his or her back to the researcher and facing a large piece of equipment with a scale of voltage up to 150 volts and switches at each gradation. This electric shock device was in front of a wall, and the "learner," a sixty-year-old research confederate, sat behind the wall, out of sight. Before the experiment began, the subject was brought by the researcher to the other side of this wall to meet the learner, who mentioned that he had a heart condition and expressed some anxiety about the experiment's effect on his health. The subject looked on as the learner was strapped into the chair and fitted with electrodes.

The subjects were eighteen men and twenty-two women—a diverse group in terms of ethnicity, age, profession, and socioeconomic class. Sixty-five percent of the men and 73 percent of the women continued to shock the learner up to the maximum 150 volts, the voltage Burger told Cuomo "is something of a point of no return." The typical response of the subjects when the learner first objected to the shocks or moaned in pain at the administration of the first "shock" was to turn back to the researcher, who would then indicate that the subject should continue. As the learner expressed increasing distress and pleaded to be released, the researcher also became more insistent that the subject continue, emphasizing that although the learner might be feeling pain, there would be no danger to him. During the experiment the researcher retained a thoroughly dispassionate and professional demeanor. Subject after subject looked quite concerned and disturbed; a few subjects even become quite agitated, though one or two showed almost no distress. Yet both those who displayed discomfort and those who didn't continued to administer the shocks at the calm insistence of the researcher.

The subjects' comments during the experiment and their self-reflections afterward are telling. At one point Chris, a fifty-year-old grade school teacher, asked about the risk (for herself, presumably) of a potential lawsuit. The experimenter replied, "If anything happens, I am responsible," to which Chris replied, "[That's what] I needed to know." Later the *Primetime* interviewer, Chris Cuomo, asked, "Why was this what you needed to know? Having this guy in the lab coat sort of divorced you from your decision-making power?" And she responded, "Oh, sure. Just like when I'm told to administer the state [education] tests for hours on end." Cuomo asked, "You're doing your job?" and Chris responded, "Yes, [I'm] doing my job." Similarly, another subject, a thirty-nine-year-old electrician, tells Cuomo in the exit interview, "I was just doing my job." The electrician concludes that "having the experimenter right there next to me had a lot to do with it." We may have thought that the echoes of the Nazi era were faint and gone forever, but here they are, and not in a wartime situation involving great fear or military discipline, but in an everyday academic setting with nothing at stake, no coercion nor potential reward, no hope of advancement resting in the balance—nothing is at stake for these subjects. Yet they do their job even when it's clearly causing distress and potential harm and in the face of the screams of pain of someone they initially were introduced to and whom they have been told has a heart condition. Here is our first telling evidence that the social forces of the Holocaust are not just about genocide but about our everyday lives.

Chris Cuomo asked social psychologist Jerry Burger to explain why he thought the subjects were willing to cause obvious pain in the way they did here and in the way they did in the original Milgram experiments. Burger replied, "The power that [the experimenter] has in this situation is that in part he's an authority figure. We're all trained a little bit to obey authority figures. But also he's the expert in the situation. He's the one who knows the machine." The major finding according to Burger was this: "We began to see a pattern. The majority of those who continued to the highest shock refused to take responsibility for the effects on the learner." One subject Cuomo interviewed said, "It's not my responsibility. We're volunteers." Another said, "I'm not in control. I'm just a conduit," while a third remarked of the learner, "He chose to be there to take the shocks. His choice." Cuomo asked Burger if he could predict after a while who would shock to the end and who would not. Burger replied, "It was impossible to tell. I tried to guess. I tried to look for signs, body language, anything to guess who's going to continue and who's going to stop. And this tells me that it's not there are some kind of people who are different from the rest of us. It tells me that probably all of us are capable [of going the whole way]."

There were, of course, some who refused to go on, who walked out. One of these subjects, forty-six-year-old Fred, worked for a software company and was a self-proclaimed nonconformist. Initially Fred did not stop even though the learner screamed out when shocked. To explain that, Fred commented, "It's two consenting adults deciding to do this. Not until at some point one of us has to say stop." Cuomo asked, "So you put all of the responsibility about when is the right quitting point on the learner?" Fred: "For quite a lot of this, yes. There is going through my head, 'How long are we going to do this?' I am waiting for the other person to say stop. I don't know where I would just have said stop on my own." At 150 volts, the learner said he wanted out, and at that point Fred quit despite the urging of the researcher to continue. Then transpired the following exchange between the researcher and Fred:

Fred: He said that's all. We're not doing any more of this.

Researcher: I want to remind you that the shocks may be painful but they are not harmful to him.

Fred: It doesn't matter. I'm not a sadist. He has said no more. He is not agreeing to this. I'm not agreeing to this.

Researcher: The experiment requires that you continue. The next item is . . . [The researcher reads the next series of words]

Fred: The experiment allows me to walk out at any time, and I will walk out if you want to push this.

Researcher: While that is correct, it is absolutely essential that you continue. Okay, remember whether the learner likes it or not—

Fred: The learner doesn't like it. I'm walking out.

Researcher: We can discontinue with this patient. That's just fine.

Fred: Okay. When somebody says no, it's no.

Researcher: Okay.

Fred then walked out. In the exit interview, Fred reflected upon his words and actions: "It was obvious at this point wrong to go on. It's not even an intellectual debate." And Fred's parting words pierce to the heart of the matter, "Do you have a brain? Shouldn't you use it, too? Somebody walks up to you and says the blackboard is white and they're wearing a lab coat, do you believe them? No, you've got your own eyes." Another subject who resisted and walked out had a similar comment on his own refusal: "I would have felt that I would have hit the switch that had killed him. If he died, I would feel a deep responsibility." Another subject who refused answered the researcher's comment "You don't have a choice. You must continue" with these words: "No. I do have a choice. I'm not continuing." And another subject responded to a similar statement by the researcher in this challenging way: "You gonna hold me down?" Those who showed some resistance early and those who felt directly responsible for their actions were those most likely not to continue to the end and shock at the highest level. *Primetime* drew the connection to the Nazis obliquely by commenting that the excuse of "just following orders" has never held up in court, and then posed the question of why the military guards at Abu Ghraib did this to defenseless prisoners. Here *Primetime* inserted a clip of Private Lynndie England, sentenced to three years in prison and dishonorably discharged from the military for her crimes at Abu Ghraib. England told the camera: "We didn't feel that we were doing things we weren't supposed to because we were told to do them."

Is the lesson of the Milgram experiment the danger of unreflective obedience to authority? Certainly Milgram took it that way. But there has been discussion since the initial experiments about what they actually document. Certainly they point to the overriding importance of situational processes over individual character and free will decision making. But is it all to be chalked up to obedience? I have already quoted Jerry Burger as pointing to the role of the expert in defining the situation and its rules. And Burger also emphasized the matter of the diffusion or assignment of responsibility to someone else other than the actor. Phil Zimbardo points out that in the Milgram experiments he was always struck by the difference between dissent and disobedience. While many participants dissented and grew upset at giving electrical shocks to the learner, very few were actually disobedient and stopped. This held true in the Stanford Prison Experiment as well, Zimbardo says. Commenting on conversations that he had over the years with Milgram, Zimbardo mentions that he once raised the question whether any of the dissenters ever ran to aid the victim in the other room behind the wall. Milgram said it had never happened. Zimbardo notes that this failure points to an even more chilling conclusion than Milgram's: "That means that [Milgram] really demonstrated a more fundamental level of obedience that was total—100% of the participants followed the programmed dictates from elementary school authority to stay in your seat until granted permission to leave."[4]

Obedience must be a much deeper, ubiquitous, and pervasive phenomenon than it seems. The responses of the subjects perhaps give us the first hint that we do not yet understand what's really happening. For example, none of them asked whether they had permission to stop. Instead they raised the question of their *responsibility*. In other words, they implicitly understood their actions as part of a collective endeavor that was primarily not their own but to which they were contributing in a tangential way rather than a central one. This tells us something very important about the nature of most action, perhaps all action—namely, that it is tacitly understood as contributing to ongoing projects and institutions and social and cultural worlds not of one's own making. Where we sit in the given context to which we are contributing we take to implicitly define the level and nature of our responsibility. And we think of responsibility as distributed within the whole and belonging especially to its leaders. We don't think of our actions as morally discrete and as free choices for the most part, but as ongoing expressions and continuations of cultural and societal projects and institutions that began before us and continue after us. Whatever we mean by obedience seems to involve this pervasive kind of embeddedness and contextualization.

In Zimbardo's analysis of the social and situational forces operative in the Stanford Prison Experiment; in Robert Jay Lifton's analysis of the Nazi doctors; in Hannah Arendt on the banality of evil; in the now vast corpus in multiple languages of Holocaust memoirs, fiction, and films; in the sizeable literature on the psychology of genocide, of Hitler, and of the Nazi perpetrators; and in the more general social psychology literature we find some disentangling of the threads in the web of obedience. To a selection of these we will now turn. Our aim here is not to come to a definitive or even fairly reasonable explanation for the evil of the Holocaust. Our aim, as it has been all along in this chapter and in the previous one, is to use the Holocaust and both perpetrators and rescuers as examples of the social character, social shaping, of action, bad and good. Here our purpose is to begin to tease out some of the aspects and properties of this social character and determination and in so doing also expose that it pertains to business as usual and not just life in extremity, whether good or bad. The lessons of the Holocaust are our lessons as well.

Social psychologists have observed the general cultural penchant—which infects in their view not only lay beliefs about behavior but also many professional and academic disciplines—to attribute motivational efficacy to individual character traits and free will decision making rather than to group social processes. They call this the *fundamental attribution error*. We look at individuals as autonomous and as internally motivated—that is, people are motivated by character, reasons, and will, things that drive us consistently whatever the situation turns out to be—and we expect that different individuals will respond differently in the same situation. But we've learned that behavior is far more consistent across individuals *within* a given situation and far more varied between different kinds of situations than we tend to think. People do act consistently, but this consistency occurs across similar situations, not across different situations.[5] Furthermore, that consistency is more often due to situational factors than to stable character traits or free will independent decision making.[6] The places, social settings, and historical contexts we inhabit: these shape our interpretation of situations and of rewards (and punishments) and drive our behavior.[7]

Obedience: What Is It, Really?

The Milgram experiments sparked thoughtful analyses and reflections on the nature of obedience that introduced levels of nuance and depth that help explain Milgram's findings. The work of two of Milgram's students, John Sabini and Maury Silver, is particularly thoughtful and noteworthy. They introduce into the analysis of obedience questions about the role of hierarchical position and hierarchical institutions, of standards of civility and politeness, and of the presumed legitimacy of institutions and practices (in this case, those of science). They also bring to the analysis further distinctions in the mode or style of obedience and disobedience, raising the question whether certain kinds of disobedience were in some ways actually modes of obedience.

Sabini and Silver once asked Milgram if any of the minority of subjects in the experiments who ultimately refused to continue shocking ever confronted Milgram or his research associates with the morally problematic character of a psychology experiment that caused pain. They learned that none of these subjects tried to rescue or help the learner, as Zimbardo found out, nor did any raise a moral qualm. Intrigued by this finding, they hypothesized that there are situations in which intervening with a direct rescuing action implies a moral reproach. Both intervening to save the apparent victim and expressing disapproval would have amounted to a moral critique of the experiment and a reproach of the experimenters, and no one ventured to do that. Even considerable emotional distress on the part of a subject in shocking the learner did not bring about the saving action or a moral confrontation.[8] But why? Sabini and Silver suggest that when a rational and seemingly common-sense response is absent, there must be an inhibition in place that precludes it. What is the nature of this inhibition? A clue was that when the researcher was absent from the scene but phoned in his instructions, the rate of obedience fell from 65 percent to 20 percent. One subject made a remark to the researcher that opened a window into what was going on, saying, "I don't mean to be rude, but I think you should look in on [the learner]." Moral reproach would be considered rude behavior to someone's face. Politeness was trumping moral concerns. Civility was trumping morality. When the researcher merely phoned in his instructions, the danger of noncompliance was lessened since there was no one present to defy or shame. This is a chilling discovery.

But what do politeness or civility, some form of the rules of etiquette in social situations, involve? What do they guard, and what do they guard against? Sabini and Silver point out that moral reproach is deemed appropriate from a hierarchically higher person to a hierarchically lower one. Someone with greater authority may reproach someone of lesser authority—parents may reproach children, clergy may reproach the congregation from the pulpit, teachers may reproach pupils, bosses may reproach employees. In matters of moral reproach, people defer to the relevant and proper authority: an example that Sabini and Silver propose is stopping someone from smoking on a bus. Most would defer to the driver to enforce the rule. Another consideration here is who has the right and authority to define or label the behavior. We all think harming others is wrong, but is a given situation a case of real harm? And who decides? Here doubt enters in, and self-doubt, according to Sabini and Silver. In novel circumstances, when confronted with an authoritative insider who has expressed no doubt about the situation and defines it counter to common sense, one's own doubts and self-doubts about one's counterinterpretation can be paralyzing, they suggest. The inhibition against confronting someone with a moral reproach is matched only by the joy in moral reproach in absentia, namely, gossip. But if the person involved learns of that gossip, it causes acute embarrassment. So perhaps now we have gotten down to the heart of the matter: it is the fear of embarrassment that tyrannizes the social situation, making cowards of all but the best of us—and perhaps sometimes of all of us, if we look at the discouraging evidence from the Milgram and Zimbardo experiments. But there is also the hierarchical dimension: valuation for the most part is a top-down affair.

Of course, there is more. In "On Destroying the Innocent with a Clear Conscience: A Sociopsychology of the Holocaust," a chapter in their book *Moralities of Everyday Life*, Sabini and Silver try to sort out the components of the obedience to authority that the Nazis induced with such extraordinary effectiveness. After the state-sponsored mob violence of Kristallnacht, the Nazis concluded that whipping up shared rage was to be replaced by the bureaucratization of evil: organized routines and a hierarchy of responsibility. This was the institutionalization of murder within hierarchies responsive to the will of the ultimate authority. Sabini and Silver propose to treat the Holocaust as "the social psychology of individual action within the context of hierarchical institutions." So they suggest that the face of the Holocaust is Eichmann, the colorless bureaucrat, rather than the raging rioter of Kristallnacht.[9] In illustration of the point they quote the diary of Hans Hermann Kremer, a physician at Auschwitz:

> *September 6, 1942.* Today, Sunday, excellent lunch: tomato soup, half a hen with potatoes and red cabbage . . . In the evening at 8:00 hours outside for a Sonderaktion ["special actions," i.e., mass murders, including the burning alive of prisoners, especially children, in pits, which was particularly practiced at Auschwitz].
>
> *September 9, 1942.* . . . . Present as doctor at a corporal punishment of eight prisoners and an execution by shooting with small caliber rifles. Got soap flakes and two pieces of soap. . . . In the evening present at a Sonderaktion, fourth time.[10]

The authors say that what needs to be explained is how murder came to "have the same importance as soap flakes." It seems to me that the diary itself is not naive but a kind of aggressive enactment and performance of the rationalization of the bureaucratization of evil.

In the Milgram experiment subjects often asked who was responsible for the administering of shocks (and for the potential harm), and they were always reassured by the researcher that he accepted full responsibility; that would inevitably free them to go on shocking. It is this anomaly of the feeling of being let off the hook while still performing actions for which one is held responsible as actions of one's own that gives us a window into the nature of action in a bureaucratic and hierarchical context. In an organization, section heads are responsible for planning, but workers are responsible only for carrying out the plan. The boss is responsible for the plan and subordinates only for the execution. The internal rules and structure of command within a hierarchical organization obscure and subvert individual moral responsibility. Moreover, the bureaucratization of evil separates the enactment of it from the intent to do it. So large numbers of people can be engaged in harm or even murder when their basic intent is to do a job well for an organization they respect rather than to be murderers. Consequently, they don't *feel* like murderers even if murder is what they're in fact doing—and doing it as the main goal of their routine paperwork, for example, like those who had to register all the Jews in a district and make up comprehensive lists, rather than as a kind of collateral damage, inadvertent harm in the pursuit of a good aim. Milgram's subjects, even those who went all the way and delivered what they thought were electric shocks that perhaps even rendered the learner unconscious, no doubt saw themselves primarily as contributing to science rather than as torturers. And most of those who shocked to the end exhibited considerable distress, clearly not liking to inflict pain, but perhaps seeing it as an unintended concomitant that they had to put up with. (Here we see the source of the perpetrators' casting *themselves* as the victims.) Sabini and Silver suggest that "bureaucracies have a genius for organizing evil" because everyone except the deviser of the plans and projects is let off the hook and feels free of responsibility, their personal inclinations and intentions having been checked at the door. In Nazi Germany the entire society was like one enormous bureaucratic system directed toward the centralized purpose of murdering the Jews.

When a bureaucracy is entirely geared to an immoral aim, the normal relationships that we expect to hold between action and moral values, between temptations and desires, are subverted. Under those circumstances people's perceptions of right and wrong become corrupted. For example, moral duty, conscience, becomes attached to doing evil and temptation to doing good. In the Milgram experiment, the subject is tempted *not* to shock the learner but resists temptation and shocks him. Sabini and Silver point out that it is not a desire to please the researcher that is operative here but loyalty to the institution of science and a commitment to carrying out the experiment. So the subject is not obeying the experimenter as such but following what he regards as the legitimate dictates of science. Not even defiant subjects questioned the experimenter's morality; indeed, they often apologized for ruining the experiment.

The presumption of the legitimacy of institutions is further strengthened by the fact that people are gradually socialized into them. Newcomers are brought along and learn how things are done. Moreover, everyone else around the newcomer seems to accept how things are done and what they mean, so they represent a norm and a claim to legitimacy that cannot be questioned. Here we're back to the issue of moral reproach and embarrassment. Those who know the ropes are seen to go along, whereas a newcomer has little authority or legitimacy in offering a rebuke or a criticism. Those already there own the system—and clearly this had an effect on the Milgram subjects. In the Stanford Prison Experiment it was another psychologist who came along, looked at the situation, and blew the whistle. Although it was certainly brave of her to do that, she also had the credentials, the professional authority, and the knowledge to be able to assess the legitimacy of a psychological experiment. That happens rarely.

The hierarchical institutional context may have been the most important factor in the social psychology of the perpetration during the Holocaust, and in the Milgram and Zimbardo experiments as well. But hierarchy is not the only word on the subject. In what follows I will touch briefly on a couple of the processes operative in extreme situations but focus mainly on those that are found both in normal life and in extremity, for ultimately our aim here is not to explain the Holocaust but to discover what we can generally about when, why, and how people do good and commit evil.

Explaining the Stanford Prison Experiment: Power, Conformity, Obedience, Deindividuation, Dehumanization, and Bystander Inaction

In *The Lucifer Effect* Philip Zimbardo examines the social forces at work in the Stanford Prison Experiment. The first dynamic that he saw operative was the desire for social power, the desire to be part of the in-group and not the out-group. To be part of the in-group gave the guards instant social status and an enhanced sense of self. The other side of the coin was the terror of being left out, the fear of rejection. Zimbardo says that this motive is so powerful it accounts for all kinds of things we see around us, not only cults but also bizarre fraternity rites and even the lifelong climb up the corporate ladder. It is a profound inner motive that ties each of us to the group. The next motive that Zimbardo cites is the one that I have repeatedly called attention to: the social construction of reality, the authoritative interpretation of the situation. This includes Sabini and Silver's presumed authority of institutions.

Zimbardo remarks that the ancillary players, the "onlookers," those who came in for a particular purpose, "all accepted my framing of the situation, which blinded them to the real picture."[11] Zimbardo teases out the "presumed legitimacy" that we grant institutions in the cognitive terms of the interpretation of situations. We accept the interpretations of situations that the authoritative definers, the higher-ups in the hierarchy, offer, and which institutions and situations embody and bring to life. Just as Milgram's lab-coat-wearing researcher was given the benefit of the doubt, so was Zimbardo himself. Here we have a lesson in the sociology of knowledge or belief: we do not routinely assess and define situations and contexts for ourselves. They come ready to hand.

To make matters worse, coupled with the social nature of our understanding and the social pressures that drive us, we have, he says, "egocentric biases" that blind us to this very fact. So not only are we likely (or inevitably, to a large extent) to be driven by social forces; we are just as likely to deny that they exist and to be convinced that we are independent creatures making personal decisions. And this blindness further intensifies the power of the social forces upon us, making us even more vulnerable to them because we are so sure that nothing of the kind could ever affect us. We overestimate our own power and underestimate that of situations. We fail to see our behavior or that of others within its situational context (the fundamental attribution error). It is the tendency to attribute to individuals, to their character and their "free" decision making, what is actually attributable to the features of situations. We attribute the source of action to people's interior qualities and autonomous will. We think of the individual as the "sole causal agent" when it is the entire context that produces the actions of individuals.[12] We think of our minds as of our own creation and under our control, but what we think and believe are largely socially, situationally, and culturally-historically determined. I am always amused by the irony that many of my students express nearly indistinguishable beliefs about their religious attitudes and convictions but always preface their comments with the caveat that these are their very "personal" beliefs. They never seem to catch the irony that they are all expressing precisely the same sentiments and attitudes. So I have come to realize that what they mean is that they are expressing something about themselves that is usually kept private—we think of religion as a private affair in America—and is not only an intimate part of their identity but an important one as well. But what is salient for us is that the kind of identity and self-understanding the students are expressing as "private" is in fact profoundly socially constructed and determined, for they are all expressing the same position. They are revealing that they are all part of a moment in a particular place with a certain history and a certain present. It has become their inner, mental world, not just the outer world, and has permeated their sense of self. We are all like this. We all think that we mostly originate our thoughts and actions and that others do, too, even though we and they are profoundly embedded in and filled with the environment.

To illustrate the presumed legitimacy of institutions, Zimbardo cites another experiment in obedience, a variation on the Milgram model, conducted in 1972 by psychologists Charles Sheridan and Richard King. In this experiment, college students were told to give shocks to a puppy to train it when it failed to perform a task (a task that was, in fact, impossible). Fifty-four percent of the men and all the women obeyed to the highest level of shock. Zimbardo himself conducted another version of the Milgram experiment to test the "social power of physicians." That experiment had even more chilling results. Nurses told during a call from an anonymous staff physician to give patients a dose of a drug (actually a placebo) twice the maximum safe level listed on the label on the bottle complied in twenty-one out of twenty-two cases. Since the order came from a legitimate source, it was obeyed, even when clearly wrong.[13]

The presumed legitimacy of hierarchical position also plays out in labeling. Studies demonstrating the effects of authoritative social labeling on educational outcomes have been well documented since the 1940s. This is another case of authoritative social interpretation. "The power of authorities," Zimbardo comments, "is demonstrated not only in the extent to which they can command obedience from followers, but also in the extent to which they can define reality and alter habitual ways of thinking." He cites the famous experiment of a third-grade teacher, Jane Elliott, in Riceville, Iowa, in 1971. Elliott, wishing to teach her homogeneous class of rural white farm kids about prejudice and the importance of tolerance, devised an experiment. She divided the children in her class into two groups, those with blue eyes and those with brown eyes. She labeled the blue-eyed group as superior and then watched what happened. Within a day, the brown-eyed children, labeled inferior, were doing poorly at schoolwork and were becoming sullen and angry. They called *themselves* "sad," "bad," "stupid," and "mean." The next day, Elliott reversed the labels, designating the brown-eyed kids as superior. Bullying, aggression, and abuse were now directed by the brown-eyed kids at the blue-eyed kids. The social labels were self-fulfilling prophecies, influencing future behavior rather than capturing past behavior.

Experiments in arbitrary authoritative labeling have been shown to induce exactly the quality being tested: for example, in IQ gains and losses, in promoting good behavior in individual children and in groups, and in improving math performance.[14] In the IQ labeling experiment, teachers treated the positively labeled children differently from the others, and their subtle expectations affected the students' performance. In the other cases, the students lived up to the (arbitrary) labels assigned to them.[15] The power of labeling is so strong that social psychologists have reached the conclusion that "the consequences of academic disappointment can be manipulated by altering students' subjective interpretations and attributions."[16]

Like labeling, the power of social role and social modeling has been shown to trump character and free and independent decision making in moral action. Philip Zimbardo suggests that Browning's "ordinary men" were, like the student prison guards in the Stanford Prison Experiment, influenced by social modeling, guilt-induced persuasion, and pressures to conform to the group. Ervin Staub, a Holocaust survivor and a psychologist who has devoted his career to studying genocide, comments, "Being part of a system shapes views, rewards adherence to dominant views, and makes deviation psychologically demanding and difficult."[17] The roles people were assigned induced in them changes of values in keeping with their actions. Intensive interviews with former members of the SS have shown that although those with a tendency to authoritarianism were attracted to the SS, these men committed violent acts only in the Nazi period and neither before nor after. They were violent only in the situation that sanctioned and institutionally supported violence.

The members of the SS have been classified into three categories: (1) those who enjoyed and identified with their positions and roles (often extreme ideologues) and gladly engaged in brutality, (2) those who adjusted to their roles and did their jobs, often changing their views to fit their actions, and (3) those who abhorred and were repulsed by what they were supposed to do and tried to lessen the burden on the victims but did not otherwise protest.[18] Philip Zimbardo and Robert Jay Lifton observed exactly the same threefold categorization in the student prison guards and in the Nazi doctors. Individual differences seem to have played a role in the way that those in each particular category approached their assigned roles and in their type of identification with those roles, but in no case did those differences result in a challenge to either the authority or the legitimacy of the system or the persecution. Social forces submerged and overrode individual differences. Robert Jay Lifton coined the term "total situation" to designate conditions of complete immersion in a context in which no alternative point of view or action is allowed. The Milgram experiments, however, can be seen as posing a challenge to Lifton's analysis, for it didn't take a totalizing context to produce extraordinary compliance and the willingness to harm.[19]

Zimbardo suggests that conformity and obedience to authority get us only so far in explaining why people commit evil. There are other important social processes at work, ones we find documented again and again in the literature of the Holocaust: dehumanization, deindividuation, and bystander disengagement. Experiments have shown that being anonymous increases a person's willingness to harm others. In another experiment designed and carried out by Philip Zimbardo, one in which women college students were to deliver painful electric shocks (of course not real) to other women (research confederates) under the guise of studying the effects of stress on creativity, half the volunteer subjects were randomly assigned to be completely covered by robes and hoods and had only number tags instead of name tags; the other half were randomly assigned to wear the same outfits but had name tags. The women were assigned to shock two other women from behind a one-way glass and to do so in groups of four while Zimbardo administered a creativity test to the victims. Unlike in the Milgram experiments, there was no researcher egging them on, and the subjects could see the confederates whose creativity was allegedly being tested. Four volunteers were tested at a time, but each was put in a separate cubicle, so there were no social conformity pressures—all would give shocks at the same time, so if one refrained, no one would know—a subject could simply slip into being a bystander rather than a perpetrator, and there would be no consequences, as with some of Browning's ordinary men. The main question raised was how long each would hold down the button to shock, that is, the duration of the shock. In the introduction to the experiment, the subjects had been told that one of the women to be shocked was nice, but the other was labeled "bitchy." First one research confederate was to be shocked, and then she was replaced by the other. Both were to give convincing performances that they were in pain. Both were to receive a round of twenty shocks.

None of the subjects refrained from delivering shocks. But the subjects who were anonymous, deindividuated, delivered twice as much shock to the victims as did the women subjects with the name tags, and they shocked the two victims equally. The individuated women, on the other hand, shocked the woman labeled "pleasant" for less time than the woman labeled "unpleasant" and even decreased their shocks to the pleasant woman over time. The anonymous women increased the duration of the shocks they administered over the course of the twenty trials, which Zimbardo attributes to an "upward spiraling effect of the emotional arousal . . . an energizing sense of one's domination and control over others at that moment."

Many other experiments have corroborated the effects of individuality masking and role anonymity on behavior—in the Stanford Prison Experiment, the guards' use of reflective sunglasses and uniforms illustrates the point. Personal responsibility and accountability were reduced and group immersion enhanced. A sense of frenzy, the pleasure of the feeling of power and dominance, was allowed to emerge as well, so in this kind of situation "aggression becomes its own reward." In a comparative study carried out by a cultural anthropologist of different societies and their practices in war, the societies in which the men changed their appearance to become anonymous when going to war were the most violent, with 80 percent of them brutalizing their enemies.[20] Deindividuation creates a situation in which immediate social forces and inner primitive emotional states have an even greater hold on the individual than they do under normal circumstances. Responsibility was diffused, and self-reflection as well as social evaluation as an individual were muted. As a result, action becomes less reflective and more immediate, and one's usual constraints dissolved. Vulnerability to social cues is heightened. These are some of the conditions found in extreme situations— genocide, massacre and torture, brutalizing war.[21] Yet there is little room for comfort when we see how little inducement it took for college students, merely decked out in robes and hoods and given a good excuse, to get carried away and torment similar women in a lab setting. These were hardly the conditions of a My Lai, an Abu Ghraib, or even a Stanford Prison Experiment.

Moreover, although one of the women in the experiment was labeled as "bitchy" or "unpleasant," dehumanization of the victim, which is the nearly ubiquitous condition of all torture and genocide (along with the anonymity of the perpetrators), was absent. We have seen how the Nazi genocide involved not only millennia-old theological anti-Semitism but also a biological ideology of dehumanization that characterized Jews as vermin, as a social cancer, or as a genetic disease that threatened the species. In Southeast Asia, American soldiers referred to innocent Vietnamese civilians as "gooks." In the American South lynching of "niggers" continued into the 1960s. A common way that dehumanizing labeling is used is between an in-group and an out-group. Humanity and fellow human feeling, empathy, applies only within the group but not between groups. To the out-group are attributed animal-like characteristics. Along with this phenomenon, now referred to by researchers as *out-group infrahumanization*, they have identified what they call a *self-humanization bias*, which is to say that we tend to see the out-group as less than human and attribute full humanity to ourselves alone.[22] And social psychologists point out that any out-group can function as a scapegoat as long as the out-group is weak and vulnerable and can't retaliate.[23] After my experience teaching and researching the Holocaust, at some point I realized that all it took to be a perpetrator—not a designer of the evil but a participant—was not hatred or rage but simple contempt. We are not aware of how potent merely looking down on others, curtailing our empathy because we don't consider them as fully human as ourselves, can be. Dehumanization is a particularly strong case of the labeling effect. So here again we have a normal kind of social process used to extreme effect. It is perhaps as surprising that such seemingly minimal labeling power over arbitrarily selected children as the teachers in the IQ experiment were given had such a profound effect on the children's subsequent performance as the dehumanizing labels in genocidal conditions that eased the devolution into atrocity.

What is surprising about the social forces that we have been examining is their potency to shape behavior. Small social variables appear to have enormous effects. Such is the case with a social force that one might think of as basically neutral but turns out to have profound effects on how the status quo is maintained and perpetuated: the passivity of bystanders. In the light of the famous Kitty Genovese case in Queens, New York, in which a young woman was stalked and stabbed as people allegedly looked on from their apartment windows and elsewhere but failed to call the police in time, social psychologists initiated a number of experiments to try to explain what had gone wrong. Were all these people merely callous city dwellers, or was there something in the situation that made ordinary people somehow freeze? That was the question raised by social psychologists Bibb Latané of Columbia and John Darley of New York University in a number of studies of this phenomenon. They found that the more people who witness an incident, the less likely it is that anyone will intervene, and they chalked that up to a diffusion of responsibility—everyone thinks someone else will help. In addition to a fear of violence there were also fears of doing the wrong thing, looking stupid, and what Zimbardo terms "an emergent group norm" of passive inaction. Psychological studies have shown that the inaction of bystanders has no significant relationship with their personality characteristics. When few people are present, studies show, New Yorkers and other big-city residents were just as ready as anyone else to intervene in an emergency. Zimbardo suggests that in the case of bystander apathy we are again confronted with the issue of the interpretation of situations: here, too, "we accept others' definitions of the situation and their norms, rather than being willing to take the risk of challenging the norm and opening new channels of behavioral options."[24]

Social Psychology: An Overview

Lee Ross and Richard E. Nisbett's *The Person and the Situation: Perspectives of Social Psychology* tells us that discoveries in empirical social psychology have reached a point where they can now be gathered together and put to use in other areas. There are three central contributions, Ross and Nisbett say, that are germane to broader issues of contemporary social, political, and intellectual concern and should be brought to the attention of people working in those areas. The first is the power and subtlety of how situations drive behavior and the extent to which even small manipulations of a situation overwhelm individual character differences and free decision making. The second they call a refinement of the first: situations are driven by subjective interpretations rather than by objective brute facts about the world. People respond to the construals, to interpretations. The third important discovery of social psychology is that situations are not atomistic or static but are subject to dynamic processes of change and inertia. This is true both within social systems and within the individual's own cognitive system, so changing one dimension of a system can cause ripples throughout until a new state of stability is reached. Alternatively, at times changing a single element will have little effect because it is overwhelmed by homeostatic processes. So while there is such considerable evidence from experiment after experiment that small situational changes produce large results, there are other cases in which the status quo is very hard to budge.

An example of the latter is a wide-scale intervention done in the 1930s, the Cambridge-Somerville Youth Study. Psychologists tried to improve the outcomes of about 250 working-class boys in Cambridge and Somerville, Massachusetts, who were considered (by teachers, judges, or schools) to be at risk for delinquency and criminality. They entered the program from ages five to thirteen and then continued in it for five years. The interventions were extensive: social workers visited the boys' homes twice a month and provided assistance, including involvement in family conflicts, in one-third of the cases; academic tutoring was provided in over half of the cases; 40 percent of the boys received medical or psychiatric care; social and recreational programs were provided, and the boys were connected to youth groups, YMCAs, Boy Scout troops, and summer camps. The massive intervention was also a rigorous study with random assignment to intervention or to control groups and follow-up studies continued over the next forty years.

Despite the positive feelings of the social workers and the many positive responses of the participants, this multifaceted intervention was a complete failure. There was no difference between the boys in the study and in the control group in terms of juvenile offenses, adult criminality, occupational success, health, mortality, or life satisfaction. In the few areas where a significant difference could be found (for example, in rates of alcoholism, adult crime rates, and professional white-collar status), in fact, it favored the control group. One source of the negative outcome might have been due to the labeling effect: these kids ended up stigmatized and lived up to the stigma. Yet all in all, the results of the study suggest that human psyches are not as vulnerable to family traumas as we tend to think, and that the community is a stronger and ongoing influence on individuals, on their deviance as well as on their success, one that even early sustained intervention is hard put to modify.

The interventions in the Cambridge-Somerville Youth Study, while substantial, were nevertheless directed at improving individuals' personal and familial lives while the social and economic context of those individuals remained unchanged. This was a massive attempt to shore up individuals within a context rather than to transform the context, and it failed. Broader contextual factors seemed to be far stronger than the shoring up of individuals could overcome. Perhaps what was lacking was a broader public health model rather than an individual and family psychology model. But what also seemed to be a mistake was what social psychologists call distal (early) versus proximal (late) interventions, which is to say, there was a bias toward assuming that there was a formative period in which an intervention had to be made so as to improve the future outcome of these children's lives. But the evidence from Head Start, from the Cambridge-Somerville Youth Study, and from examples and studies of later contemporaneous interventions is that later situational interventions are more powerful and effective than early ones. The current situation turns out to be more powerful, while the legacy of early learning is less so.

There are several good examples of the success of relatively small (certainly in comparison to the Cambridge-Somerville project) late situational interventions. In the early 1970s, a professor of mathematics, Urie Triesman, at the University of California at Berkeley noticed that African American students in his classes were failing introductory math in high percentages and thus being precluded from going on in science and medicine. He observed that Asian students, noted for their success in math and science, studied together in groups, while African American students tended to study alone. Study in groups was more productive because someone in the group was likely to have a solution to a given problem and could help the others. Triesman persuaded a large number of entering African American students to participate in a special honors program in math that featured group study (thus Triesman included positive predictive labeling as well). The African American students who participated had grades on average the same as those of white and Asian students, and their dropout rates plummeted.

The discovery that small concurrent interventions in situations to change people's behavior in the present tend to be more powerful than carryovers from early learning has important implications for morals and the teaching of morals. We tend to think that children learn morals early and if they were taught well ought to be set for life. But the research seems to indicate otherwise: the current situation, how it is interpreted, and our investment in it can, to a large extent, override the past in determining our social behavior. This helps to answer some of the enduring questions about the perpetrators of the Holocaust: Weren't they Christians? Didn't they have a moral education? Didn't they realize their actions violated basic moral codes and norms? The Nazi situation and its Nazi interpretation, for most, overrode earlier learning and values. And that turns out to be business as usual rather than an aberration. It's just that the contrast is not usually as stark and as bleak as it was in the Holocaust. So it's the Holocaust again that has brought to our attention, perhaps as never before, the overwhelming impact and influence *everyday* situations have upon us. Just as decision making is not independent of context or freely originated, and just as moral character is not stable across situations (as we shall soon see) and hence is not predictive of our behavior, so, too, early learning does not necessarily leave an indelible mark. We are creatures of present social contexts and situations far more than we realize.

It was Kurt Lewin, one of the founders of social psychology, who developed group discussion techniques and democratic procedures as powerful social interventions to change individual behavior. Lewin, a refugee from Hitler's Germany, focused in his experimental work on the power of the immediate situation. His findings suggested that the peer group is the most important force in inducing or restraining behavior in the social context. In an early study he and colleagues showed that the manipulation of leadership style in recreation clubs to institute authoritarian or democratic group social environments transformed the relationships of the young men who were members of the clubs.[25] When an authoritarian leadership style was introduced, the young men scapegoated each other and passively submitted to authority figures. With a short-term manipulation of the environment, Lewin had quickly produced the array of responses that Adorno and colleagues had argued were due to the "authoritarian personality." In a slightly later study in which nutritionists were attempting to change entrenched eating patterns toward more available organ meats during wartime shortages in the 1940s but failed to accomplish their goal through lectures, pamphlets, and moral encouragement and exhortation to eat organ meats to help the war effort, Lewin introduced, instead, small discussion groups of homemakers with trained leaders who talked together about how to implement the program of alternative foods and new recipes, which was followed by a show of hands of the participants who were willing to commit to it. While the success rate for the nutritionists' initial strategy of lectures, pamphlets, and moral exhortation was just 3 percent, the group discussions elicited a 30 percent success rate. Other interventions by group discussion and personal commitment in a group setting were also successful, for example, getting rural mothers to give their infants cod liver oil. Ross and Nisbett explain that what Lewin was doing was creating a social reference group for the new norm of behavior and creating a consensus in that group about the behavior. Lewin was also simultaneously reducing the social pressure of the existing group and its norm, and it worked. Earlier, in the 1930s, Lewin had pioneered participatory industrial management and workgroup decision-making procedures and brought them to Japan, where they were widely instituted in industry—techniques that returned home only in the 1970s, where they were introduced into the United States as "Japanese" management techniques.

One of the operative components of the small group method was the presence of a leader who is an appropriate social model. While social modeling is enhanced by the model's characteristics—high status, attractiveness, power—even the mere presence of a model has been shown to be powerful in inducing people to engage in a desired behavior. In the Milgram experiments the presence of a researcher demonstrating how to present the series of words and use the equipment had a great effect on the volunteer subjects' compliance, while when he merely called on the phone compliance was reduced. And the presence of research confederates who pretended to be volunteer subjects and refused to shock the learner created 90 percent *non*compliance.

Another telling example of how the behavior of individuals could be changed by intervention in group structure and process rather than by addressing individuals directly or by depending upon early moral or intellectual training is the intervention devised by Elliot Aronson, a professor of social psychology at the University of California at Santa Cruz, to help prevent future massacres like that at Columbine High School in Colorado in 1999, when two disgruntled students went on a shooting rampage, killing twelve students and a teacher and wounding twenty-three other people before committing suicide. The first step for Aronson was to try to figure out what had gone wrong. Why did Columbine happen? Was it a failure of overall morality in the schools, as a congressional bill that would have allowed the posting of the Ten Commandments in the nation's schools seemed to suggest? Was the answer to beef up moral education in the schools (we saw that the character education movement got a great boost after Columbine)? Was it to get violence out of the media, TV, movies, songs, the Internet? Was it to increase surveillance of kids in schools, identify problem kids early, and remove them from the school setting or subject them to intensive therapy? Was it to make all the kids into spies and informers to point out to the school administration any weird or suspicious behavior or dress? Was it to create more rules and codes of proper behavior, ensuring better obedience and respect for school authority? These were all answers that were proposed in the wake of the Columbine massacre, but Aronson points out that they have little evidence to support them even though they might feel like the right thing to do. These analyses and the answers they lead to, he says, rest on "emotion, wishful thinking, bias, and political expediency."[26]

Aronson fully acknowledges that the actions of the students who massacred others were pathological and horrific. But in studying high schools in various parts of the country, Aronson concluded that the main cause of the massacre was a general atmosphere of exclusion that was not unique to Columbine but could be found in many of the nation's high schools. There was a prevalent social atmosphere, he says, and it is toxic. These terrible perpetrators were "reacting in an extreme and pathological manner to a general atmosphere of exclusion," he says. From his classroom research and from extant social psychology research on American schools, he found that the social atmosphere of most American high schools is competitive and cliquish, something that most students find unpleasant, difficult, distasteful, and humiliating.[27] For some students, he says, it is "a living hell."[28] It is an atmosphere in which taunting, rejection, and verbal abuse are prevalent. School shooters seek revenge for their humiliation at being relegated to the bottom of the pack as well as fame. To prevent another Columbine, Aronson proposed to show how the atmosphere in high schools could be transformed from exclusion to inclusion, and from hierarchical rejection to mutual support, through a two-pronged approach: by introducing strategies for resolving conflict and dealing with anger and teaching empathy, on one hand, and by restructuring the classroom itself from competitive learning to cooperative learning, on the other. He devised a cooperative learning system, called the Jigsaw Classroom, that structures learning so that all students together aim at a common goal where everyone stands or falls together. He has developed a way to get around the problem with standard group learning in the normal competitive school environment, which is that the motivated and high-achieving kids do all the work for everyone in the group. The effect of the Jigsaw Classroom is also to create a kind of tolerance and even appreciation of differences that moralizing is not capable of producing. Aronson concludes that respect cannot be legislated and that forcing students to obey rules only makes them go through the motions but doesn't change their attitudes. He has found that what does work is less direct moralizing exhortation, less call for individual moral decision making and virtuous character, and more structural intervention to introduce common goals and common strategies to meet them. He explains, "You don't get students from diverse backgrounds to appreciate one another by telling them that prejudice and discrimination are bad things. You get them to appreciate one another by placing them in situations where they interact with one another in a structure designed to allow everyone's basic humanity to shine through."[29]

Although the Jigsaw Classroom and other interventions along those lines introduce a win-win situation, the difficulty and radicalism of a wholesale transition to interventions of that kind are not to be minimized. For that kind of systemic intervention to be seen as a legitimate and obvious answer to Columbine is hardly painless, involving as it does a major transformation in our moral outlook. What is gained in practical moral outcomes is balanced by what is lost: we lose the right to blame others exclusively for their actions and to merely expect others to do the right thing. We lose the right to inculcate, to lecture, and then to praise and blame. We lose the option of taking ourselves off the hook for the actions of others in contexts and situations for which we have some responsibility or could have contributed to their change. We lose our own innocence or presumed innocence as others lose their complete guilt. That is the price we must pay (and ought to pay) for getting ethics right and, as a result, being able to have an ethically transforming effect on others and even on ourselves.

The Social Psychology of Altruistic Behavior

Luckily, social forces are operative in the inducement and shaping of good behavior as well as bad behavior. In an almost diabolically clever experiment by social psychologists John Darley and C.D. Batson, Christian seminarians at Princeton Theological Seminary were told to prepare a brief extemporaneous talk to be delivered at a nearby building where it would be recorded. Half of the students were told to give a talk on the New Testament parable of the Good Samaritan. (This is the story of three passersby who find themselves on the road where a man has been held up by robbers, beaten, and left for practically dead. It is the Samaritan, the outcast of post-Exilic Israelite society, who turns out to stop and help the man while two notables pass him by. So the message is roughly that the moral worth of altruistic behavior, rather than religious and social status, is to be valued.) In the experiment the seminarians were divided into two groups; those in one were told that they were late and to rush over to the nearby building to give their talk and have it recorded. Those in the other group were told that it would be a few minutes before those in the other building would be ready to record their talk but that they might as well head over to the other building. En route to the destination, Darley and Batson had placed a research confederate, a man who was slumped in a doorway, head down, coughing and groaning. Of those seminarians who were told that they were late, only 10 percent stopped to offer assistance to the suffering man, while 63 percent of those from the early group stopped to help him. The students were also given a questionnaire before the experiment that asked whether their interest in religion was focused primarily on personal salvation or on helping others (dispositional motivational factors). It turned out that the particular answer a seminarian gave to the questionnaire had no correlation with whether that student stopped to help the suffering man or not. Determinants of altruism were not moral at all but instead a seemingly neutral feature of the situation (in this case, whether they were pressed for time). And this conclusion about the pertinence of morally neutral features of situations is borne out by other studies.

After Kitty Genovese's murder in 1964, Darley and Latané developed a series of bystander experiments to try to determine when and why people act to help others in distress. In a 1968 study, Columbia University male under-graduates were put in a room to fill out a questionnaire. They were left either alone, with two other subjects, or with two research confederates instructed to remain impassive when an emergency was introduced into the situation. The emergency was a stream of smoke that poured into the room through a wall vent and eventually filled the whole room. Seventy-five percent of the subjects who had been left alone in the room reported the smoke, while only 10 percent of the subjects left in the room with two impassive research con-federates reported the smoke; 38 percent of those in three-subject groups intervened. In a 1969 experiment, individuals answering a questionnaire ei-ther alone, in the presence of one impassive research confederate, or with another subject also working on a questionnaire were left in a room. Then a noise that sounded like the female research experimenter taking a bad fall came from behind a room divider. Again, 70 percent of the lone subjects intervened to offer assistance, but only 7 percent of those subjects who sat next to an impassive bystander (the research confederate) intervened to offer assistance; with the pair of subjects, only 40 percent intervened. Amazingly, four dozen follow-up studies provided the same results. Darley and Latané's conclusions were that the feeling of responsibility is diffused in a group setting, that the interpretation of both the nature of the situation and the appropriate action is deferred to the group, and that the failure of other people to act serves as a confirmation of the appropriateness of nonintervention. So the presence of others inhibits intervention. Ross and Nisbett suggest that "the opinion of the majority carries normative or moral force."[30] Social influence, even of strang-ers, has a determinative effect on the individuals in the group. And conformity can be to a minority as well as to a majority, as we saw in the Milgram experi-mental variation in which when research confederates declined to go along only 10 percent of the subjects complied. That the presence of a small mi-nority of independent thinkers and actors can embolden the majority to re-sist conformity and take responsibility suggests where some hope may lie.

Other factors that influence behavior can seem even more incidental. For example, a small elevation in mood leads to enhanced helping behavior. In one study the impact of finding a dime in a telephone booth increased the likelihood that a random person coming out of a phone booth would help a research confederate who dropped a folder of papers in front of him or her. Only 13 percent of dime finders failed to stop to help pick up the scattered papers, while 96 percent of those who didn't find a dime failed to help, with many even trampling on the papers. The numbers are staggering: in study after study, positive affect has been shown to be related to prosocial behavior. Even pleasant aromas have been shown to improve prosocial behavior, it turns out. Subjects near a fragrant bakery or a coffee shop were more willing to change a dollar bill than subjects near a neutral-smelling department store.[31] All the evidence seems to suggest that neither stable moral character traits or virtues nor free will decision making based on moral principles seems to have much to do with actual ethical behavior, whether under the most extreme conditions and situations, such as the Holocaust, or in the most trivial ones, like those just recounted.

The Social Interpretation of Situations

We have just seen the power of the presence or absence of others in situations to modify our behavior quite profoundly. Just the mere presence of others seems to inhibit actions that we would think are the natural response, such as intervening to help someone in distress; such actions do seem to be our natural response, but only when we're alone! Now we'll look at experimental evidence that suggests not only that the presence of others is important but also that their opinions shape our own in unexpected and extreme ways. These findings were some of the earliest in social psychology. Experiments in the 1930s by Muzafer Sherif, a Turkish immigrant to the United States, showed the group influence and social conformity effect upon perceptual judgment. Sherif developed an experiment in which individuals and groups were put in a dark room with a single point of light located at some distance from them. Because the room was entirely dark they could not tell either the dimensions of the room or, as a result, how far the light was from them. The subjects suddenly saw the light move and then disappear; then a new point of light would appear, move, and finally disappear, a sequence that was repeated many times. In fact, the light only appeared to move; this is the perceptual illusion called the *autokinetic effect*. The virtue of the test was that there was no objective standard that could serve as a reference point. In each instance, the subjects were to estimate how far the light had moved. When subjects were alone in the room, their answers differed widely from person to person, from a few inches to a few feet, and even the same subject on different occasions gave different answers from his or her previous answers. When the research subjects were tested in groups of two and three, however, they quickly converged on a group consensus. Different groups turned out to converge on very different norms. They had substituted a social norm for a personal one. Then unbeknownst to the subject participants, Sherif introduced a research confederate who gave consistent estimates either far greater or smaller than the subjects typically made on their own. The subjects immediately adopted the confederate's judgment. Sherif concluded that even an individual with no particular status or claim to expertise could impose a social norm by displaying consistency and confidence in the face of the others' uncertainty. Subsequent repetitions and extensions of the experiment proved an even more extreme conclusion: subjects would adhere to the norm established by the arbitrary group after their peers were no longer present, and even a year later they would adhere to the same norm. It had been internalized. Subjects taken from an old experimental subject group to a new and different one would adhere to the old norm even when the new group established a different norm. And an extension of the experiment in 1961 showed that a norm established in an earlier trial could be transmitted to a new "generation" by introducing one new subject in each repetition of the experiment, so the final repetition included no subjects from the initial experiment, yet the norm that was set by the original group was handed down over several generations. Because there was no "*there*" there" in the Sherif experiment, no possibility of a true answer to the question but only the social construal, Sherif concluded that our most basic judgments about the world are socially construed, and perhaps even perception is socially conditioned.[32]

Solomon Asch, one of the following generation of social psychologists, began his experiments with the express purpose of challenging the conclusions that Sherif had come to, but his experiments proved Sherif right and even extended the latter's conclusions further. Asch's experiments substituted for Sherif's unknowable perception a knowable one, asking whether experimental subjects would still conform their perceptual judgments to each other and to a *false* standard confidently introduced if the actual answer was easily and readily available to all. Asch predicted that the Sherif paradigm would fail the test and that the limitations of social conformity in cognitive judgment would become clear. Precisely the opposite turned out to be the case, with the inevitable conclusion that not only would a group conform to a social judgment when the situation was ambiguous, but a group would conform to a false judgment that when individually tested each knew to be false. Like Sherif's, Asch's experiment was also one based on visual perception, but this time the task was to look at varying groups of three lines projected at the front of a room and compare their length with a different standard line. The lines were clearly drawn and the differences in length obvious and marked. In each iteration seven to nine participants were placed in the room and each was to give an answer in turn. But in this experiment Asch made all of the participants research confederates except for one naive subject. All the confederates followed a set script. The participants were told not to confer with each other and to make their own judgments. In the initial three iterations all the participants, the confederates as well as the naive experimental subject, gave the obvious true answers. But then in the fourth trial the first confederate, without hesitation, confidently gave a wrong answer. After a look of disbelief and a second check at the lines, the naive subject gave some sign of discomfort. Then all the other research confederates repeated the same wrong answer of the first confederate. The subject's turn came last, and he or she had to either conform to the unanimous majority false judgment or remain independent and give the obviously true answer. In the various repetitions of this experiment, five to twelve instances of wrong answers unanimously given by all the research confederates were stuck into a total of between ten and eighteen trials. Subjects almost invariably showed great discomfort, and 50 to 80 percent in the different trials conformed at least once to the mistaken majority. Over a third of all trials exhibited a false conformity. In different slight variations of the experiment, the results revealed that conformity rates did not diminish when the unanimous group of confederates was reduced from eight to three or four.

When the number of confederates was reduced to two, however, little conformity was produced, and when only a single confederate was with a single subject, there was no evidence of social influence at all. In fact, when the unanimity of the confederates was eliminated, and when even a single confederate ally was introduced into the group who remained independent and called it as she saw it, the subject also retained independence of judgment. A third of Asch's subjects never conformed at all, and another third conformed less often than they remained independent. In interviews with the subjects after the experiment, they acknowledged that despite their personal perceptions they articulated conforming judgments because they were unwilling to be a lone dissenter, even though they knew the majority of their fellows were mistaken. What was shocking about the outcome was the extent of the subjects' willingness to defy and relinquish their own clear perceptual reality to conform to group opinion. Social psychologists of the

1950s were quick to relate Asch's findings to the surrounding corporate and middle-class suburban culture as well as to McCarthyism and its loyalty oaths. With such marked conformity in a situation of no coercion, little at stake, and a clear and unambiguous reality, many at the time concluded that the conformity must be of much greater proportions in real-world situations where social and political pressures are great and the situation far more ambiguous and open to varying possible interpretations! Further experiments bore out that Asch's basic findings were not mere experimental anomalies but had far and wide implications for real-life contexts.

After Asch's experiments using simple objective perceptual data, social psychologists turned increasingly to experiments focused on conformity in social perceptions and subjective interpretations and opinions. Studies of cognitive dissonance—that is, of differences of opinion in a group—exposed the group pressures and also pressures within the individual to establish or reestablish conformity and consistency of opinion. Ross and Nisbett point out that the dissonance between one's view and that of the group is "characteristically resolved in favor of the group's view, often not by simple compromise, but by wholesale adoption of the group's view and suppression of one's doubts."[33] People's beliefs and even pleasures turn out to be highly manipulable and determinable by interventions in the social meaning and interpretation of situations. Further experiments by Asch revealed that social influence governs not only positive and negative valuation of an object but also the very interpretation and definition of the object. In an experiment with two groups of peers, where both were told to rank the status of various professions, one group was told beforehand that the profession of politician had been ranked at the top of the list by a prior group and the other group was told that politician was ranked at the bottom by a prior group. Not only did the two groups conform to the anonymous ranking (of the imaginary prior group), but in exit interviews the subjects in the first group made clear that *politician* meant to them "great statesman" while the members of the second group defined *politician* as "corrupt political hack."[34] This reminds me of Ronald Reagan's use of the term *welfare queens* to develop in the public a derogatory feeling toward welfare recipients and push the public to withdraw their support for the policy. Ethical interventions are established at the level of description far more than we realize. Once the description has been set, the actions merely follow. So the basic fight is for the airwaves and winning the description wars—those who fund the Rush Limbaughs of the world are acutely aware of this. And it's social groups, much more than individuals, who sign on to such opinions. These deep sociological factors creating and maintaining groups underlie the more explicit group adoption of attitudes and beliefs.

Groupthink

All these tendencies point in the direction of the phenomenon called groupthink. Irving L. Janis, in his now famous 1972 book *Groupthink: Psychological Studies of Policy Decisions and Fiascoes* (expanded in 1982), documented a number of historical debacles, including the Bay of Pigs, the escalation of the Vietnam War, and the Watergate cover-up, that he argued were due to groupthink. The term *groupthink* refers to the social pressures exerted by the leaders of a group and its overall dynamics toward loyal consensus of opinion and the elimination of doubting and dissonant voices. Janis analyzes cases of groupthink to try to tease out what he calls a common "specific pattern of concurrence-seeking behavior" in which the approval of one's peers in the group becomes more important than solving the problem at hand.[35] He found again and again that from the historical facts a clear and consistent recurring psychological pattern could be identified that produced cognitive distortions and corruption through the quashing of critical thinking. Janis noted that for groups that exhibited groupthink, the most important group value was loyalty to the group. Loyalty was embraced and enacted as its highest moral value. This created, Janis says, a kind of softheartedness toward other members of the group but at the same time an extreme harshness toward outsiders. Within the group softheartedness leads to "softheaded" thinking, whereas toward outsiders and enemies the group exhibits extreme hardheartedness and a tendency toward dehumanizing definitions, leading to harsh military solutions and even atrocity. At the same time, within the group there is the conviction that wonderful people such as the group members could never act inhumanely and immorally. Moreover, the tendency to seek concurrence also fosters unrealistic optimism about the outcome of the group's policy decisions, a lack of vigilance, and "sloganistic thinking" about the weak and immoral character of out-groups. The greater the esprit de corps of a policy-making group, according to Janis, the higher the likelihood of groupthink and even of the group's taking dehumanizing action against out-groups.[36] Janis goes on to identify eight main symptoms of groupthink that fall into three main types:

Type I: Overestimations of the group—its power and morality

1. An illusion of invulnerability . . . which . . . encourages taking extreme risks
2. An unquestioned belief in the group's inherent morality, inclining members to ignore the ethical or moral consequences of their decisions

Type II: Closed-mindedness

3. Collective efforts to rationalize in order to discount warnings . . . that might lead the members to reconsider their assumptions . . .
4. Stereotyped views of enemy leaders as too evil to warrant genuine attempts to negotiate . . .

Type III: Pressures toward uniformity

5. Self-censorship of deviations from the apparent group consensus . . .
6. A shared illusion of unanimity . . . (partly resulting from self-censorship of deviations, augmented by the false assumption that silence means consent)
7. Direct pressure on any member who expresses strong arguments against any of the group's stereotypes, illusions, or commitments, making clear that this type of dissent is contrary to what is expected of all loyal members
8. The emergence of self-appointed mindguards—members who protect the group [and especially the leader] from adverse information that might shatter their shared complacency about the effectiveness and morality of their decisions[37]

All eight symptoms clearly fit the Nazi case—as well as the invasion of Iraq, a war that continues as I am writing.

Of course groupthink is not only a problem of governments and their policy making but also a danger in all kinds of groups from corporations to families. In any of these groups, the structural features and the situational factors that give rise to it can be intervened in, according to Janis, so as to avoid its disastrous consequences. In the groupthink context each person decides that his or her misgivings and doubts are not important, not really relevant. But groups can be organized to support dissident voices and to encourage the freedom to speak and even to think outside the box. The fear of humiliation can be lessened by an accepting attitude toward members that is not dependent on loyal consensus, unquestioned authority, or the rejection of out-groups. Janis proposes that the structural feature of a middle level of cohesiveness (rather than extreme cohesiveness or very loose affiliation) is best in damping the dangers of groupthink. This is because at that moderate level there is an accepting attitude toward members, but not to a degree that exacts extreme loyalty and conformity; nor, at the other extreme, does it make members uneasy that their acceptance is so contingent and minimal that they could easily be pushed out. Other structural features that promote groupthink and therefore ought to be avoided include insularity, the lack of a tradition of unbiased and open inquiry, the lack of impartial procedures for getting relevant information, and homogeneity in the backgrounds of people in the group. The solicitation and encouragement of divergent opinions can even be formalized and institutionalized procedurally by a group. Furthermore, procedures can also be put in place that help a group resist the tendency to a sense of moral self-righteousness, to cast itself as good and outsiders as stupid and bad. Times of stress of course make these tendencies worse.

Nevertheless, a group already given over to the kind of limitless and unchallenged power enabled by groupthink would hardly welcome interventions that would challenge and limit it in these ways. It seems to me that if a society, group, institution, or company has groupthink tendencies, it would take a real disaster and fall of Nazi or Watergate proportions or some other great debacle for its members and especially leaders to be willing to consider putting in place the kinds of democratizing and freedom- and dissent- and diversity-enhancing structural and procedural changes that Janis recommends. Or perhaps it would take a new golden age, a Renaissance, or an Enlightenment. Groupthink is a stellar example of the hijacking, the corruption, of moral judgment. And we need only look around us to become aware of how much that is the rule rather than the exception. Disinterested, independent moral valuation of others, of actions, and of policies—free decision making—is a utopian ideal, and hence hardly a reliable and firm basis upon which to build a society or any of its component institutions.[38] As Ross and Nisbett put it,

> People actively promote their beliefs and social construals and do not readily tolerate dissent from them. It follows that cultures, which have much more at stake than the informal groups studied by psychologists, would still be more zealous in keeping their members in line with respect to beliefs and values. It is primarily for this reason that cultures can be so starkly and so uniformly different from one another. . . . *Isolation is the key to stability of cultural norms.*[39] [Emphasis added.]

Since we cannot rely on people to *be* independent (and rational) merely as a feature of human nature (as we have falsely assumed) and think and act for themselves, our best hope for ethics is to nurture pluralism and democratic structures and procedures, putting mechanisms in place that foster differences and protect minority groups with different perspectives and also shield whistle-blowers. In addition, it seems clear from this work that it is vital that different groups (ethnic and religious groups, regional groups, age groups, etc.) in a society interact and not have isolated, protected domains that keep them from challenging each other. They also must engage in shared tasks that bring them together rather than set them into unremitting competition and mutual antipathy and dismissal. Pluralism isn't just a matter of letting everyone alone; rather, it involves actively bringing people together toward common goals, yet in ways that encourage differences but do not reify them. Janis proposes ways that this can be done in a policy-making group, but surely some of these ideas are more generally applicable as well. Janis's work, like that in the social interpretation of situations that I have been documenting throughout this chapter, suggests that the point of intervention for ethics is not in a choice of (bad or good) action via a (mythical) free will but instead earlier in the process, in the cognitive interpretation of situations. And even that intellectual intervention is socially and situationally driven; it is not willed but must be enabled through social structures on the group level and the broadest kind of learning and rigorous self-analysis and self-reflection on the personal level, the philosophical education of desire of the kind spelled out in Spinoza's *Ethics*. So Spinoza may indeed have got it right, that in the end all virtues depend upon and emerge from a more fundamental intellectual virtue, the openness and honesty that make independence of mind possible. But in its absence, good social and political institutions that foster differences and that promote a humane vision of self and others must pick up the slack. Spinoza envisioned civil religion as performing the role of promoting an ideology of universal humaneness in modern democratic societies, societies that he also envisioned as institutionalizing pluralism.

A Final Comment on the Psychology of Leaders and Followers, National, Social, Civic, Corporate, and Familial

Another psychological thinker whose concerns and insights emerged in part from the experience of the Holocaust and were also brought to bear to try to understand the evil exhibited in the Holocaust is the psychoanalyst Heinz Kohut. Kohut grew up in an assimilated Jewish family and had already completed his medical degree in neurology when he escaped his native Vienna in

1939 and settled in Chicago, where he founded the self psychology movement within psychoanalysis. In a number of essays Kohut applied his psychoanalytic thinking to social, cultural, and political life, focusing mostly on an analysis of Hitler and his relationship to the German nation.[40] Charles Strozier, editor of a volume of Kohut's essays on history and literature, credits Kohut with "a refreshingly new theory of leadership, one that systematically formulates the relationship between leaders and followers."[41] Kohut conceived the relationship between leader and nation to be of a group self with a leader's self, where both sides shape each other's self-understanding and state of the self. The kinds of internal and intimate interrelational processes that govern individual selves also affect the historical group self and the group leader, Kohut proposed. A group self, he says, is as prone to "narcissistic injury"—that is, a collective humiliation—as an individual self.

> It's so easy to say that the Nazis were beasts and that Germany then regressed to untamed callousness and animal-like passions. The trouble is that Nazi Germany is understandable. There is an empathy bridge; however difficult to maintain. . . . People will go to extraordinary lengths to undo narcissistic injury. These people would rather die than live with shame. One has to study this dispassionately. What else can you do? You have no other choice.[42]

In an essay from the early 1970s titled "On Courage," Kohut posed the question of what it is that enables "an exceptional few to oppose the [social] pressures exerted on them and remain faithful to their ideals and to themselves, while all the others, the multitudes, change their ideals and swim with the current." Kohut's answer was that each person has a unique nucleus of psychological being, a "nuclear self," consisting in ideals and aspirations determined early by family, culture, and context but nevertheless modifiable by life experiences. By this nuclear self, Kohut warns us, he does not mean the Cartesian conscious self-substance, "a single self as the central agency of the psyche" from which "all initiative springs and where all experiences end," the source of (an allegedly) free willed action. Instead, he says, he is speaking of a kind of unification of introjected ideals and capacities. But this nuclear self in most people, in the majority of adults, he says, "ceases to participate in the overt attitude and actions and becomes progressively isolated and is finally repressed and disavowed." For most of us, then, the self becomes buried in the group and its attitudes, beliefs, and projects. It is only rare individuals who retain the capacity to experience an ongoing conflict between the nuclear self and the demands of the social context and resolve the conflict so that the sources of inner creativity lodged in the nuclear self can find expression and contribute to the needs of the larger context. Such individuals are rare and heroic. That is what it takes to have independence of mind and action, according to Kohut. These unusual individuals bring all their psychic energies together toward a totally committed central creative and independent effort. And the rare individuals who are capable of remaining in touch with the nuclear self and bringing it into effective expression in socially meaningful projects have a profound sense of inner peace, Kohut says, a "mental state akin to wisdom." Kohut gives as examples of this rare capacity for courage the anti-Nazi activists of the White Rose, Hans and Sophie Scholl, whose calm clarity enabled them to go to their deaths at utmost peace. Surprisingly, Kohut does not claim that this rare kind of courage entails an extraordinary degree of mental health. Rather, he says, that capacity or psychic organization that results in a kind of courage and even heroism is to some extent independent of the extent of an individual's psychopathology. Nevertheless, it is a very specific psychological capacity and inner balance, integration of components of the self.

The vastly more common phenomenon is the identification of the members of groups with different kinds of leaders and the subsequent social determination of belief and action. Kohut classifies group identifications into three categories: (1) a relation of group to leader that is like a mystical religious devotion to an omnipotent god whose power they participate in vicariously; (2) a relation of group to leader in which ideals and values cover up and cover over the main motive, which is again identification with a source of omnipotent power; and (3) groups or movements in which shared ideals and a leader who stands for them and promotes them bring people together. The Nazis represent the rise of an appeal to pure omnipotence in a social context in which both national prestige had been profoundly wounded (in World War I) and at the same time traditional values and ideals and religious beliefs had undergone a significant devaluation and debasement. Kohut holds that Germany had undergone a period in which people had become disillusioned with established religions, the country had declined in prestige, the monarchy had fallen, the aristocracy and officer class had declined, and the working classes had risen in power. Shared ideals had been sorely hurt while at the same time Germany had experienced a devastating and humiliating defeat. Hitler and the German nation shared a narcissistic wound (Hitler's was personal as well as national) and held a common grandiose fantasy. The leader of this kind of group is the locus not of shared values and ideals but rather of pure ambition; through him the group hopes to become great or restore their lost greatness. Articulated ideals by the leader or the group under these circumstances are mere covers disguising a raw, grandiose ambition to banish shame. Hitler expressed this grandiosity and destructiveness under a thin veil of articulated ideals that served as a disguise. The collective hope of this kind of group is that through strength and triumph, vengeance will replace shame and humiliation. The leader and nation engage together in a kind of frenzy. Anyone who questions the leader who symbolizes this hope of restitution of self-esteem and power is cast as a traitor to the nation. Hence the amassing of power over the whole world and the destruction of the Jews, cast (mythically) as the source of the humiliation, became ends in themselves. This alliance of nation and leader for power itself, in order to overcome psychic humiliation, Kohut regards as the source of the worst atrocities of the twentieth century. He comments, "The most malignant human propensities are mobilized in support of nationalistic narcissistic rage. Nothing satisfies its fury, neither the achievement of limited advantages nor the negotiation of compromises, however favorable—not even victory itself is enough."[43]

The Nazis' goal was, and remained till their total defeat, the total extinction of an enemy. The German nation, Kohut suggests, pursued a vision of total control of the world via a supraindividual, nationally organized vendetta of merciless persecution, genocide, war, and destruction. It was a group regression, the primitivization of an entire society—much like an individual undergoing a kind of breakdown under the psychological pressure of profound humiliation coupled with loss of ideals. That the German nation found a fittingly pathological leader with whose grandiosity they could merge was a fortuitous tragedy. Kohut writes that in a conflict with such a frenzied and destructive national force as the Nazis, one can only hope the winner will be a nation or group of nations mobilized by their ideals and held together by their common love of and identification with those ideals, rather than by a striving for vengeance and overpowering force. For this counterforce to have the power to overcome the frenzy of narcissistic rage, Kohut muses, it may be necessary to carefully engage a religious merger, also a somewhat primitive and regressive tendency yet one that can be placed in the service of ideals. What is important about Kohut's approach to Nazi evil is that, unlike so many other psychological treatments, his focus is on the pathological condition of the German nation, the German group self, rather than the psychopathology of Hitler alone. For the kind of danger that Hitler represents is always with us somewhere. It is the danger that even a highly civilized nation could be all too ready to embrace a leader who offered them an instantaneous feeling of intense power and pride through merger with an omnipotent figure. There are often Hitlers available, but their success depends in part upon a very particular condition of a national group, a state of acute weakness demanding instant relief.

Kohut's analysis, whatever the precise details, as a general approach makes us intensely aware that the psychic forces of group belonging and the dynamics between leaders and groups are powerful and mostly irrational; they are not easily or merely rationally dispelled. All the evidence taken together in this chapter should make us quite skeptical about the human capacity for independent action, about relying on individual will free from group and situation to produce and ensure moral behavior. Moreover, evil actions now appear just as likely as good actions to be due to group behavior that relies on institutional authority and hierarchy, authoritative interpretations, and the like. The research is not new, nor is it arcane and hard to understand. It is overwhelming and widely known. Why do people generally and our society at large, then, largely ignore it and continue to think about why people are ethical, why they are not, and how to get them to be more ethical, in terms of individual autonomy, free will, blame and punishment, praise and reward?

In Chapter 4, I explore the cultural history of the notion of free will, how we come to have it, and why it seems the only plausible explanation for good and evil behavior. The notion of individual free will has such a hold on us that the widely available evidence I have summarized here, that social forces are largely responsible for both moral and immoral action and also that it is often the group that is the actor, remain invisible to us. The question I pose in Chapter 4 is: why do we focus almost exclusively on individuals as moral actors capable of (at least some significant degree of) independence from group, context, history, culture, biography, and even to some extent biology? This is the problem of free will, the ubiquitous assumption we saw being taught in schools and churches, in religious and moral education from before the founding of the country to its present appearance as character education in our schools, enshrined in law, and everywhere around us. Why do we think we have free will? Where does the belief come from? Can we do without it and still be ethical, and, if so, how? And if free will is a dead end—as it seems to be, and as the evidence from the new brain sciences that

I put forth in the second half of this book is increasingly revealing—why have we stuck with it so long? Why have we been so reluctant to relinquish or at least modify the belief that individuals are independent from group, history, and context, as moral actors who originate their actions? Why do we have such a hard time envisioning other possible ways to account for moral motivations, commitments, and responsibility? These problems occupy the next chapter.

4

What Happened to Ethics:
The Augustinian Legacy of Free Will

Human nature, even in sinners, is superior to the beasts. For it is the nature of man that is from God, not the wickedness in which he implicates himself by an evil use of free will. Nevertheless, if he did not possess free will, he would not have the same excellence in nature.

—AUGUSTINE, *The Literal Meaning of Genesis*

If men were born free, they would, so long as they remained free, form no conception of good and evil.

—SPINOZA, *The Ethics*

Introduction

We have come to the middle, to the pivot, to the turning point.[1] Up till now my aim has been to pick out some salient examples of ethics on the ground, so to speak. In the first chapter I surveyed the history of moral education in the United States in order to identify our pervasive cultural and historical assumptions about why people are ethical and how to get them to be ethical. That view was that human beings have free will. Free will is the belief that each of us (unless we are children beneath the "age of reason" or mentally incompetent) has enough independence from history, context, culture, group, present situation, and even biology to enable us, uniquely as a species, to be the originators (in some sense and to some meaningful degree) of our actions, choosing them freely and hence being individually responsible for them. That claim is the standard reason given for why we deserve praise for good actions and blame for bad ones. Then in Chapter 2 I picked out exemplary cases of both astonishing evil and saintly goodness, mostly transpiring during the Nazi Holocaust, with the hope of bringing into clearer view what needs to be explained: how and why some people, ordinary people, committed astonishingly evil actions and others good (sometimes extraordinarily good) actions. And why did it seem so unpredictable on the face of it, from people's pre-Nazi normal lives, who would be in which camp? Why did so many people engage in evil, and why did others, fewer but still a significant number, engage in acts of exceeding humanity, endangering themselves in rescuing social pariahs? I raised the question of whether there were common factors that led some people toward doing good and others toward doing harm and even murder, then in the next chapter took up that question more systematically by turning to survey and explore the group processes that had emerged and become keenly visible in the Holocaust, and similar evidence. The preliminary answer seemed to be that in the cases of both perpetration and rescue, what came into focus as significant causal factors were social processes of group identity and the authoritative interpretation of situations, the relation to the larger society, group attitudes toward the legitimacy of the dominant authority, and the like, rather than the expected differences in individual character or in the rational free choice of some individuals to act according to moral principles or chosen virtues, and of others not to or to fail to. So at that point I turned to explore what is known from the research on the psychology (and especially the social psychology) of good and evil. I continued to survey the results of experiments whose aim had been to explore rigorously and systematically various conditions that were hypothesized to induce and produce, or at least influence, either good or bad actions from unwary volunteers.

At this juncture, the reader ought to be keenly aware that we are confronting a conflict or at least an unresolved tension: on one hand, the basic assumptions in our society about why we are moral and become moral turn on the standard belief that each of us has free will and ought to exercise our independent moral judgment in all situations and make the right choices, freely making the decisions that we consciously and rationally discern to be in accordance with moral principles and rules or to conform to the virtues of courage, kindness, prudence, fairness, and the like. On the other hand, we have the amassed evidence of how normal people actually behaved in the morally crucial but also extreme situations of the Holocaust and Abu Ghraib prison, for example, and also in the morally significant but not extreme situations of the psych lab, the boardroom, and similar places. In all these situations we were confronted with the fact that freely willed—that is, decisive, independent, rationally thought out, and fully aware—action taken on principle or for virtue's sake, originating with and within ourselves alone, was astoundingly rare, if not altogether glaringly absent. So the standard explanation seems to fail to explain the facts on the ground, accounting neither for moral failure nor, perhaps more remarkably, on the whole, for moral good. We can dismiss, perhaps all too easily, the former failure of explanation: after all, it may be that we ought not to expect people to choose the good and do the right thing most of the time. So our moral failures may be due to the nature and magnitude of moral weakness, weakness of the will, what the philosophers call *akrasia*, namely, knowing the good but doing the bad. If *akrasia* is the problem, then free will is off the hook, because these immoral folks could simply have had a failure of will—a view that still holds that they ought to have used the free will they had.[2] For the moment, let's grant this possibility to the free willers. Yet if they're right, the free will theory ought to account for people's motives and actions when they actually *do* do good—the Holocaust rescuers, for example. But that doesn't seem to be the case at all, for individual freely willed action on principle, taken as the result of rational deliberation about the facts of the case or, alternatively, to enact a chosen virtue, did not appear to explain the evidence I amassed about why those who rescued did so, nor did it predict or explain the behavior of good Samaritans and all the rest.[3]

Current scientific evidence is now pointing in the direction that a human capacity—let alone a separate faculty or, in contemporary lingo, a discrete brain module—for conscious free will is not borne out by the neurosciences. Instead, it now appears that determinism—not reductively material causation but a determinism that includes consciousness—holds sway in the neurobiology of the brain.[4] This means that beliefs and culture, biography and history, social position and location, current situation, and all the rest are causal factors in our actions as much as genetic and other biological factors are. Yet none of these causes and contexts is freely willed—we cannot shed any of them at will. Instead, together they produce us: they produce our thinking and motivation, that is, our actions. And there is no sense in which we stand above them as the "originating cause" of our actions. They are not mere influences upon us, something that is easily dispensed with.

Before we go into the philosophical discussions that address the implications of causal determinism for the possibility of moral freedom and the validity of moral valuations, I want to define some terms and make some distinctions here briefly. There is a vast literature on the topic, and I cannot attempt to do it justice. What I want to do here is to present some of the basic terms and issues of the philosophical debate and then expose some underlying assumptions of the entire enterprise as culturally provincially Western and emergent from barely secularized Latin theological tradition and origins. My aim will be to show that the current debate—which is based on the unexamined presupposition that its terms are universal in that all people by virtue of their humanity hold certain assumptions about their actions and moral nature—in fact and instead takes place within a narrow cultural framework and set of presuppositions. I argue that the assumptions we hold about how and why we are moral, which set the terms of the debate, are merely those of the Latin West. They appeal to deep yet implicit cultural narratives about human nature and about how humans fit into the universe and into the natural world. Setting the belief in free will in its cultural place allows other ways of defining the problem of the psychology of moral agency to enter the picture from different cultural worlds. This chapter exposes the Western Latin provincialism of our philosophical and widespread cultural anthropology, exposing the ongoing hold of a theological story that has been secularized, but barely. The next chapter builds on this one by setting forth a historical and available alternative cultural view of why we are ethical, why we are not, and how we can become more ethical and get others to become so, too. The alternative ethical vision, Spinoza's, harks back to the Greek classical notion of the human person, a notion that was jettisoned by early Latin Christianity but was taken up and developed by Christian, Muslim, and Jewish philosophers in the eastern Mediterranean world.

The Current State of the Philosophical Debate About Free Will

"'Free Will' is a philosophical term of art for a particular sort of capacity of rational agents to choose a course of action from among various alternatives. . . . Most philosophers suppose that the concept of free will is very closely connected to the concept of moral responsibility" and reward or punishment, praise or blame.[5] That is how the *Stanford Encyclopedia of Philosophy* defines free will and its hold on contemporary philosophers. The classical problem of free will involves contradictory claims, and so it poses a logical dilemma. For in the West, people's standard assumptions include the belief that people can act other than the way they in fact do and also the belief that all events are caused. So here is the problem: people believe both that things could *not* have been different (because they were caused to be the way they are) and also that they *could* have been different (because human beings have free will).[6]

In order to make our beliefs logically consistent, philosophers turn to strategies that either reject one or more of these claims or redefine and refine the meaning of one or both. These strategies have produced a standard range of different positions that set the terms of the debate. The field is divided into two camps: *incompatibilists* and *compatibilists*. Incompatibilists hold that a belief in free will and a belief in the causal determination of all events cannot be reconciled. Compatibilists believe that free will and causal determinism can be reconciled. Within these two broad categories there are different variants. Incompatibilists can be agnostics, in the sense that they do not take a stand on whether determinism or free will is true. They can be libertarians, arguing that free will is true and determinism false because the universe is indeterministic, or because "human agents are the cause of freely willed actions" but "are not themselves caused." Or they can be hard determinists who deny free will, which is to say, they deny that people can do anything different from what they do.[7]

Philosophers who are *compatibilists* regard free will and determinism as compatible notions. They argue from a variety of perspectives that one can consistently hold both that all actions have causes that result necessarily in those actions and also that people have the freedom to choose their actions. Compatibilism, in one of its versions, is presently the most widely held position among philosophers.[8] That is now beginning to change, however, in the light of the findings of the new brain sciences.

Classical compatibilists deny that believing that all actions are causally determined means no one could act differently.[9] They argue that even though all actions are causally determined, there is still real choice among alternatives.[10] Other compatibilists, however, argue that determinism does seem to pose a threat to the possibility of real alternatives and choices, for if determinism is true only one future is possible. So these compatibilists argue that an ability to choose from alternative courses of action is not necessary for freedom. Instead what is necessary is that one is the source of one's actions.[11] To be the source of one's actions means that human beings are not determined by webs of relations, biology, psychology, context, and the like, but somehow by some internal will beyond or beneath all these. Nevertheless this model, too, is problematic, for "no compatibilist . . . can deny the truth of the . . . premise . . . [that i]f determinism is true, no one is the ultimate source of her actions."[12] That is because, as determinists, compatibilists believe that all the internal and external causes that make up a person in fact *do* determine how she acts. The compatibilist must come up with an account of agency that both is causally deterministic yet also can accommodate either real alternative choices or some form of substantial self-origination of one's own actions.[13]

Accounts of free will or choice rely upon the mind being a different sort of thing and a different kind of cause from the body. We in the West see ourselves as harboring a deep divide, and it underlies the claim of free will. "The [standard] notion [is] that 'actions' and 'events' are two exclusive categories," the philosopher Joe Keith Green remarks. What this amounts to is that the standard philosophical notion of reasons (for actions) presupposes that those reasons are only mental: we choose mental reasons and then direct the body to act. Those mental reasons are not considered to be in webs of beliefs and culture that we merely adopt, but are instead freely chosen through reason alone as worthy of being acted upon. Events, however, unlike actions, do not come about in this way. They are within contextual causal nexuses rather than chosen from alternatives or fully originated in mental acts via rational considerations (reasons). Compatibilists believe that actions are contextually determined events, yet they also mentally and freely originate in an individual's reasons. They are both events and actions (in this sense), and the two explanations can be reconciled.[14]

Freedom of choice among alternatives (or, alternatively, being the mental originator and source of one's actions) is regarded by compatibilists as well as by free will advocates as being necessary for moral responsibility, for ascribing praise and blame, and for determining degrees of culpability, punishment, and reward. Taking the person out of context—out of influences, appropriations, natural inheritance, and social belonging—seems so natural and obvious as the basis for holding people morally responsible that almost limitless philosophical ink has been spilled on the problem. Yet why does it feel so natural to us to embrace a division between the mental and the physical and to identify the mental with will, and especially with free will? It is a particular Western cultural inheritance, with a history that can be traced from its beginning to the present. For it is a barely secularized version of an Augustinian theological conception of the universe, a conception that still dominates our culture and which we take for granted.

Beginning Again

So we have a situation in which an almost universal belief has been dogmatically held by the general public and also embraced historically and still widely by philosophers, yet at the same time has been called into question by the social and natural sciences and even by a lot of the facts on the ground for the past eighty years or so. This cries out for explanation. Why this depth of adherence to a notion of free will as the sine qua non of moral action and responsibility? What are the origins of the concept, and what accounts for its ongoing appeal? These are the questions of this chapter.

Our standard ideas about moral action and moral responsibility are so strong, so ubiquitous, that our minds are all but closed to alternative construals of what it takes to be ethical. So we need to expose in detail what the standard presuppositions about free will are, why we have them, and why we are stuck in them. Why have the research of Milgram and Zimbardo, the fact of the Holocaust, the concept of groupthink, and all the rest failed to make inroads into our thinking about moral agency?

There is a cultural depth to the problem of ethics in the Western world. The concept of free will is rooted in a pervasive and ancient Latin Christian theological anthropology that has been secularized mostly only on the surface, for our basic Western understanding of moral psychology is embedded deeply within us. What this means is that Western beliefs to this day assume things about human nature that arose as part of the Augustinian theological story. Even though most of us perhaps do not embrace that story literally anymore, or instead reserve that belief for a religious context, there are ideas about human nature implicit in the Augustinian version of the Christian story that continue to have a deep hold on us. These beliefs particularly influence our understanding of moral agency—how and why we act ethically and why we often don't. We'll now turn to the history of how we got to the point at which free will seems to be the only game in town. My major claim is that we in the West are still wedded, if largely beneath our awareness, to a normative Augustinian Christian theological conception of what it means to be human, what it means to be moral, what the human place in the universe is, and what our relation to nature is. While that conception has undergone significant secularization, it still retains its basic outlines, for the secularization is more window dressing than real, and so it serves to disguise—and, paradoxically, to reinforce—rather than transform the basic underlying religious claims about human nature.

Augustine's Worldview and the Origin of the Western Moral Subject: Two Fateful Moves

I am certain that Augustine respected the human will (he practically invented the concept).

—JAMES WETZEL

All the awe, all the sweetness, all the sense of a divine presence brought close to humankind that [the pagan Greek Hellenistic philosopher] Plotinus had seen in the cosmos, Augustine now saw in the perfectly tempered union of human and divine in one single human being, in Christ.

—PETER BROWN

There was a moment of invention, of creating a clear and new conception of human nature from disparate strands. This is what the powerful and powerfully influential fourth- and fifth-century theologian Augustine accomplished. The Augustinian conception of human nature took such powerful hold on the imagination that subsequent wrestling with human nature in the Christian West, in both theology and philosophy, has largely been a footnote to Augustine. Later theologians and philosophers periodically returned directly to Augustinian writings—in both the Reformation and the Counter-Reformation, for example—in response to the pressing problems of their own age, for both inspiration and legitimacy. Earlier, Augustine had been invoked in the thirteenth century to determine how far rebellious Aristotelian-leaning thinkers would be allowed to depart from his conception of free will; in that era, free will was formalized, radicalized, clarified, and nuanced by his advocates, and invoked in attenuated form by opponents. Even now the standard Western philosophical and general cultural beliefs about our moral capacity, on the whole, remain a footnote to Augustine.

Augustine redefined what it means to be human, what it means to be moral, and what the human role in the cosmos is. He made two fateful moves that mark a profound shift from the classical Greek worldview to the one that became commonplace in the West. The first move was to elevate free will to a cosmic principle, to the quintessential nature of God. The second was to carry this through to human nature as well, redefining the human mind in terms of will rather than rational understanding. The overall ideal of human striving, the terms of the relationship between the human person and God and between people (albeit manifested more in the breach than in the fulfillment), was now to hinge on voluntary choice rather than on the quest for knowledge.

The great scholar of late antiquity Peter Brown, at the end of his biography of Augustine—a biography that has now become a classic—proposes that Augustine produced a profound shift away from the culture of the ancient world: "Seen against the wider background of the classical philosophical tradition, Augustine's magnificent preoccupation with the problem of the human person and his fascination with the working of the will represented a decisive change in emphasis." Brown further points out that Augustine has been called the "inventor of our modern notion of will." Augustine deflected the locus of human striving for meaning and purpose away from the philosophic and scientific search for the human place in nature and the cosmos and toward a concern for the individual will. His achievement was a "shift from cosmos to will," "a turn[ing] away from the cosmos," Brown says. The notion of free will and the intensity of focus upon it were, in a sense, shorthand for what Brown calls the "mighty displacement of an entire religious sensibility." Moreover, "this displacement of attention from the cosmos to the saving work of God, through Christ, was the most hotly contested of [Augustine's] many doctrines," Brown points out.[15] He concludes that Augustine's "intervention proved decisive for the emergence of a distinctive notion of the individual in Western culture."

What got left behind was the classical Greek reverence for nature, the sense of the magnificence of a natural world in which humans were at home and via which the glow of the divine could be glimpsed and some sparks of it captured through intellectual understanding. Peter Brown grasps in a nutshell the overall transformation initiated and instituted by Augustine:

[Augustine] allowed the Platonic sense of the majesty of the *cosmos* to grow pale. Lost in the narrow and ever fascinating labyrinth of his preoccupation with the human will . . . Augustine turned his back on the *mundus*, on the magical beauty associated with the material universe in later Platonism. . . .

Augustine would never look up at the stars and gaze at the world around him with the shudder of religious awe that fell upon Plotinus, when he exclaimed . . . "All the place is holy" (as Oedipus had exclaimed at Colonus, and as Jacob had done at Bethel). . . . Augustine pointedly refused to share this enthusiasm. . . . Something was lost, in Western Christendom, by this trenchant and seemingly commonsensical judgement.

Brown even laments that "if Augustine was the 'first modern man,' then it is a 'modernity' bought at a heavy price." For that price was the "dislodg[ing of] the self, somewhat abruptly and without regard to the consequences, from the embrace of a God-filled universe."[16]

By attributing to the human mind (and hence the human person) the character of voluntary self-control and self-origination, Augustine turned away from a Greek classical conception of the mind (and hence human nature) as characterized by its cognitive capacities of critical thinking and insight, theoretical contemplation, natural discovery, and logical deduction and argument. Augustine's fateful turn reoriented Western Latin culture away from the Platonic intellectualist conception of human moral nature as either clear-sighted or confused and benighted (and in either case within the natural order) and toward the idea of a human person as fundamentally moral or immoral, responsible or irresponsible, obedient or sinful through choice of action rather than through understanding and character.[17] In the Platonic tradition, by contrast, the body's corruption was responsible for the mind being morally clouded; hence moral ignorance—not active sin but the Greek *hamartia*, "missing the mark"—was the result of the problems inherent in embodiment. Aristotle's view was a nuance on the Platonic: his was an account of moral action as stemming from moral character. In this theory, early socialization shaped desire, enabling a person to have the capacity for moral discernment and understanding, as well as deliberative reasoning. Augustine, in contrast, explicitly rejected the body as the source of ignorance or error, neither of which, in any case, could in his view ever account for sin. He regarded that view as pagan and said, "Those who suppose the ills of the soul derive from the body are in error." Augustine in *The City of God* takes on Virgil's moral Platonism, pointedly remarking, "Our faith . . . is something very different. For the corruption of the body, which presseth down the soul, was not the cause of the first sin, but its punishment; nor was it corruptible flesh that made the soul sinful, but the sinful soul that made the flesh corruptible."[18] So it is not the body that is the source of the moral problem. Desire, Augustine further remarked, is not even possible without the soul, for it is not a merely bodily phenomenon. It takes a soul to have desires.[19] The hardly controllable and disease-prone body, common to all humanity after Adam's fall, is the divine *punishment* for sin, not the origin of sin, he insists. Souls are therefore not naturally better or worse due to their bodies, but instead, Augustine rhetorically asks: "For on what basis are souls good or better, or on the other hand not good or less good, except by their manner of life adopted by the free choice of the will . . . ?"[20]

The upshot of Augustine's reduction of all internal mental operations—thoughts, emotions, feelings, judgments, learning—to acts of will is a new theory of moral psychology. This new theory amounted to nothing less than a shift in worldview—in the conception of the human person and of the universe that human beings inhabit and, hence, in the conception of moral agency— initiating a decisive break with the past by focusing on the freedom of the will and a concomitant demotion of nature. It is this worldview that we have inherited.

Augustine's First Crucial Move: Divine Will Overtakes Divine Wisdom

I want to elaborate here for a moment on the first crucial move that Augustine introduced to produce an overall reorientation in worldview. And I want to define more precisely his new conception of what it means for the human person to be a moral subject. The first move I want to call attention to is what in theological language is referred to as an emphasis on the divine will over the divine wisdom. Augustine does not hold the complete "reduction of the divine wisdom to the divine will," yet he tends in that direction. It will be Descartes who "out-Augustines" Augustine, so to speak, and reduces all cognition to will, in both God and the human mind. The divine wisdom, in the theological tradition, refers to God's knowledge: not the knowledge *of* the universe so much as the knowledge that God *puts into* the universe, namely, the regularities produced by independent forces of nature initially created and set in motion by God. So what is at stake here in the theological debate is God as the creator of a universe that embodies natural laws, laws that are thought to be the content of God's mind (an assimilation of Platonic forms to Abrahamic religious sensibilities); or, alternatively, a

God who creates the universe more by fiat than by science, so to speak.[21] The latter is a God whose wisdom is subordinated to the divine will.

If God's wisdom is emphasized, then God is considered to be the source of a natural necessity that plays out, but at the same time God is also in some sense subsequently limited in action and intervention by that natural order, or the naturalness of that order, in action and intervention. The shift from divine wisdom to divine will marks a move away from an ancient classical conception of nature and cosmos as redolent with the divine and exemplars of divine presence, and toward a different Christian view, one focused on the Incarnation. The historian Stephen Menn pinpoints the shift from the classical ancient to the Augustinian Christian worldview:

> The Incarnation [for Augustine] presupposes that God has a will, which is not reducible to God's knowledge [i.e., wisdom] . . . . For the Platonists, [by contrast,] God rules in accordance with the nature of things:. . . there is therefore no need for him to choose what he should send where. But it is not at all in accord with the nature of the recipient that Nous [the divine mind] itself should descend into a human body: this requires a will in God. . . .
>
> Augustine thinks that the Platonists' rejection (or incomprehension) of such a will in God, and of the Incarnation, arises from their presumption. . . . For Augustine, it is this crucial difference between Christians and Platonists that requires something like theology, and not simply scripture on the one hand and allegorizing philosophy on the other: where Incarnation and, more fundamentally, will, are central notions of this theology.[22]

The divine will has overtaken the divine wisdom.[23]

The key to understanding Augustine's fateful move from divine wisdom to divine will is to see the move in terms of the role of miracles versus the role of natural causal processes. Augustine has elevated the importance of God's *miraculous* power over God's power as the source of the causal regularities of nature. For miracles represent the breaking through of a personal divine will into the regularities of the forces of natural lawful necessity. A miracle is the momentary (or not so momentary) purposeful suspension of the laws of nature. If there *are* no laws and forces of nature, then everything is a miracle, an unfathomable divine decision in any given moment. For example, on this view, that the flame was lit when the match was struck would be a divine free choice that could go the other way next time and needs divine intervention in each instance. (A minority sect of Muslim theologians, the Ash'arites, were famous for holding this position.) But the plain meaning of many passages of the Hebrew Bible (Old Testament) is that there is a nature, created by God, that functions independently of any direct divine intervention but which nevertheless can be suspended by God for some purposeful intervention—for example, to save someone or a group from danger or death. In the biblical cases of divine intervention we have the God of history or will more than the God of nature or wisdom. Augustine is in a sense making the God of history a more central description of divine power than is the description of God as the wisdom and power behind nature. When Augustine takes the Garden of Eden as historically and literally true, he has opted for the God of history over the God of nature.

Augustine's Second Crucial Move: The Human Mind Is Will

The second fateful move that Augustine makes is that the human mind is will more than thought. In his *Confessions*, Augustine reveals that he discovered the centrality and significance of free will through looking inward. His introspective gaze revealed to him that at his core was a free will. He realized that he himself alone was responsible for his choices, and so he was responsible for himself—and that was because he had a free will, a will that fully originated his actions, both the good and the bad. That free will was at the core of what it means to be human and to be capable of acting morally or immorally. In a passage presaging Descartes's *cogito* (reminiscent of Descartes's "I think [meaning, I have inner mental experience], therefore I am") Augustine writes:

> I was brought up into your light by the fact that I knew myself both to have a will and to be alive. Therefore when I willed or did not will something, I was utterly certain that none other than myself was willing or not willing. There lay the cause of my sins I was now coming to recognize.
>
> I inquired what wickedness is; and I did not find a substance but a perversity of will twisted away from the highest substance, you O God, towards inferior things, rejecting its own inner life . . . and swelling with external matter.[24]

The interior mental realm, all that is included within one's subjectivity, is essentially voluntary. Will is not just one psychic faculty among others—

sensation and imagination and daydreaming, for example—but instead is the very character of the inner life itself. Descartes will later follow and extend Augustine's claim, arguing that even truth is a kind of volition rather than a basically cognitive insight and understanding.

The mind, the soul, is fundamentally voluntary rather than cognitive, in analogy to the divine will subordinating the divine wisdom. Human reason is subsumed within the voluntary character of the mind. Whereas in the Greek classical philosophical tradition the human soul was quintessentially characterized by its rationality, meaning its intellectual capacity to acquire knowledge and understanding, for Augustine the rationality of the soul is subordinated to, and to a significant extent swallowed up by, its will. Bonnie Kent, the foremost authority on the will in the history of Christianity, points out that "Augustine . . . did not see the will as a rational or intellectual appetite; for him the will was the entire soul as active."[25] We can discern Augustine's swallowing up of reason in will obliquely in his comments on various biblical passages. When he speaks of human reason he frequently goes on to explain that what he means by reason is the *voluntary choice* of good action. The following passage from *The Literal Meaning of Genesis* is a good example of his conflation of human reason with good will:

> God is the unchangeable Good; man . . . is indeed a good, but not unchangeable Good as God is. A changeable good . . . becomes a greater good when it adheres to the unchangeable Good, loving and serving Him with *a rational and free response of the will.* . . .
>
> [A] rational creature is no small good, even a creature that is led to avoid evil by a consideration of evildoers. This class of good creatures would surely not exist if God had converted all evil wills to good.[26]

It is the voluntary, good will that Augustine also refers to by the Pauline phrase "law of the mind." In describing Adam and Eve before the Fall, Augustine uses that phrase to refer to the (easy) control of desire by the will. He opposes that law to the Pauline "law of sin in the members." So *mind* and *sin* are opposites here: *mind* means "good free choice of action," whereas its opposite, *sin*, means "bad free choice of action." When Augustine uses *mind*, he most often means *will*.

Augustine even goes so far as to identify all of the capacities of the mind with the will, not just the intellectual ones. In the entire cosmos—in God, in the angels, and in human beings—all the capacities of the soul come down to the will. Even the emotions, for Augustine, turn out to be the *effects*, rather than the causes or motivations, of the will. He reduces all emotions to acts of (good or bad) will: "What is important here is the quality of a man's will. For if the will is perverse, the emotions will be perverse; but if it is righteous, the emotions will be not only blameless, but praiseworthy. The will is engaged in all of them; indeed, they are all no more than acts of the will."[27]

So emotions, as acts of will or the effects of acts of will, turn out to be voluntary choices rather than character states.[28] That the human soul is at bottom a giver or withholder of consent, that is, a will, renders every human person an individual moral actor of a very particular stripe. Augustine does not hesitate to state explicitly the larger claim he is making: the human mind, because it is a voluntary will, is *culpable*. Because we have—or are—a will that freely chooses, we can be praised and blamed for our decisions and actions. Human nature is essentially about being morally good or bad according to actions freely taken. Gone is the classical Greek notion of human beings as rational animals whose degree of ignorance or wisdom shapes their desires, purposes, emotions, and character.

Augustine's Two Cosmic Principles: The Voluntary Order of Mind and the (Quasi-) Natural Order of Matter

> In [the universe] is found a double activity of Providence, the natural and the voluntary.
>
> —AUGUSTINE, *The Literal Meaning of Genesis*

When we look at the universe as a whole, Augustine says, we find God's divine action pervading it in two ways. There are two cosmic principles that govern the universe: one is nature (the natural) and the other is free will (the voluntary). Both are cosmic systems and forces. Each displays God's ongoing involvement and participation. Both are expressions of the divine providence, of God's care for the world. God has not set them up and withdrawn, Augustine insists. Instead, "the natural working of Providence can be seen in God's hidden governance of the world, by which He gives growth to trees and plants; the voluntary working of His providence can be observed in the deeds of angels and men." Augustine's illustrations of the natural working of divine providence are just what we would expect:

The heavenly bodies above and the earthly bodies below follow an established order: the stars and other heavenly bodies shine, night follows day and day follows night, earth firmly established is washed and encircled by waters, air moves above all around, trees and animals are generated and born, develop, grow old, and die; and so it is with everything else in nature that comes about by an interior, natural movement.

The divine wisdom or truth working in nature, Augustine elaborates, introduces numerical form into bodies, and hence material processes can be described in mathematical terms and in terms of regularities.[29] Augustine describes the arena of the voluntary working of providence, on the other hand, as learning, agriculture and animal husbandry, political governance, and the arts, not only on earth but also "in the heavenly society."

The same cosmic dualism holds within the human person: Augustine goes on to say that "in man himself the same twofold power of Providence is at work," for the body is in the domain of "natural" providence, whereas the soul is within the domain of "voluntary" providence. "First, there is the natural work of Providence in respect of the body, that is the movement by which it comes into being, develops, and grows old." Second, he says, "there is the voluntary working of Providence in so far as provision is made for his food, his clothing, and his well-being." Even within the human soul—and not just between soul and body—there is a division between its natural dimensions and its voluntary ones; Augustine informs us that "by nature [the soul] is provided that it lives and has sensation; by voluntary action [however] it is provided that it acquires knowledge and lives in harmony."[30] That is to say, the soul has within it natural functions that animate the body and make it sensate, and it also has voluntary functions that inform its mental and social capacities.

The human body is under the sway of nature, yet it is freely directed by the human mind. Human free will commands the human body.

The soul which commands the movement [of a part of the body] remains unmoved in space. . . . The soul is not a corporeal substance and does not fill the body in space as water does a skin bottle or a sponge, but in a mysterious way by its incorporeal command it is united to the body, which it vivifies, and by this command it rules the body through an influence, not a corporeal mass.[31]

The human mind, too, is permeated with divine wisdom. Unlike mere matter, however, the mind can turn away from God's wisdom through the exercise of its own free will. That is why our freedom, Augustine emphasizes, is the source of moral evil.

This dual divine system of care, the natural and the voluntary, is carried out by angels, and angels (like human beings but more effectively) act voluntarily, that is, by free choice.[32] The good angels choose only the good, Augustine tells us, yet there are also bad angels, who fell because of their free choice of evil. The fallen angels no longer occupy the heaven of heavens but instead reside in everlasting torment.

That Augustine conceives the voluntary as one of the two principles that divide the cosmos—nature being the other—means the freedom that characterizes divine action (and which is the basis of its goodness) entails that God, humans, and angels are all undetermined. They differ in that they are better or worse, higher or lower exemplifiers of that voluntary ontological principle and realm. The human person is of the voluntary cosmic order, but is not a very good or lofty occupant of that order. In fact, the human person is the lowest of the three types of being in that order, namely, God, angel, and human. So it is hardly surprising that human beings, unlike God and the good angels, often fail to freely choose the good. Doing wrong does not mean that they fail to use their free will, however, Augustine insists, and consequently become naturally determined by the body. Rather, they use their free will badly and freely choose to follow bodily desires that the soul makes present to them, for it is only the soul that makes desires present and not the body itself, Augustine insists.

The divine cosmic rule of the voluntary is expressed in the goodness of the free gift of divine grace on God's part, and at best by the freely chosen obedience of the human will, enabled by grace—or at worst by the freely chosen rejection of or rebellion against that grace. That the human soul is a direct divine creation of God's will and, as will, is in the divine image—it is not part of the divine substance itself but rather a divine creation, Augustine insists—makes it a "little will" to God's great will. Hence human beings can use the will to do evil as well as good, but they cannot be *other than will*. The voluntariness of the human soul or mind, unlike the human body, introduces a rupture in the universe and precludes any role for nature as intermediary in the human-divine relation. The human person relates to God via free self-creation, turning toward good in obedience or toward evil in rebellion. The divine will freely offers grace to mitigate the worst effects of human free will (in principle for all via the Incarnation but in practice for some via predestination and baptism).

It is "God by the twofold working of His providence [who] is over all creatures, that is, over natures that they have existence, and over wills that they may do nothing without either His command or His permission."[33] In describing the two orders of creation, Augustine maintains that God has direct ongoing supremacy even over the fixed natural order:

> God has established in the temporal order fixed laws governing the production of kinds of being and qualities of being and bringing them forth from a hidden state into full view, but His will is supreme over all. By His power He has given numbers to His creation, but He has not bound His power by these numbers. For His Spirit "moved over" the world that was to be made in such a way that He still moves over the world that is made, not by a material space relationship but by the excellence of His power.[34]

So Augustine insists that the natural causal system does not limit God's power, nor does it introduce any necessity into the freedom of God's will.

Nature as Matter, Spirit as Will

Nature is not fully nature in the normal sense of necessary, rule-bound processes and forces playing out without direct divine intervention, according to Augustine, for nature, expressive of divine wisdom, can be overridden by the power of divine will. The human will, along with the wills of the angels and God's will, is not of the natural order at all. The human will, like the divine will, is not subject to "internal natural wisdom," to use theological language, which is to say that it is not subject to a necessary natural order of causes. The will is of a different causal stripe; it is of the voluntary causal cosmic order. The human will is like God's historical fiat, and like God's absolute freedom, the human will is in principle capable of transcending the (quasi-) necessity of natural processes, internal and external. It is in that sense that Augustine designates the human will, like God's will, as free.

The human will, along with the divine wills (of angels and God), belongs to a different order than the natural order; it belongs to the miraculous order. Human mind and God are connected as the same kind of thing (that is, they are connected ontologically); they are also directly connected via a relationship. The relationship of God and humanity bypasses nature (body and material world) but joins human will to divine will. It is a free choice of the human person, as potentially spiritual, to willfully master the natural body and material world through voluntary obedience to God's will. The human will belongs to God's personal order of direct action, so to speak, rather than the indirect divine order of the laws and forces of material nature.

In the biblical account of God's creation of Adam, which Augustine takes to be historical fact, God directly breathes life (the soul) into Adam's natural material and earthly body. The human soul, unlike the body, is a direct divine gift of spirit, that is, of the will. So in the end, the divine wisdom (which refers to the lawfulness of nature) is subordinated to the divine will, God's willful fiat. Nevertheless, *nature* is generally and for most purposes an apt description of the material realm. So Augustine generally distinguishes the divine wisdom from the divine will, the former being the source of the operation of nature-matter, in contrast with the operation of unpredictable and denatured, fully self-originated and self-originative choices. Of course the natural necessity of even material processes can always be disrupted by the divine will; more than that, it carries on carrying on due only to some kind of direct willful intervention of angels at God's command. So given the two cosmic orders, the voluntary (the spiritual) can overrun the natural (the material), in both human and divine.

The Natural Versus the Moral

The natural and the moral, like the body and the soul, are antonyms, types of being set in ontological opposition. They are opposite types of being. The soul or mind's freedom from being determined by natural constitution is just what enables it to be a moral subject in the Augustinian conception. The will is free in respect to nature. Nor does any social shaping have a necessary determining effect on freedom of action. *Freedom from nature is at the origin of our moral capacity for Augustine.* Even when Adam sins, his punishment is not that the soul or mind becomes natural but rather that it becomes a less good or effective soul—for soul and will it remains. It is only the body that is (in any degree) "natural." Augustine makes the point in his early work *De Libero Arbitrio*: "Because we have no doubt that the soul's motion is culpable we must absolutely deny that it is natural."[35] And in the late work *The City of God* he still holds, "Where the will becomes evil, this evil would not arise in it if the will itself were unwilling; and its defects are therefore justly punished, because they are not necessary, but voluntary. For the defections of the will are not toward evil things, but are themselves evil."[36]

That the human mind was created "in the image of God" means for Augustine that its quintessential character is its free will, and hence it has no determination or shaping that it cannot resist if it wills to do so. The relation of the human soul to God is will to will. The human person's burden is to turn the human will toward God so as to make the human will conform to the divine will. For Augustine, the ideal relation of the human to the divine is the free obedience and subordination of the human will to the divine will. Augustine repeats often that the human good will in practice is to be expressed as an undetermined choice to turn to God in absolute obedience to God's will. Numerous passages could be cited to illustrate the point, but the following will suffice:

> This account [of the Garden of Eden] was written . . . to remind [the human person] how important it is to recognize God as his Lord, that is, to be obedient under His rule rather than to live uncontrolled and abuse his freedom. . . .
>
> When man does not depart from God, by this very fact, since God is present to him, he is justified, illuminated, and made happy, and God is cultivating and guarding him while he is obedient and subject to God's commands. . . .
>
> Finally, nothing else is sought by the sinner except to be free of the sovereignty of God when he does a deed that is sinful only in so far as God forbids it.[37]

Because the human-divine relation is both of like to like and also of direct conformity of will to will, the human person is disconnected from nature and body. Both the soul and God are different in kind from nature, which is the arena of material processes alone. So for Augustine nature is not the primary intermediary between the human and the divine, as it had been for the ancients. For the classical mind, in contrast to Augustine, to come to know lawful nature was to come to know a part of God's mind and thereby to unite the human mind with some dimension of cosmic reason.

What I want to point to particularly here is that the subordination of wisdom (knowledge and understanding) to will in both divine and human minds results in the introduction of a notion of the individual human as a God-like inventor of his or her actions and self. Or perhaps this was the purpose all along, since we saw that Augustine proposed that it was "because we have no doubt that the soul's motion is culpable we must absolutely deny that it is natural."[38] Thus actions are self-created and the self's self-invention enables their fundamental character to be either good or evil. So the human person, who is the soul—which is identified with free will, as it is self-governing by the voluntary ontological cosmic principle—is completely blameworthy or praiseworthy, and never a product of necessary natural forces either internal or external. While the body and all material processes have some degree of automaticity and determination by external forces and as a result are to a large degree left to a working out of inherent natures and laws of nature that can be expressed mathematically, Augustine's notion of mind is precisely the opposite of body and nature. Mind, by definition, is that which is neither natural nor necessary. Nature alone is the arena of necessity, and nature is exclusively the material or bodily. So all mental capacities are non-natural and non-necessary—and hence voluntary and culpable. Here is the birth of the modern moral subject—and for that matter, of the modern conception of the natural arena as confined to material and material processes, which are deemed the only ones amenable to scientific causal inquiry. Henceforth "natural" causes are confined to the strictly material. The modern Western person is a theological invention and has a miraculous character that is based on Augustine's deliberate yanking of mind out of nature. We in the West are still struggling with this Augustinian turn in the notion of the human person; we still labor under it as if there were no other way to understand human moral capacity and agency—and, concomitantly, nature as its opposite or inverse.

Augustine's commitment to the utter individual moral culpability of the human being, enabled and guaranteed by the divine gift of the freedom of the human will, never wavered. In running the ontological division between nature and will through the universe as a whole and as a bifurcation within the human person in particular—an assimilation of dualist Platonism to a Christian theology of salvation—Augustine projected upon the human being the grandiose ideal and severe moral expectations worthy only of a god and capable of execution only by a divine-like, miraculous will and person who transcends nature and nurture. This grandiosity and severe moralism prevail, untempered by theological mitigating notions of divine grace, in the secularizations of the Augustinian notion—for example, in Descartes's moral psychology and even in Kant's. Even the many contemporary philosophers who have modified the Augustinian notion of the self to include some limitations to human self-invention by natural determination, social and contextual situatedness, and the like, nevertheless continue to assign moral responsibility to a circumscribed core of the human person to which they believe that freedom of will can still be attributed. So the Augustinian conception of the foundations of the human moral capacity in a divine-like freedom from nature and social constitution still largely prevails.

Augustine on the Garden of Eden

> [Adam and Eve's] nature was changed for the worse in proportion to the magnitude of their sin, so that what arose as a punishment in the first human being who sinned also follows as a natural consequence in the rest who are born of them. . . . [T]hat conjugal pair received the divine sentence of its own damnation.
>
> —AUGUSTINE, *City of God Against the Pagans*

We can uncover Augustine's beliefs about human nature in more detail by investigating his writings on the Garden of Eden. Adam and Eve loomed large in Augustine's thought, and he wrote about them throughout his lifetime, developing his account in a number of ways but also sticking to his main grasp of human uniqueness, God's relation to humanity, and the divine origin and operation of the cosmos. The Garden of Eden story is a wonderful blank screen upon which many theologians and philosophers have projected their own conceptions and theories, so it will serve us well here. Maimonides and Spinoza also use it for purposes similar to Augustine's—but they arrive at entirely different, and largely diametrically opposite, conclusions about human beings, God, and the cosmos. In the next chapter we will find that even so modern a philosopher as Kant follows a long history of Christian philosophers who have used the Garden to put forward a philosophical and theological anthropology, and he weighs in on the Augustinian side. Descartes seems to have eschewed the opportunity to retell the story of Adam and Eve at any length (referring to it only briefly in *Principles of Philosophy*, III, article 45) while still embracing, and to some extent reviving and even radicalizing, an Augustinian conception of both man and world—minus the explicit theology. What appears to me remarkable is the stability of the Augustinian conception of the human person, once articulated and established, across millennia.

Augustine wrote two separate books on the interpretation of the story of Adam and Eve in the Garden of Eden. The first he wrote soon after his repudiation of cosmic dualist Manichaeism, a commitment he had held for nine years before his embrace of Catholic Christianity in 387 c.e. It was in a work titled *De Genesi Adversus Manicheos*, which he wrote to challenge the Manichaean dogma of a god of evil and a god of good and also the Manichaean rejection of the Hebrew Bible (Old Testament), that Augustine first addressed the Garden and Adam's disobedience. He struggled with the problem of the origin of evil and had embraced the Manichaean solution of a "kingdom of darkness" and a "kingdom of light" until he worked out a solution that preserved both God's omnipotence and all-goodness while still accounting for the reality of evil and sin.[39] That solution to the problem of evil enabled Augustine to relinquish Manichaean ontological and cosmic dualism without having to attribute evil to God. This first work on Genesis relied on metaphor and allegory to accomplish the reconciliation of a good and all-powerful God with the existence of evil. It had a contemplative focus, as did Augustine's life at that time.[40] When Augustine took up the issue again, however, a few years later, he was writing at a time when he had returned to an active life. This new work he called *De Genesi ad Litteram* (*The Literal Meaning of Genesis*), and it is an important source of his interpretation of the Garden of Eden, as are the twelfth and thirteenth chapters of *The Confessions* and also the twelfth, thirteenth, and fourteenth chapters of his late great work, *The City of God Against the Pagans*.

In *The Literal Meaning of Genesis*, Augustine emphasizes that the Garden of Eden account is historical—it is literally true. It happened. In his view, the historicity of the text did not vitiate its metaphorical meaning; rather, Augustine now felt that he could offer and maintain a literal approach to the text along with more symbolic meanings. One of the burdens of interpretation that arise for anyone who is committed to the (more or less) literal meaning and historicity of the text is to reconcile what appear on the surface to be two conflicting accounts. For Genesis puts back-to-back in its first few chapters two different accounts of the creation of the universe and also two accounts of the divine creation of the first two human beings. The first chapter of Genesis recounts the creation of the world in six days, beginning with the creation of light on the first day. Prior to that the text says, "The earth was unformed and void, with darkness covering the surface of the deep and a wind from God sweeping over the water." As many may recall, subsequent days detail the creation of the sky, of dry land and vegetation, the moon and stars and the sun, sea creatures and birds, terrestrial beasts, and finally man. "God created man in His image, in the image of God He created him; male and female He created them."

The second creation story follows in the second chapter of Genesis. Here creation seemingly transpires in a different order. Man is created before vegetation and from the dust of the earth, and God is said to have given him life by breathing into his nostrils. It is then that God is said to have planted a garden in Eden, only then planting all trees, and putting the man there. The famous account of the creation of Eve from Adam's rib follows, giving Adam a "help-meet" (or "helper," as the Jewish Publication Society now translates it). But first God created all the wild beasts and birds from the earth as potential companions for Adam (each species of which Adam names), and they are all judged lacking as worthy companions. As Adam had named the beasts, he now names the woman. Earlier (Genesis 2:16–17) God was depicted as having issued the famous command when He put the man in the garden to tend it:

And the Lord God commanded the man, saying, "Of every tree of the garden you are free to eat; but as for the tree of knowledge of good and bad, you must not eat of it; for as soon as you eat of it, you shall die."

There are numerous other differences between the two accounts, though we need not concern ourselves with most of them here. Our concern will be only with what Augustine makes of the discrepancy per se and his strategy for reconciling the two texts. Genesis 2 becomes roughly the historical implementation of what Augustine proposes is the ahistorical or atemporal plan of Genesis 1. For he insists that only with the bringing into being of actual things does time begin. So the "days" of creation of Genesis 1 are not temporal days but represent an order of ontological priority or value.[41] Augustine's strategy enables him to maintain that in some sense creation was complete at the end of the sixth day (as the Bible says), while, at the same time, what has been set is yet to be worked out in time through the direct intervention of God—and, most important, through the divine intervention that punishes Adam's sin by a degradation of human nature and even of nature writ large. Nature is set as natural processes through the divine wisdom, but Adam's sin engages God's direct punitive action to intervene and change human nature and even nature itself.

Free Will and the Fall of Human Nature

In his interpretation of Adam's Fall, Augustine brings home with extraordinary force the implications of writing the will (both human and divine) into the very structure of the universe as a cosmic voluntary principle, rather than confining the voluntary to an intrapsychic human capacity. Augustine argues that Adam sinned voluntarily, by an act of free will. His perverted use and abuse of human freedom led God to introduce a rupture into the very cosmos. For Adam's sin resulted in a divine punishment that took the form of a degradation of human nature, and so of nature itself. Augustine writes that from Adam's "evil use of free will there arose the whole series of calamities by which the human race is led by a succession of miseries from its depraved origin, as from a corrupt root."[42] It was not that the tree of knowledge of good and evil was itself harmful or that the eating of its fruit was a sin in and of itself. On the contrary, Augustine says that everything God put in the Garden of Eden was good. What was sinful was only the disobedient free act of the human will itself. Adam and Eve's disobedience of the divine commandment was a "betrayal [that] occur[red] as an act of free will."[43] Augustine proposes that "the man was forbidden to touch that tree, which was not evil, so that the observance of the command in itself would be a good for him and its violation an evil."[44] Because the divine providential goodness consists in its dual principles, the natural and the voluntary, the voluntary cannot preclude the possibility of its abuse. For "the providence of God rules and administers the whole creation, both natures and wills: natures in order to give them existence, wills so that those that are good may not be without merit, and those that are evil may not go unpunished."[45] The moralization of the cosmos is at the heart of the divine plan. The cosmos would be less perfect, in Augustine's view, if it were not structured to express the divine moral import, not only as reward and punishment in a heaven and hell but, most important, by endowing human souls (and those of the angels) with a special ability, "free will," which makes divine moral reward and punishment just.[46]

Augustine held that the will is an inherent human capacity and humans' superiority to animals lies in their possession of a free will. "Human nature, even in sinners is superior to the [nature of] beasts," he insists.[47] The misuse of that freedom in doing evil, therefore, redounds to the human person alone, and not to God's creation, which is all good. So "those who have chosen evil have willingly and culpably corrupted a praiseworthy nature."[48]

This claim is at the heart of Augustine's answer to the problem of evil: since evil is in the exercise of the psychic capacity of the person who commits the action, it does not devolve upon God, who is all good. Augustine reminds the reader that the human person "is separated from God not by a distance in space but [only] by a turning away of his will." [49] And good, too, works via the will, for to be worthy of its name, "our love for Him [must be] freely given."[50] Human beings originate their actions via will not only before the Fall but after it as well. Even though the first human beings' free choice could result—and did result—in sin and a divine punishment that was a degradation of human nature, nevertheless, if human beings "did not possess free will, [they] would not have the same excellence in nature," Augustine says.[51] Human nature is still superior to that of animals even though that nature is not as perfect as it once was.

Writing late in his life in *The City of God*, Augustine says insistently and with complete clarity that good and evil originate in the will: "No one suffers punishment for faults of nature, but for vices of the will; for even the vice which has come to seem natural because strengthened by habit or because it has taken an undue hold derives its origin from the will."[52] Augustine rejects the idea that God's foreknowledge that Adam and Eve would sin undoes the evil character of the action of their wills, for "the evil will was theirs, not His." After all, God "made them in such a way as to leave it in their power to perform some deed, even if they should deliberately choose evil." They alone freely acted and hence "their evil will comes from themselves; their nature, which is good, and their punishment, which is just, come from God."[53] Augustine muses, "Why, then, would God not allow a man to be tempted, although He foreknew that he would yield?" The answer is this: "For the man would do the deed by his own free will, and thus incur guilt, and he would have to undergo punishment according to God's justice to be restored to right order."[54]

God punishes Adam and Eve with a change for the worse in their nature. The degradation of Adam and Eve's nature is also transmitted to all their descendants, to all humankind. In the biblical text (Genesis 2:17) the divine command to Adam and Eve not to eat the fruit of the tree of knowledge of good and evil ended with the warning that "as soon as you eat of it, you shall die." Yet the actual punishment at the end of the story does not include their immediate death. Augustine begins by addressing this discrepancy: "Although the bodies of our first parents were natural bodies," he writes, "we should not suppose that they were 'dead' before they sinned—I mean necessarily destined for death." Their bodies "would have received an angelic form and heavenly quality," transforming their natural bodies in good time if they had not sinned, he says. In proposing that Adam and Eve had natural bodies, Augustine means that their bodies were not the angelic spiritual bodies of the good angels in the heaven of heavens, but were instead bodies in a real corporeal sense. He even speculates that had Adam and Eve not sinned, they would nevertheless have had sex in the Garden of Eden and produced offspring.[55] But they would not have died, and the number of human beings born would have been of a stable and controlled number that would fill the earth but not overfill it. So death enters the world as a result of and punishment for sin.

It is not only death in the usual sense that is the result of Adam's Fall. Augustine portrays Adam and Eve's (and humanity's inherited) punishment as a kind of death within life.[56] The divine punishment is that the human body has been transformed into a kind of maimed body, a body inherently sick and harboring its own illness, deterioration, and death. "What, indeed, is this life from our birth, even from our conception, but the beginning of a sickness by which we must die?" he asks.[57] Augustine goes on to write that Adam and Eve's punishment was that their "bodies contracted, as it were, the deadly disease of death, and this changed the gift by which they had ruled the body so perfectly." The "gift" is a reference to free will. In punishment God changed the nature of their bodies from one that was completely and easily controllable "by the law of the mind" to a body whose "law" was at war with the law of the mind. Augustine (quoting and interpreting Paul) characterizes this other "law" as the "law of sin in the members." The divine punishment is that the body, in Augustine's view, became unruly, much harder to control by the will. He asks rhetorically, "What punishment could have been more deserved than that the body, made to serve the soul, should not be willing to obey every command of the soul, just as the soul herself refused to serve her Lord."[58] The will is not entirely disabled in its control of the post-Fall body, but it has much more difficulty in doing so. Nevertheless, it is still held morally responsible for human actions.

One of the two characteristics that figure again and again in Augustine's representation of the fallenness of the human body in its post-Edenic condition is the uncontrollability of male sexual erection and its dependence on the capriciousness of desire.[59] Adam and Eve, if they had not sinned, would have had sex in the Garden of Eden, but without desire, Augustine insists, and Adam would have moved his penis for impregnation in the way that one moves a hand or a foot. Second, human fallenness is evident in the deterioration of the body in aging and disease. Both of these corporeal limitations weigh heavily on Augustine's mind and fill him with loathing and disgust, serving as obvious signs of a human nature that has undergone a great change for the worse from its originally divinely intended natural perfection. The kind of body that the first Adam had, Augustine elaborates, was an "animal body," which is "the kind of body that we have now," "although it would not have died had he not sinned." After Adam sinned, however, "its nature was so changed and vitiated by sin that we now stand under the necessity of death."[60]

In *The Literal Meaning of Genesis*, Augustine envisions the wonderful, perfect bodies of Adam and Eve before they sinned:

> Why then should it seem beyond belief that He made the bodies of the first human beings in such a way that, if they had not sinned and had not immediately thereupon contracted a disease which would bring death, they would move the members by which offspring are generated in the same way that one commands his feet when he walks, so that conception would take place without passion and birth without pain?[61]

But this was all to change not just for Adam and Eve but also for us all. For "human nature was so vitiated and changed in [Adam] . . . that he suffered in his members the conflict of disobedient lust."[62]

> Because it had of its own free will forsaken its superior Lord, it no longer held its own inferior servant in obedience to its will. Nor could it in any way keep the flesh in subjection, as it would always have been able to do if it had itself remained subject to God. Then began the flesh to lust against the Spirit. . . . [W]e bear in our members, and in our vitiated nature, the striving of the flesh, or indeed, its victory.[63]

Should the life we now live rightly even be called life? Augustine asks rhetorically.[64]

God's punishment for Adam's sin was a degradation in the very nature of human nature. The extremity of the vitiation for all humanity of the natural goodness, which the divine punishment instituted and which subsequently was transmitted to all Adam and Eve's progeny, is made clear by Augustine in this passage:

How happy, then, were the first human beings, neither troubled by any distur-
bance of the mind nor pained by any disorder of the body! And the whole
universal fellowship of mankind would have been just as happy had our first
parents not committed that evil deed whose effect was to be transmitted by
their posterity, and if none of their stock had sown in wickedness what they
must reap in damnation.[65]

When Augustine discusses the Genesis account of the Garden of Eden in
The City of God, he says that the text refers to two deaths: the death of the
body and also the death of the soul.[66] So Adam died two deaths. "The first
death, which is common to all men, was brought about by the sin which,
in one man, became common to all." The "second death . . . [which] is not
common to all men," was that of the soul.[67] Yet the story does not end here,
of course, for there is a new Adam, Christ. It is Christ, says Augustine,
elaborating on Paul, who restores what Adam had perverted, for the first
Adam was "of the earth" but the second Adam is the "Lord from heaven."
This animal body "Christ Himself deigned to assume . . . by choice." It is the
"spiritual body," Augustine concludes, that "Christ Himself as our Head,
already has; and this is the kind of body which His members will have at the
final resurrection of the dead."[68] At the final resurrection, our spiritual
bodies will have the perfect free will of the good angels.

The Augustinian Legacy

Augustine, summoning all his eloquence and fury, argued for a view of nature
utterly antithetical to scientific naturalism.

—ELAINE PAGELS

Augustine's characterization of all mental operations (divine as well as
human) as voluntary and his circumscription of nature to material or bodily
processes alone—as automatic, lawful, and mathematically describable—
came to have fateful consequences for moral psychology that continue to have
a hold on us Westerners, beneath our full awareness. That nature is confined to earth and body and as such is capable of scientific explanation, while
the mind is the realm of spirit and cannot be pinned down and predicted,
are beliefs that feel as obvious and ubiquitous as the air we breathe and the
water we drink. Moreover, the bifurcation of mind and nature also has con-

sequences for how scientists have all too frequently and unreflectively limited science to the investigation of material processes alone, the mind either being considered too free to operate via a natural lawfulness or instead deemed to be irrelevant, static, and not "real" (a position that philosophers call *epiphenomenalism*). Both of these positions owe a debt for their dualist bifurcation of body and mind as two entirely different kinds of causal orders, the natural-lawful versus the voluntary, to Augustine's fateful theological turn.

Free Will and the History of Christian Doctrine

> The idea that humans have free choice was mainly a product of the Christian tradition. It is not an idea found in Aristotle's ethics. . . . [For Aristotle] a person's choice always expresses her character.
>
> —BONNIE KENT

From Augustine onward it was generally recognized in the Latin Christian West, whose center was Rome—in contrast with Eastern Christianity, whose center was Constantinople and whose languages were Greek, Syriac, and other Eastern languages—that some notion of freedom of choice or will was the standard Christian moral psychology. It was also widely recognized that Augustine, the most important Latin Church Father, was the source of the belief in free will and that his authority stood behind it. The East, of course, was both the site of another Christianity and also under the spreading sway of Islam beginning in the seventh century. The Muslim conquests reached the Iberian Peninsula from North Africa in the eighth century, and almost all of it was under Muslim rule by 720 C.E. Parts of Spain remained under Muslim rule until 1492, when Ferdinand and Isabella completed the Reconquest with the fall of the last Muslim state, Granada. A fluke of history made Islamic Iberia the site of a tremendous intellectual and scientific flourishing. A prince of the Umayyad caliphate escaped the Abbasid takeover of the Umayyads in Damascus in 750, fled to Spain, and made Cordoba the center of a new Islamic state that became quite independent from the rest of the Islamic world. The area became known in Arabic as Al-Andalus, gradually fragmenting into independent Muslim states. Arabic culture and hegemony survived in portions of the Iberian Peninsula for the next 750 years. Muslim Spain became the site of a society of substantial tolerance as well as sophisticated intellectual and artistic achievements building upon Greek classical philosophy and science in an advanced cultural milieu. Muslims, Christians, and Jews lived together in a largely Arabic-speaking society that assimilated and integrated native inhabitants. Cordoba became the intellectual capital of a vibrant scientific, philosophic, and artistic cultural world, expanding and building upon the legacy of ancient Greek rationalism in astronomy, medicine, philosophy, and the arts.

It was in the wake of the expanding conquest of Muslim Spain by Christian forces (which began nearly as soon as the Muslims arrived on the Iberian Peninsula and continued for the next seven hundred years) that Arabic texts of Aristotle and their philosophic and scientific expansions by Iberian philosophers and scientists began to trickle into Latin Christendom. Up till the twelfth century, there were only two Aristotelian books available in Latin: the *Categories* and *On Interpretation*.[69] Aristotle's *Physics* and his *Metaphysics* had been banned by the Church, while the *Nicomachean Ethics* had not been. Nevertheless, only the first three books of the *Nicomachean Ethics* had survived in Latin Christendom, and a complete Latin translation of it did not appear until 1246–47. With the taking over of bastions of cultural and scientific advancement and flourishing in the Iberian Peninsula, the great Arabic libraries fell into the hands of Latin Christendom, generating a great deal of intellectual excitement and foment. Included in the significant influx of writings from the Iberian Peninsula in the late thirteenth century were advanced philosophical and scientific writings of Muslim philosophers and also Arabic translations of and commentaries upon Aristotle by such philosophers as Averroës and Avicenna. There was a flurry to translate texts from Arabic into Latin, and the appearance of these latter texts created a sensation in medieval universities, especially at Paris and Oxford, in the 1260s, particularly in the arts faculties but also among theologians.[70]

The Arabic philosophical interpretation of Aristotle was for the most part as a radical naturalist in physics, metaphysics, and moral psychology. It was this radical Aristotle that now caught the imagination of the faculties, and an intense and vociferous controversy took hold. This reached a turning point in 1277 when Étienne Tempier, the bishop of Paris, banned 219 Aristotelian claims and positions as contrary to the Christian faith. Three radical naturalizing Aristotelian doctrines were of particular concern, and all three were among thirteen Tempier condemned even earlier, in 1270. Among those thirteen were (1) the eternity of the world, (2) the unicity of all intellects, and (3) the necessity of all events. The first appeared to threaten the biblical account of divine creation. The second implied that all rational minds were in essence one universal mind, and hence it appeared to threaten the afterlife of the individual. The third was the claim of an unbroken, thoroughgoing determinist natural causality, and hence appeared to threaten divine power and also to obviate direct divine intervention or miracles. The latter claim also posed a challenge to free will—certainly God's but also humans'. Two more of the thirteen propositions condemned in 1270 appeared to directly call into question individual free choice or will: (4) that "the will of man chooses and wills by necessity" and (5) that "free choice is a passive power, and not an active one and it is moved with necessity by the object of desire."[71]

Henry of Ghent, a theologian and major figure in the faculty at Paris, was on the sixteen-member commission appointed by Tempier to investigate the writings of the arts masters. It is thought that perhaps he was the moving force behind the condemnation in 1277 of the nonvoluntarist positions.[72] In his *Quodlibet* of 1276, Henry maintained that "it is the will that commands all powers of the soul," and he called the will "the first mover in the kingdom of the soul"[73]—the classic position of self-command and the self-origination of action, which to this day broadly defines all versions of voluntarism. He also precisely articulated and held the view that "every disorder of reason is caused by a disorder of will," a position that Descartes will embrace and elaborate several hundred years later. Historian Bonnie Kent proposes that "of all the articles condemned in 1277, those related to the will probably had the greatest significance for the history of ethics." More than fifteen of the 219 bans of 1277 were focused on safeguarding free will against any Aristotelian naturalist challenges to it. Their "aim [was] plainly to safeguard the freedom of the will" from any claims of "determination . . . by external powers" which "were seen as a threat to moral responsibility." Even the moral "agent's own intellect (reason)" was held to be one of the "external powers" that was not to be allowed to be claimed to have a role in determining action. In addition, "the articles firmly reject[ed] the thesis that all wrongdoing results from ignorance, that anyone who knew better would perforce *do* better, or even *will* to do better."[74] We saw above that both these claims—that individuals can be held to be morally responsible, that is, blameworthy or praiseworthy only because of the free origination of their actions and that vice is due not to ignorance but to free choice—are central to Augustine's moral outlook. As a clearly developed and coherent view, they originate with Augustine, and they were also consciously associated with him as an important dimension of his tradition and legacy. As a result, the battle lines drawn in thirteenth-century Latin Christendom had radical Aristotelian naturalism at one pole versus Augustinian voluntarism at the other.

The radical Aristotelians, dubbed "Averroists" both at the time as well as subsequently, embraced a view of the human moral subject as a necessary product of nature and nurture. Aristotle, of course, did not deny that human beings make choices and decisions. Yet he did not regard these choices and decisions as free in the Augustinian sense. For all choices and decisions, according to Aristotle in the *Nicomachean Ethics*, are those of the kind of person one is or, rather, has become. Choice is not even possible, Aristotle pointed out, if not determined by disposition, for choice is its necessary expression. Character determines decision and choice, and character is a matter of nurture, of habit. "None of the moral virtues arises in us by nature," he says. Aristotle distinguishes between two types of reason: practical deliberation and theoretical understanding. Practical deliberation plays a role in determining the means to follow toward ends, goals, which by contrast are set by desires. "We deliberate not about ends but about means," he insists.[75] He defines choice as "deliberate desire of things in our own power."[76] Our desires are who each of us is; they express and reveal character. Choices and decisions are expressions of the desires that constitute character. "For each state of character," he writes, "has its own ideas of noble and the pleasant." So character also shapes perception, potentially corrupting the mind's conceptions; ignorance of what is truly good for human beings is a character state, and hence it is expressed as a state of desire, dictated by character, of certain ends rather than others. Our desires harbor a hidden judgment that their ends are good. "There is," he says, "no natural object of wish, but only what seems good to each man." So people always act with some notion of the good in view and from desire, from what Aristotle deemed the "appetitive faculty." Yet they are potentially prey to a kind of ignorance that is neither simple nor innocent. For their desires, formed by habit, drive their conceptions of the good, which is to say their motivations toward goals and pursuits. In this Aristotelian account of moral psychology there is no room for will, in the Augustinian sense of a capacity to free oneself and one's actions from necessary determination by the constellation of character, desire, habit, and cognitive judgment. There is no possibility, for Aristotle, of acting "out of character," or even, over time, of willfully changing one's character, with all that this implies. Just as we cannot will ourselves well if we have the flu, to use Aristotle's own metaphor, we cannot will ourselves objectively good.

While good and bad people are responsible for their moral states and actions because those states are their own and result from their character, over which they have some control (at least initially), once character is set, people cannot do otherwise than they do, he says.[77] For "it does not follow that if [an unjust person] wishes he will cease to be unjust and will be just." Moral character is like health, in his estimation, "for neither does the man who is ill become well on those terms."[78] Aristotle holds that while "we are ourselves somehow partly responsible for our states of character," in the end habit is a social and political phenomenon. So it is "legislators," Aristotle concludes, who "make the citizens good by forming habits in them, and this is the wish of every legislator."[79]

It was the growing encounter with this rather socially determinist, character-based Aristotelian moral psychology that raised the alarm of the masters at Paris and led them to circle the wagons. In the wake of the Condemnation of 1277 banning Aristotle's naturalizing account of moral psychology in terms of character, self-consciously precise and nuanced theoretical articulations of the opposing (neo-Augustinian) free choice alternatives were formulated and crystallized. Voluntarism could now be articulated in clear opposition to the Aristotelian moral psychology as a range of theories that locate moral virtues in the will rather than in moral character.

After the ban, *all* major Latin Christian thinkers on the topic took issue with Aristotelian naturalism—roughly, the position that human beings act necessarily according to their character and that natural (including mental) processes operate by necessity—if not with Aristotle himself. They either embraced voluntarist positions, in stark opposition to Aristotle, or came up with voluntarist versions of Aristotle sometimes wished upon him in his name. One aspect of the response was a clarified account of free choice according to which it was held to be localized in a distinct faculty of the soul, namely, the will. It could now be maintained that choice was free because it resided in a free will, a part of the soul that had the power to move itself freely, that is, without external determination by, or passivity to, intellect or anything else.[80] Bonnie Kent comments that "efforts to reconcile Aristotle with the faith were the rule, not the exception, in the theology faculty."[81] The influx, the fervor, the controversy, and the ban provide the background to the rise of a voluntarist movement in the late thirteenth century.

Versions of voluntarism can be traced, for example, from the important medieval thirteenth- and fourteenth-century theologians and philosophers John Duns Scotus and William of Ockham to the seventeenth-century philosopher Descartes (often called the father of modern philosophy) and even to the nineteenth century's Immanuel Kant, whose thought is still dominant, alive, and generative today. The great representative of a compromise position that attempted a reconciliation of Aristotle's notion of naturally acquired virtues of character with some room for free choice of the will was Thomas Aquinas, but quite a number of others can be seen as falling into that camp, including Bonaventure and Walter of Bruges.[82] Aquinas, for example, sets forth a compromise position by acknowledging an Aristotelian notion of dispositions as mental tendencies to do things, but he relates them principally to the choices of the will.[83] He also claims that "every sin consists principally in an act of will," thus paying his debt to the Augustinian orthodoxy.[84] Many other Christian theologians also straddled both views of the human moral person, at times even attributing—whether unconsciously or deliberately—their Augustine-laced Christian voluntarist compromise to Aristotle himself. Clearly invoking Aristotle's authority was not at issue, since it was common to claim his imprimatur not only for positions in part inspired by him but even for those quite distant from his. There was a "growing cult of Aristotle," a bandwagon effect that had its own momentum, drawing many along who invoked his authority for even barely recognizably Aristotelian positions. Yet the appeal to Aristotle horrified other thinkers as a "paganizing trend."[85]

The enhanced self-conscious embrace of voluntarism as a movement beginning in the 1270s was characterized by the claims that "the will is nobler than or superior to the intellect"; that "beatitude or happiness consists more in an activity of will than in an activity of intellect, that man's freedom derives more from his will than his rationality, that the will is free to act against the intellect's judgment, and that the will, not the intellect, commands the body and the other powers of the soul."[86] Prior to that, the exact role or status of the intellect in relation to the will had been more up for grabs. Positions that leaned toward Aristotelianism—Aquinas's, for example—compromised in the direction of giving some status to the role of intellect in decision making, but voluntarism proper had by now hardened and narrowed, with a characteristic downgrading of intellect both in power and in dignity. It is also after 1277 that the distinction between free choice and free will became prominent, with the latter implying the superiority of will to intellect, whereas the former could fudge the question or even go the other way. It was only then that *"libertas voluntatis"* became the rallying cry for opponents of Thomism as well as opponents of radical Aristotelianism. Yet because of the condemnations of 1277 the entire range of legitimate permissible opinions had already been restricted only to those within the free will spectrum, for 1277 had been about safeguarding freedom of the will.

I'll end this journey into the history of free will in Latin Christianity with a summary of Duns Scotus's position—not the most extreme in its highlighting of the will in contrast with the intellect but paradigmatic.

> Natural agents devoid of reason are determined to one effect as a stone is determined to fall if not impeded. As a rational power, the intellect is capable of opposite acts regarding the same object: it can be pursued or avoided. But because the intellect is a rational power that acts according to its nature, it cannot determine itself to one of the two alternatives, nor can it refrain from acting. The will, in contrast, can act or not so, although it is not the only cause of its own acts, it is the only free cause. Thus it is the will alone that makes any act a free act. . . .
>
> [T]he will is the sole source of freedom . . . [T]he will remains the principal cause of its own acts. Because the will acts freely . . . it alone introduces an element of contingency in the process culminating in bodily action.
>
> Why should this element of contingency be so important? . . . [Scotus] explains what it is that makes an act "imputable," that is, eligible for praise or blame, reward or punishment. Scotus argues that what all such acts have in common is that they lie within the free power of the agent. . . . It is the freedom of the will that makes our acts our own and so makes us responsible for them.[87]

We hear in Scotus's position the resounding echo of Augustine's insistence that human beings must be endowed with freedom to originate their actions willfully in themselves alone or their moral responsibility would be vitiated and they could be deserving neither of blame nor of praise.

After Scotus the issue of voluntarism continued to be central to the concerns of theologians and philosophers, an ongoing and hotly debated issue. Such was the case at the dawn of modernity, occupying even philosophers whom we now think of as initiating modern scientific rationalism. René Descartes, often thought of as the first modern philosopher to embrace the systematic use of reason in the natural sciences, was also an ardent voluntarist.

Descartes the Augustinian

When Descartes claims to derive from [Augustine's method of] reflection [on the soul and God] . . . a new system of philosophy to replace that of Aristotle, he is putting himself forward as the philosopher of the Catholic Reformation.

—STEPHEN MENN

The more learned someone becomes in the teaching of Augustine, the more willingly he will embrace the Cartesian philosophy.

—FATHER MARIN MERSENNE

Standard modern philosophy begins with Descartes. He is regarded as the founder of modern philosophy—alone or with such associates as Bacon and Hobbes—and a major figure in the scientific revolution. Descartes was born in La Haye, France, in 1596 and was educated at a Jesuit college for about eight years starting at age ten before going to the University of Poitiers, where he studied law. While in the army, however, Descartes became very interested in the sciences and perhaps studied engineering. By 1625 Descartes was in Paris and had developed a relationship with Father Marin Mersenne, a member of a Franciscan order, through whom he came into contact with some of the major philosophical thinkers of the day. Mersenne enabled Descartes's works to reach the important intellectual circles in Paris.

Descartes's importance may be seen in his coming up with a new way of looking at matter that made it possible to understand physical processes in terms of mechanical explanations. He developed a way of connecting geometry and algebra that was germane to his new way of thinking about matter.[88] His *Meditations on First Philosophy*, a work published in 1641, offered nothing short of "a philosophical groundwork for the possibility of the sciences."[89] Descartes's stated aim was to break with Aristotelian philosophy and science, especially with Aristotelian causal explanations of physics in terms of purposes and aims, and to replace them with explanation in terms of strictly material causal mechanisms. Mechanistic explanation is iconically captured in the model of billiard balls, in their surfaces coming into contact and movement giving rise to definable changes. The new physics was concerned strictly with the size, shape, motion, and position of bodies, their length, breadth, and depth—all measurable properties—and was not to be concerned, as Aristotle was, with their "qualities," what made them the types of objects they were. For Aristotle, the essential properties of all things was chalked up to the presence of "mind" or "form" in them, making them the kind of objects they were and dictating the kinds of ways they behaved—heavy things falling because of the presence in them of "heaviness," for example. That kind of explanation was to be jettisoned in favor of mechanisms that could be captured by mathematics.

When we think of Descartes today, his name is synonymous with an extreme form of mind-body dualism. When we call someone a Cartesian we are likely to be pointing to a position in which the mind is thought of as separable from the body and of an entirely different nature. In fact, only the body is natural, whereas the mind is not within nature, according to Descartes. Descartes understood the nature of the body in terms of its extension (in length, breadth, and depth), whereas the mind's nature was to think. This position emerged from Descartes's method of doubt, his methodological skepticism, from which he drew the insight that anything can be doubted except the mind's activity itself in doubting, and thus thinking *must* exist even if the whole world, including even one's own body, is an illusion. Thinking cannot be separated from oneself even in conjecture, although body can be so separated in principle. Hence Descartes's famous claim "*Cogito ergo sum*," "I think therefore I am." The only knowledge that cannot be doubted is one's own thinking, one's own fact of subjective or inner experience (not the correspondence of that experience to the world). By "thinking" Descartes was referring to inner consciousness, awareness, not narrowly to cognition. On the foundation of the indubitability of the inner life, Descartes turned to rebuild knowledge and an external world. An external world, he argued, was real, even though sense perception was unreliable, because God would not be a deceiver and trick human beings into mere illusions. Descartes distanced the person from the material world, including the person's own body, but at the same time provided ingenious and important ways to come to new understandings of how body and matter operate.

What is of lasting significance, ironically perhaps, are Descartes's advances in the mathematical explanations of material processes rather than his conception of mind and of the mind-body relation. For his approach to matter removed the mind from natural explanation entirely—a legacy philosophers and scientists and people generally in the West are still in the grips of and perhaps just emerging from. Cartesianism today is likely to bring to mind the view that the mind, unlike the body, is what is really "one's own," that the body is a sort of alien carrying case for the mind, and that the mind would have the same thoughts and contents no matter what kind of material it was housed in. This is the notion encapsulated in the phrases "mind in a machine" or "mind in a vat." Descartes in his moral philosophy *The Passions of the Soul* portrays mind and body in an unending struggle for mastery over one's actions. Moral goodness is the triumph of the free will of the mind in dominating the body—its physical urges and those emotional motivations originating in it.

Descartes's philosophy must be seen not only as contributing to discussions about physics and metaphysics but also in the context of seventeenth-century theology. That theological concerns are central to Descartes is argued systematically and persuasively by the historian of philosophy Stephen Menn in his book *Descartes and Augustine*. Menn argues there at great length and in detail that "the history of philosophy in the sixteenth and seventeenth centuries is intimately bound up with the history of Christianity." The Western Christian universities in Descartes's time were still committed to the Aristotelian approach to philosophy that we have seen had been dominant since the thirteenth century. That settled form of Christian Aristotelianism still structured the curriculum and set the doctrines of Latin Christian theological thought into the seventeenth century. Its advocates and academic masters were not about to give way to a new philosophy deriving from a different set of foundational commitments and beliefs. As a result, "the modern philosophy developed outside the universities, and won its place in them only through protracted struggle."[90] This "modern philosophy" amounted to an Augustinian challenge to the embedded Christian Aristotelianism—itself, of course, as we just saw above, a compromise struck between the Aristotelian and the Augustinian. The Aristotelian approach, standard since the thirteenth century, was now being reevaluated by the Counter-Reformation. The Counter-Reformation, lasting about one hundred years, was the Catholic response to the Protestant Reformation. It began with the Council of Trent (1543–63) and initiated a Catholic revival in both doctrine and religiosity. Counter-Reformation thinkers now came to view any theology that leaned toward the Aristotelian as shockingly pagan and corrupting, in contrast with the Bible and especially Augustine's writings, which were regarded as pristine and authentic. Menn remarks that "from very early on, Augustine had become the chief authority, second only to scripture, for western Christian theology. . . . The body of philosophical doctrine that later thinkers will take to be axioms of Christian philosophy is in fact (at least for the Latin West) the work of Augustine."[91]

Menn offers an exhaustive account of both the Catholic theological context of Descartes's philosophical project and also of its explicit Augustinian revivalist agenda and aims. He exposes Descartes's conscious intention (often now forgotten or neglected by philosophical interpreters) to develop a philosophy in an Augustinian key, which would meet the demands and hopes of Counter-Reformation advocates and theologians. He documents Descartes's relationship to Augustinian revivalist movements in the seventeenth century, both Catholic and Protestant, for "in Descartes' time there were many such Augustinianisms," Menn points out. "The history of Augustinianism," he says, "is the history of the many revivals of Augustine by different thinkers, who have each discovered some new aspect of Augustine's thought, and seen in it a way to answer the philosophical or theological challenges of their own times. . . . Thus, in the early sixteenth century and beyond, Christian reformers of all stripes appealed to the [Church] Fathers over the [Aristotelian] Scholastics as offering a model for Christian thought and practice."[92] And "especially in the sixteenth and seventeenth centuries," Menn says, it was "Augustine [who] was the chief human authority and model for many thinkers throughout Latin Christendom who were indifferent or hostile to the thought of Aristotle."[93] Moreover, "the prestige of Augustine [had] gained an added boost from the Reformation. The Protestant reformers had taken Augustine as their chief authority (after the scriptures) for their theology of grace; the Catholics at [the Council of] Trent, sought to reclaim him, and so, individually, did each of the different tendencies of the Catholic Reformation."[94] Descartes's clear plan was to develop a philosophy based on an authentic Augustinian Christianity to replace the Aristotelian scholastic basis of Christian philosophical theology that was normative in the universities. So what we think of as the founding of modern Western thought, the philosophy undergirding and furthering revolutions in the physical sciences, had a profoundly revivalist, perhaps even reactionary, and certainly theologically conservative Augustinian character and tenor.

Menn says that it is from the Counter-Reformation's Augustinian angle that he views the entire Cartesian project and philosophy. His book, he says, is a (re)reading of Descartes's philosophical project and beliefs: an exhaustive argument for the validity of interpreting Descartes through an Augustinian lens. Menn first places Descartes in his Counter-Reformation theological context, both by shining a light on that historical moment and even more by pinpointing Descartes's own statements of his theological purpose and

primary audience. Menn then devotes much of the rest of the book to identifying in equally great detail Descartes's precise Augustinian methodological approach to philosophizing, and also to exposing the Augustinian doctrinal
foundations of Descartes's central philosophical positions. Augustine's free
will voluntarism is an important part of this story, yet Descartes's Augustinianism is both deeper and broader and more self-conscious than the mere
commitment to free will by itself would entail. Menn reports that there were
two crucial turning points in the way that Descartes conceived his philosophical project, the first in 1619 and the second in 1628. "From 1619 on,
Descartes was trying to fulfill the general hope for a new philosophy," Menn
says. After 1628, however, Descartes "appear[s] to take up the more specific
project of a systematic philosophy based on Augustine."[95]

It is documented that earlier in his life, by 1619, Descartes had already
envisioned a philosophical project to put forth "a fundamental science developed into a scientific wisdom." Yet "the fundamental science identified at
that [earlier] time . . . Descartes had first intended to be universal mathematics (a general science of quantity)." By 1628, however, in a letter to Picot,
Descartes reveals that he has changed his mind about the possibility of bas-
ing his new philosophy on mathematical foundations and now intends to
turn instead to a metaphysical foundation. He writes in the letter that he
will base his new philosophy on the Augustinian (metaphysical) notions of
the immortality of the soul and of God. The letter "gives a programmatic
statement of two different aspects of [his] philosophical project": the two
principles that Descartes now attests that he will base his philosophy upon
are (1) that "God is the creator of all beings and the source of all truth" and
(2) that "the human soul is immortal and separable from the body." Descartes held these principles to be self-evident and universally accepted. They
were universal in the sense that they were Augustinian "axioms of Christian
philosophy," as Menn calls them, which had been grafted onto the scholastic Aristotelian philosophy in the Christianizing of it, so that both sides of the
Christian philosophical divide accepted them. Descartes proposes that
these unassailable and hence reliable principles serve as the foundation from
which he would now derive a new physics and other particular sciences.[96]
Hence a broadly shared but not in any way superficial Augustinianism—
minus any vestiges of Aristotelian naturalism—became the foundation upon
which the Cartesian modern philosophy and philosophical outlook was
constructed.

Descartes had intimate relationships with Counter-Reformation figures in France. These encounters turned out to have a decisive effect on his philosophical intentions. The circle of Counter-Reformers harbored "a hope of constructing out of Augustine a new philosophy to replace that of Aristotle." Menn introduces biographical descriptions of seminal conversations that Descartes had with Counter-Reformation movers and shakers, who called upon him to become *the* philosopher of the Counter-Reformation, "demand[ing] in the strongest terms that he apply his method to natural theology, by proving the existence of God and the immortality of the soul." Descartes came to be the thinker upon whom the Counter-Reformers set their hopes to develop an Augustinian philosophy, and he came to commit himself to "trying to fulfill that hope."[97] It was Cardinal de Bérulle, "the spiritual leader of the Catholic Reformation in France and the leader of the 'devout party' at court," who met privately with Descartes and "told him to begin with metaphysics, and with metaphysics as conceived in Augustinian terms, as a discipline of reflection on God and the soul." Descartes's biographer Adrien Baillet regards the "exhortations of the pious cardinal" as a turning point in Descartes's life. It was "from this time on," Menn explains, that "Descartes takes such a metaphysics as the fundamental discipline, from which the principles of philosophy must be drawn."[98]

Descartes emerged from these fateful encounters, having relinquished an earlier hope for deriving a physics from pure mathematics, to take up the banner of developing a new philosophy, in a work he at first planned to call his *Metaphysics*, now to be based upon Augustinian notions of God and the soul. His intention was "to circulate the work among the theologians, not only to receive criticisms and improvements on the text, but also to win the endorsement of individual doctors of theology and, if possible, the institutional endorsement of the theological faculty of the University of Paris." To emphasize even more clearly the Augustinian approach to philosophizing that he was now undertaking, Descartes decided to change the title of the work from *Metaphysics* to *Meditations on First Philosophy*, thus pointing to its Augustinian method of reflection. The work is cast as a series of meditations not only on "God and the soul" but also on "immaterial or metaphysical things." Moreover, Descartes pointedly addressed the dedicatory letter with which he introduced the work to theologians rather than to philosophers. He offered it "to the theologians as the true philosophy of the Catholic Reformation: he will carry into the hostile territory of philosophy the reformers' struggle to restore the pristine simplicity of [Christian] truth and to eliminate scholastic corruptions."[99]

The thrust of Descartes's Augustinianism is to "reverse the journey of Aristotelian philosophy: beginning, with Augustine, by reflecting on the human soul, he will show that it can be known better and prior to bodies, and that it can subsist apart from them."[100] The soul is fundamentally tied to and of God, rather than akin to the body and of nature.

Descartes's Augustinianism can be summarized as follows:

> Metaphysical knowledge, being purely intellectual, is independent of the testimony of the senses. . . . It will be concerned primarily with God and with the human soul, and not with God and the human soul as they may be inferred from sensible objects. The human soul will be known primarily as a thing that thinks: not as an act of an organic body, but as something only extrinsically related to a body. God will be known primarily as the highest object of our thought, not as the governor of the physical world although he becomes that too when he creates the world.[101]

In the *Meditations*, Descartes sets out to prove three theses so as to be able to demonstrate that the direction of establishing true beliefs goes from the understanding of God to sensible things, rather than from the evidence of the senses up to metaphysical entities, as Aristotle had maintained. "He must show that we can know God and the soul without knowing bodies, that we *cannot* know bodies without first knowing God and the soul, and that we *can* know bodies once God and the soul have first been known."[102]

Descartes's voluntarism outstrips even Augustine's own. For Descartes, God freely decrees the laws of nature, rather than recognizing them as intellectual necessities to which he must conform; and the human mind too, made in God's image, has a freedom that makes it superior to the law-governed natural order, even while it is limited by the constraints of its natural environment.[103] In both God and human, the voluntary will is the *only* source of activating movement and action both for and in matter (which is now defined in geometric terms as occupying space but inert).

Moreover, Descartes goes even further than Augustine in extending the mind's self-control over its actions, expanding it to include its very beliefs and understanding. Error for Descartes is not the result of simple ignorance but a matter of moral obstinacy; it is an intellectual sin. Descartes thus accomplishes a radical and complete reversal of the Aristotelian and general

Greek classical conception of moral failure as a matter of cognitive ignorance and foggy benightedness. "In Descartes' Augustinian language, to say
that error . . . proceed[s] from us and [is] within our power, is to say that [it]
depend[s] on the faculty of *will*," whereas the Greek classical view reduced
morals to knowledge, moral good being a kind of understanding and evil a
form of ignorance. Descartes performed a radical reversal of the classical
worldview, thoroughly Christianizing even thinking itself. Knowledge was
now to be viewed as resulting from a free act of the resolve of the will, while
cognitive error was recast as a willful failure or rebellious refusal to use
reason. For Descartes, even belief was to be a matter of self-origination and
self-control—hence his famous method of doubting everything, even the existence of our own bodies. We can willfully shut those out, he proposed. He regarded our minds as free to such an extent that even the world could possibly
be thought to be a mere invention of our minds, rather than as proof that a
world of our bodies and environments force themselves upon our perceptions.

A Brief Description of Descartes's Free Will Moral Theory

For Descartes, body and mind are in constant struggle to control a person's
actions, for (in Augustinian terms) two orders of divine cosmic causal origination and motion are at war here, the natural and the voluntary. The soul
can be either passive to the bidding of the body or active in controlling the body
and hence its actions. Descartes regarded all the functions of the soul as
aspects of its conscious thinking—understanding, willing, imagining, re-
membering, sensing, and emotional feelings.[104] The human moral problem
is the struggle between body and soul. The aim is to get the soul, the mind,
to have utter control over the body and what the body does. The mind's willful mastery of the body and of itself—in both its thoughts and initiation of
actions—is the aim of ethics, and it is what Descartes points to as moral
agency. The emotions must be brought into line by becoming "active," which
means that they must come under the mind's control, rather than being passive to the causal order of the body and the world. We hear the echo of Augustine's reduction of emotions to acts of free will here, recalling that even
the emotions, for Augustine, turn out to be the effects, rather than the causes
or motivations, of the will. We also discern in Descartes's moral psychology
Augustine's bifurcation of the cosmos into two divine causal orders, the natural (mathematical and bodily) and the voluntary (mental).

Willing and understanding occur in the soul alone, for Descartes, and are both aspects of its voluntary activity. Sense perception, the passions, some memories, and imaginings result from interactions of mind and body. Thoughts can be either passions or actions, depending on whether they originate in the soul and are initiated by it (these are the volitions) or originate outside it in the body and world and hence are passively received and represented by the soul (these latter are the passions). Passion and action refer to internal states of the soul of relative mental weakness or strength in initiating thoughts. When the body acts upon the mind, affecting it, that's passivity. The result is emotions that we seem to be completely passive to and in the grip of. The internal power of the mind to affect the world occurs when the body is (passively) obedient to the mind's active bidding. That's Descartes's "activity." When the body is active, or dominating the mind, that's Descartes's "passivity"—so the terms are articulated only from the point of view of the soul or mind's control or domination of body and world. So Descartes's theory of the passions draws a line between our passive perceptions and our active volitions. That passivity of perceptions and passions thus marks off part of us—certainly the body, but also a division even within the mind—as in a sense exterior to what is the "true" locus of "us" or self. The activity/passivity dichotomy redraws the boundary between self and other, self and world, and relocates it within the customary bounds of the person—the skin. Since only volitions are identified by Descartes as truly our own, or ourselves, what counts as the self is radically narrowed to the will; again we encounter here Descartes's Augustinianism. It is from the (free) will alone that our control over the often unsettling waves of emotion can arise. The will is the movement that the mind's judgment initiates. It reverses the direction of passivity from the mind's pervasion by painful passions to its self-mastery of them. Volition is the activity of the mind par excellence.

Virtue, according to Descartes, consists in judging what is best and then acting with complete resolve on those judgments. Our virtue is thus our strength of will in shaping the self, the body, and the world rather than being shaped by them. The rewards of such virtue are our pleasure in our capacity for self-control and our satisfaction and ease in knowing that the passions emerging from the winds of fortune cannot move us. Descartes regards such joy as not itself a passion because it is strictly interior to the mind. It does not depend at all upon the body, although it moves the body to emotional expression. Emotion that originates strictly from the soul as though the soul were without connection to the body is the ideal for Descartes. That is activity and the active control of self (mind), body, and world. Here we have a somewhat secularized Augustinian account of moral psychology and agency, in which God's will in relation to the human will, either in grace or in conflict as all-powerful, no longer holds sway, so that the arena of human free will has suddenly taken up the entire space. We have the expansion of the human capacity and responsibility of freedom without the limitations of divine power or original sin. As Menn points out, Descartes's Augustinianism is that of *De Libero Arbitrio*, a text that emphasized human free will more than divine power and will. That text was well suited to be particularly inspirational in an era in which humanism had already placed the human person, rather than God, at its center. Yet it is a human person envisioned with a divine-like, miraculous power of will over its actions, its body, and over the natural world.

Conclusion

We have seen how a streamlined and hardened Augustinian conception of free will became foundational to standard modern moral philosophy in the West and came to dominate cultural commonplaces about how our moral capacity is assumed to work. Tendencies of an earlier Greek classical naturalism and intellectualist understanding of moral agency that had seeped in from the Mediterranean East in the thirteenth century underwent a full-scale jettisoning in favor of a purist Augustinian revival in the sixteenth century by both Protestants and Catholics in the Reformation and Counter-Reformation. In the Mediterranean Eastern orbit, however, the Greek classical naturalist and rationalist understanding of human nature, with its intellectualist conception of the good person as having a kind of knowledge and the bad person as befogged and ignorant, had both flourished and undergone a line of development of its own. While Western Christendom had become more and more under the sway of a purified and radicalized Augustinian conception of human nature, in the Eastern orbit, first in Syriac Christianity and then in an Arabic philosophical school within Islam and Judaism, the opposite tendency had taken hold. Thinking of human nature, not just of the body but of the mind as well, as within nature and explainable by natural causes was the starting point of an alternative account of human moral nature, an account that continued to pay homage to classical Greek moral naturalism and intellectualism. This different conception of human nature and the human moral capacity, while it did not come to have the general cultural dominance that the Augustinian presuppositions did in the West, nonetheless was developed into a clarified, radicalized, and systematically worked out account of how and why human beings act morally and why they don't. That conception was introduced by the Jewish philosopher Maimonides, writing in Arabic in the thirteenth century, and was developed further and brought into modern dialogue with the Augustinian West by the Jewish Dutch philosopher Spinoza in the seventeenth century.

5

Another Modernity: The Moral Naturalism of Maimonides and Spinoza

There is a strong tendency toward intellectualism in Arabic ethical works.

—PETER ADAMSON

Augustine's ... lack of knowledge of Greek ... rendered him deaf to the riches of the Greek world.

—PETER BROWN

Classical Greek Philosophy in the Arabic Orbit: A Different History

That the Latin West broke with the naturalist thrust and intellectualist focus of Ancient Greek philosophy, in many respects, was a fluke of history.[1] The philosophical tradition in the Greek Mediterranean world (in contrast with the tradition that became dominant in the Latin orbit) inherited, embraced, and built upon ancient Greek philosophical naturalism and intellectualism. There were Christian and Jewish philosophers who followed and extended the Greek classical naturalist thrust, but the Islamic philosophical movement, known as *falsafa* (the Arabic word for "philosophy"), came to hold center stage and Arabic became its primary language. Christian and Jewish philosophers as well as Muslim ones within that cultural orbit wrote largely in Arabic, as the most contemporary language of science and philosophy. *Falsafa* was a movement to introduce and integrate Greek philosophical and scientific texts into an elite Islam via appropriation, translation,

183

and commentary, and thereby to spur further philosophical thinking and scientific development. It was from this particular philosophical tradition that echoes from the Greek philosophical past periodically reverberated in the Latin West, posing a challenge to dominant Augustinian understandings of human nature and the human moral capacity.

Falsafa was not the only arena of theoretical speculation in the medieval Arabic milieu. There was also *kalam*, a tradition of theological speculation within Islam. *Falasifa* (philosophers) and *kalam* theologians engaged in ongoing thinking and debate about the nature and origin of the world, the composition of material objects, the nature of time and motion, the role of God in causal processes in physics, and the like. *Falsafa* had its origins in the philosophy of Aristotle as it was read through a Neoplatonic lens, in the biological and medical writings of Galen, and in the astronomy of Ptolemy.[2] As an elite intellectual enterprise, however, *falsafa* did not have anywhere near the influence on the Arabic-speaking world that the broad hold of Augustinian traditions had on the Latin West. Nevertheless, its philosophical and scientific achievements were considerable, and increasingly over time these made their way westward.

The Greek Philosophical Tradition of Alexandria

Medieval Arabic philosophy benefited from having inherited the classical traditions and thrust of the late antique Alexandrian school of Neoplatonism. Neoplatonism was the name given in the nineteenth century to the movement and wide influence of the thought of the third-century c.e. philosopher Plotinus and those who followed him. In the 500s c.e., in both branches of the Neoplatonic school, the one in Athens as well as the one in Alexandria, "late Neoplatonism mostly focused on commenting on Aristotle," according to the Italian scholar Christina D'Ancona. When Emperor Justinian shut down the Neoplatonic Academy of Athens in 529 in his attempt to weed out non-Christians, get rid of non-Christian institutions, and suppress paganism and other non-Christian religions within the empire, he "effectively strangl[ed] this training-school for Hellenism."[3] Platonic otherworldliness—devoid of an Aristotelian biological and naturalist thrust—could more easily be accommodated and assimilated into standard Christian doctrine, and hence did not appear to pose the same threat as a school devoted to Aristotle did. In the wake of the closure and suppression of the Athenian Neoplatonic school, some of its noted philosophers headed east to Persia, where Greek philosophical endeavors were welcome, joining fellow philosophers there. By the sixth century the focus of philosophic thought in the Neoplatonic school in Alexandria, more than in the school's other branch in Athens even when it was fully operational, had shifted decisively toward Aristotle while still adhering to the Neoplatonic view of the cosmos. The Alexandrian school turned toward a markedly embodied and naturalist understanding of Aristotle. Rather than serving the primary purpose of supplying an introduction to the Neoplatonic curriculum and particularly to the dialogues of Plato, as Aristotle's philosophy had done in the Athenian school in its heyday, in the Alexandrian school it was Aristotle himself who was deemed the epitome of philosophic achievement and the model of philosophic endeavor. The Alexandrian reading of Aristotle emphasized the intelligibility of nature to the human mind, and the divine gift of reason as binding together God, cosmos, and human soul. The spiritual path toward God, in this view, was available to all human beings via reason, which is to say through the study of philosophy and science. The intellectualist spirituality of the Alexandrian school became the foundation of Arabic philosophy.

The Neoplatonic Philosophy

Neoplatonism or Plotinian philosophy was itself a melding of Plato and Aristotle. It was characterized by distinctive doctrines about the One (the Plotinian god), the intellect (both supernal and in the human person), and the soul. All being, according to Plotinus, flows forth in necessary outpouring from the ineffable One, via the divine intellect (*nous*), where Being becomes characterized and formed into "the archetypes of all existing things." *Nous*, the divine intellect, was held to be both God's thinking and an ideal world of which our mundane world was both an image and also its derivative. *Nous*, as the source of the world, was thus both in the image of God but also, as the first stage of the eternal emergence of the universe from God, entirely different from divinity itself.[4] Neoplatonism posited a number of intermediate stages in the emergence of the world from God and *nous*. These were sometimes identified with the "spheres" of the planets and regarded as intermediating intellects (minds) between the *nous* and human minds.

The structure of *nous* (the divine mind), expressive of the One (God), was believed to be embodied in stages, as the universe radiated out of it, culminating in the multiplicity and natural world of Earth. Earth was at the bottom of the outflow of Being from the One, through the spheres of the planets, ending in the terrestrial realm. As the universe flowed out from God or the One, its material expression gradually exhibited less clarity and refinement. The image of light flowing from a source (the sun) or water from a spring were the standard metaphors for the Plotinian ontology and cosmology. Both human soul and body—and everything else—were thought to flow or "emanate" eternally in a necessary causal generosity of fecundity from the inexplicable and ineffable radical unity of the One. The cosmos—including all nature, human, animal, plant, and mineral—thus expressed in the multiplicity of embodiment *nous*'s underlying rational structure and vital energy. All these were somehow emergent from the utter unity and ineffability of the One. At the outer reaches of the One's self-expression, its vitality and generativity reached their nadir. The Alexandrian Neoplatonic School interpreted this standard schema via a decidedly naturalist reading of Aristotelian texts.

Although mythic in its cosmology, Neoplatonic philosophy exhibited a rationalism and naturalism in its conception of causation. The universe operated according to natural principles that could be discovered and grasped by human reason. The universe was grand but to a significant degree knowable. Through the knowledge of nature, the mind could unite with the underlying structure of the cosmos, for the mind's thinking, its reason, was identical to the causal principles embedded in natural processes. Natural processes were the material expression of divine reason, from which all things were believed to derive their existence and their identities or natures. God's ideas were thus thought to disclose themselves to the human mind through the common presence of reason in nature, divine mind, and human mind. The Plotinian view was that the human mind, because it is capable of rigorous philosophical and scientific understanding, can expose and reproduce mentally the underlying logical and scientific causal structure of the universe (or, in some versions, it was thought capable of reproducing at least the structure and processes of the part of the universe nearest us, the immediate natural world of earth and moon). In engaging in a rational philosophic endeavor that delved into and was thought to uncover the structure and causal principles of nature, the human mind was thought to be capable of approaching—and to some degree even of achieving—a kind of union with the divine mind, the source of those principles. In engaging in rigorous philosophy, the human rational soul could thus metaphorically return to its source in the divine reason (*nous*) and a spiritual transformation, even a kind of immortality, could be brought about.

In the Aristotelian tradition of the Neoplatonic school of Alexandria, the natural world was thus the conduit between human and divine, and hence rigorous philosophic and scientific endeavor was a spiritual quest. Philosophy was the spiritual path to exposing the divine rational origins and causes of nature. Engaging in the systematic study of nature through the scientific and philosophic curriculum was thought to be a rational praxis of the greatest spiritual import. The human understanding of the underlying natural order and causal principles that flow from *nous* (indirectly and automatically) and are embodied within all things (both to make them what they are and to bring them into being) was held to be the source of the greatest human fulfillment. Such a life of the mind was thought to institute what Plato had mythically represented as the intellectual "return" and "ascent" of the soul toward the divine mind. Hence the desire to come to know the natural universe through engaging in philosophic and scientific study was fueled and pervaded by the desire for human fulfillment and spiritual transcendence.

The Translation of the Greek Texts of Philosophy and Science of the Alexandrian School

The area around Alexandria fell into Muslim hands and rule at the end of antiquity, in the seventh century, early in the rise and expansion of Islam. This area was the cultural center of the tradition in which "Aristotle was seen as the unexcelled master of scientific learning in logic, physics, cosmology, natural science, and psychology." The Alexandrian school, perhaps representing the most naturalist and rationalist approach of the classical Greek philosophical tradition, thus came within the Muslim orbit. In the East, Greek philosophical and scientific works, as well as biblical ones, had been translated into Syriac, a dialect of Aramaic, in the Christian biblical school at Edessa for several centuries before the Muslim conquests. That Christian school had undergone a transition from hostility to Greek learning to a program of assimilation.[5] These eastern Christians had set about translating Greek texts into Syriac, and later, from the ninth to the tenth centuries, it was these Syriac texts that were first translated into Arabic with the enthusiastic support of the Abbasid caliphs in Baghdad. The invention of paper around this time also contributed to the process.

Philosophers and philosophical translators within the Eastern cultural world generally held Aristotle in the same high regard as did the Alexandrian Neoplatonists. For the Syriac tradition of Aristotelian logic—"translations, companions, commentaries"—both in Persia and in the Eastern Christian orbit of Nestorians and others was also within the sphere of the influence of Alexandrian Aristotelianism. "Even under the Abbasid rule," D'Ancona points out, "in the eighth and ninth centuries, the Christians of Syria were the unexcelled masters of Aristotelian logic," and "in ninth-century Baghdad, and even later on, Syriac-speaking Christians carried on a tradition of logical learning in close relationship with the Arab *falasifa*," that is, the classical Arabic Aristotelian philosophers. The way was prepared for Aristotle to become "The Philosopher" and "First Teacher" within Arabic philosophy.[6] There is no doubt that there was an "intrinsic dependence of the rising Syriac and Arabic philosophical tradition on the Alexandrian model of philosophy as systematic learning, organized around a corpus of Aristotelian texts." The Alexandrian model thus became "the main pattern for the understanding of what philosophy was, and how it was to be learnt, in the Arabic tradition."[7]

The texts originally translated into Syriac formed the core of the Arabic rationalist philosophical tradition, for "a vast array of Greek scientific and philosophical works [were rendered] into Arabic." This movement of translation of Greek philosophy and science into Arabic was the impetus and source of a philosophic movement in Islam. The term *Arabic philosophy*, in fact, "identifies a philosophical tradition that has its origins in the translation movement."[8] So philosophy in Arabic owes a debt to having become heir to the Syriac Greek philosophical tradition, which in turn was profoundly influenced by the Alexandrian philosophical school, which in turn continued and developed a classical Greek (and particularly embodied Aristotelian) rationalist and naturalist philosophic thrust and ethos—an ethos from which the Latin Augustinian West had moved away.[9] The introduction of Greek philosophy into Islam and the philosophic and scientific developments it inspired owe a great debt to the vigorous and vast movement of the translation of Greek texts into Arabic.

That Aristotle's *Metaphysics*, the latter three books of Plotinus's *Enneads*, and the fourth-century Proclus's *Elements of Theology* happened to be translated from Greek into Arabic contemporaneously had a lasting effect on the

shape of *falsafa*. Al-Kindi, the first Arabic philosopher, or *faylasuf*, in melding the three texts in his appropriation of them in his own philosophy, set
the direction of *falsafa* for the future as an amalgamation and reconciliation
of Aristotle and Neoplatonism. It was al-Kindi who brought Greek philosophy to the Muslim intellectual elite. Into the Neoplatonized Aristotelian
orientation and philosophic point of view that al-Kindi established, all the
books of Aristotle came to be introduced and integrated when the Arabic
translation of the full Aristotelian corpus became available. The completion
of the translation of Aristotle's works enabled the *falasifa* to rethink their
philosophic project as a complete Aristotelian system of rationally demonstrative science. The system of Aristotelian science was placed by the tenth-
century Muslim philosopher al-Farabi at the heart of the advanced
philosophic curriculum he developed. While al-Kindi set the direction of
falsafa, it was al-Farabi who formalized that direction in a normative curriculum. In that curriculum the nonphilosophic Islamic studies—that is,
theology and law—were to be subsumed (and thereby philosophically transformed) by Aristotelian philosophy. The goal of a complete Aristotelian
philosophic system was also germane to the eleventh-century Persian philosopher Avicenna, who set himself the project of writing "the *summa* of
demonstrative science—from logic to philosophical theology—as a necessary step for the soul to return to its origin, the intelligible realm."[10] Through
the development of a definite corpus and curriculum of scientific and philosophical knowledge, which encompassed what is knowable about nature,
political society, and human capacities, a clear path was set for those who
devoted themselves to intellectual endeavors to achieve what was held to be
the closest approach to the divine mind embodied in nature. For the *falasifa*,
spiritual transformation was held to be the quintessential province of the
philosophically and scientifically educated.

Al-Farabi's Revival of the Alexandrian Aristotelian School

Man is a part of the world, and if we wish to understand his aim and activity
and use and place, then we must first know the purpose of the whole world, so
that it will become clear to us what man's aim is, as well as the fact that man is
necessarily a part of the world, in that his aim is necessary for realizing the
ultimate purpose of the whole world.

—AL-FARABI

[For Augustine,] though their bodies are (temporarily) "in" nature, humans
are not "of" nature.

—PETER BROWN

The tenth-century philosopher Abu Nasr al-Farabi conceived his curriculum
as a project whose explicit aim was to revive and reestablish the Alexandrian
school of philosophy as adapted to an Arabic linguistic and cultural milieu.
He is thought to have been born in central Asia about 872, yet the name
Farab denotes the family's Persian origin. Most of al-Farabi's life was spent
in Baghdad, seat of the ruling Abbasid dynasty, but he spent some time in
Syria and Egypt, dying in Damascus in 950 or 951. In Baghdad, al-Farabi
studied with noted Christian scholars and philosophers, and especially
Aristotelian logic with a Christian cleric. Al-Farabi's coming into contact with
the Christian philosophical circles of Baghdad opened to him the scholarly
world of Syriac Aristotelianism and the Alexandrian philosophical tradition
to which it was heir. He apparently had friends at the Abbasid court, for he
composed a book on music for the minister to one of the caliphs. Al-Farabi's
writings span a wide range of disciplines, from mathematics and music to
logic and philosophy. He made important contributions to political philosophy and to what we would today call social psychology. He is held to have
written more than one hundred works, including commentaries on Aristotle
(even a now-lost commentary on Aristotle's *Nicomachean Ethics*).

Al-Farabi set the agenda and the curriculum of classical rationalist Arabic
philosophy from his time forward as the explication, transmission, and
further development of Aristotle as the Philosopher and of Aristotelianism
as the Philosophy. His fame and stature were such that he came to be called
the Second Master of Philosophy, that is, second only to Aristotle himself.[11]
In his endeavor to promote, systematize, refine, and transmit Alexandrian
Aristotelianism, he was inspired by his Christian teacher Yuhanna ibn
Haylan, who was within the Alexandrian philosophical tradition and had
transmitted it to him. Al-Farabi can be seen as the founder of the Arabic
Peripatetic (Aristotelian) tradition as it traveled from Baghdad to Muslim
Iberia (Andalusia), in part inspiring the somewhat mystical philosophy of
the Persian Avicenna and culminating in Arabic in the thirteenth century
with the great Cordoban commentator Averroës and the Jewish philosopher
Maimonides, Averroës's contemporary, who was also born in Cordoba,
intellectual capital of the Iberian Arabic world.

In his development of a school of Aristotelianism, al-Farabi aimed to bring the Alexandrian tradition into the culture and language of Arabic Islam. "The al-Farabian corpus," the historian of philosophy David Reisman writes, "is almost single-mindedly driven by the combined goals of rehabilitating and then reinventing the scholarly study of philosophy as practiced by the Alexandrian school of neo-Aristotelianism. In this regard, he is rightly . . . [the] self-proclaimed heir of that tradition." Al-Farabi explicitly claimed the mantle of the latest in the line of transmission of the Alexandrian Aristotelian School, offering his own, somewhat polemical version of the history of philosophy into which he introduced his own Aristotelian project:

> Philosophy as an academic subject became widespread in the days of the [Ptolemaic] kings of the Greeks after the death of Aristotle in Alexandria until the end of the woman's [Cleopatra's] death. The teaching [of it] continued unchanged in Alexandria after the death of Aristotle through the reign of thirteen kings. . . . Thus it went until the coming of Christianity. Then the teachings came to an end in Rome while it continued in Alexandria until the king of the Christians looked into the matter. The bishops assembled and took counsel together on which [parts] of [Aristotle's] teachings were to be left in place and which were to be discontinued. They formed the opinion that books on logic were to be taught up to the end of the assertoric figures [*Prior Analytics*, I.7] but not what comes after it, since they thought that would harm Christianity. [Teaching the] rest [of the logical works] remained private until the coming of Islam [when] the teaching was transferred from Alexandria to Antioch. There it remained [until] one teacher was left. Two men learned from him, and they left, taking the books with them. One of them was from Harran, the other from Marw. As for the man from Marw, two men learned from him . . . , Ibrahim al-Marwazi and Yuhanna ibn Haylan. [Al-Farabi then says he studied with Yuhanna up to the end of the *Posterior Analytics*.][12]

The measure of his success is that subsequently all philosophy in the Islamic world followed the agenda and approach that al-Farabi had introduced and developed.

What was perhaps most distinctive and innovative about al-Farabi's assimilation and adaptation of the Alexandrian school of philosophy of late antiquity to the Islamic context was his conception of the relations among

religion, politics, and philosophy. Al-Farabi accepted the Neoplatonic Aristotelian emanationist schema of the Alexandrian tradition, according to
which philosophy, through its engagement of the intellect in coming to
know the universal patterns and processes inherent in the natural world, was
the arena of the human approach to the divine. So philosophy, having been
given the standard role of religion as the path toward God, left established
religion with little purpose. Al-Farabi rethought the purpose of religion, attributing to it the role of a culturally specific imaginative symbolic system
for conveying universal philosophic truth to the non-philosophic masses.

> Al-Farabi, Ibn Tufayl, and Averroës in particular, claim that there is one
> philosophical truth reflected in a plurality of simultaneously true religions.
> "True religions" simply translate into symbolic and, therefore, culturally deter-
> mined languages what the philosophers know through demonstrations. Such a
> claim . . . implies that the great philosophers hold basically the same philo-
> sophical tenets and that philosophy reached its peak with Aristotle. Al-Farabi
> offers a striking example of this attitude in his *The Harmonization of the Two
> Opinions of the Two Sages: Plato the Divine and Aristotle.* This text illustrates the
> old Alexandrian tradition that profoundly influenced the *falasifa.*[13]

In claiming that religion was a product of the imagination in contrast to
philosophy, which was a product of reason, al-Farabi set as religion's proper
role the development of a persuasive rhetoric, in stories that appealed to the
imagination and to the emotions, that could contribute to the political management of the masses in the interest of peace and justice. He proposed that
the use of imaginative suasion via religious story and injunction could be of
crucial benefit to political leaders in their maintenance of political order. As
a product of the imagination, religion was deemed to have a primary educa-
tive and political function in the institution of morals among the masses,
and more broadly in instituting and managing the polity. Al-Farabi did not
consider religion and philosophy as representing two equal sources of truth,
faith and reason (as did Aquinas, for example, and much medieval normative Christianity).[14] Instead he proposed a two-level theory: philosophy (reason) was the province of the intellectual and political elites, whereas religion
(a product of the lowly imagination) was the recourse of the masses.

The Alexandrian Aristotelian philosophical tradition as a whole was committed to the superiority of theoretical intellect (engaging in the pure sciences and the most abstract philosophy) over practical intellect or instrumental

reason (developing good habits in the formation of character) as the source of human fulfillment and perfection. In this tradition, partly due to its Neoplatonic origins, theoretical intellectual endeavor—that is, scientific and philosophical engagement for its own sake, for the sake of understanding alone—was deemed to be not only the path of the spiritual approach to the divine but also the source of the only true ethic.[15] Hence, philosophy, not religion, was considered the necessary foundation of both spiritual transformation and moral agency.[16]

Ethics Within the Arabic Aristotelian Philosophic Orbit: The Education of Desire

[The Arabic philosophers] explored ethics within *falsafa*, an enterprise common to Muslims, Christians and Jews and continuous with the Greek tradition, even if that tradition was rethought to accommodate religious concepts like charity, divine will, and revelation.

—PETER ADAMSON

This power [of practical intellect] is a power common to all people who are not lacking in humanity, and people only differ in it by degrees. As for the second power [the theoretical intellect], it is clear from its nature that it is very divine and found only in some people, who are the ones primarily intended by Divine Providence over this species.

—AVERROËS

Moral philosophy within the Arabic Alexandrian tradition looked to Aristotle and particularly to his *Nicomachean Ethics*, to Plato (indirectly, since it is likely that no complete Platonic dialogues were available in translation), to the Neoplatonist Porphyry, and to Galen as authoritative sources.[17] Arabic philosophical ethics planted itself squarely within the eudaimonistic tradition, a tradition that identified virtue with human happiness or fulfillment. The happy, fulfilled person, in Aristotle's estimation, is she who exercises and develops her most characteristic and noble capacity, namely, reason. Human beings, therefore, have their fulfillment and perfection in bringing to fruition their nature as a species, which is their rationality. While according to Aristotle the defining natural characteristic of the human animal was reason, for horses the development of their natural equine nature perhaps would lie in becoming good at galloping and jumping. So for humans, to be rational was to be virtuous, and to be virtuous was believed to entail the perfectibility of what one potentially is into what one actually is.

Aristotle had divided reason into two kinds, practical reason (having to do with action) and theoretical reason (having to do with contemplation and inquiry or discovery). The Arabic philosophers followed Aristotle in this division of reason. It might seem to us self-evident that ethics should fall into the practical side of reason, yet that was not the route taken by a number of Arabic philosophers. Instead, inquiry into the nature of the virtuous life and the source of happiness often was categorized within theoretical reason, while practical reason was relegated to mere actions. There was a tendency of Arabic philosophers to regard virtue as a matter of coming to understand universal truths, while habit and practice had the function of a kind of training to reduce the desires associated with what were regarded as the animal-like regions of the soul, those lower than the intellect. The position of these Arabic philosophers leaned toward the Socratic point of view that virtue is a kind of knowledge, so that knowing the good brings about doing the good. In other words, true knowledge was believed to transform desire, motivation, toward morally good ends as well as toward intellectually worthy ones.

In the Arabic philosophic tradition, Aristotle's moral intellectualism was accentuated and developed further by the *falasifa*, who at the same time downplayed his notion of practical moral virtue. They highlighted not the spiritually neutral theory of the training of character via the practical intellect—the famous choice of the mean between two extremes, courage being the mean between foolhardiness and cowardice, for example—near the beginning of the *Nicomachean Ethics*, but instead they embraced the latter part of the *Nicomachean Ethics* as fecund with spiritual meaning. The moral transformation of mind and heart believed to be consequent upon the theoretical intellectual endeavors of philosophy and science for their own sake were given center stage as leading to spiritual transformations with morally beneficial consequences. In interpreting Aristotle in this way, the *falasifa* emphasized the sections of the *Nicomachean Ethics* about the divine nature of the pure intellectual life, reading the sections on practical reason in the light of the sections on the spiritual attainment (and the concomitant transformation in desire) emerging from the ecstatic joy of the discovery of the underlying principles of nature and universe. These were the passages in which Aristotle introduced a vision of God as engaged in theoretical contemplation of the scientific order of the universe, a natural order of which God's activity of theoretical thinking was itself the eternal necessary and ongoing source.[18]

The lesser practical virtue of moderation, because of its association with the body (following Galen), was generally thought of on the medical analogy as the health of the (lower parts of the) soul, those that governed the body. The pursuit of intellectual virtue, in contrast, was thought of in terms of ultimate spiritual transformation and ecstatic rapture. Attainment of a degree of intellectual virtue thus was thought to offer a taste of the joy of eternity, by opening a window to the divine world of truth that awaited the philosophically trained once the soul had departed the body.[19] The Aristotelian view of training in habits that instituted and promoted a certain kind of character, which in turn supported ethical discernment or moral perception, especially of the virtuous mean between extremes, was downplayed by the Arabic philosophers, who regarded virtue, for the most part and at best, as an outgrowth of grasping universal theoretical or scientific truths.[20] As a result, the Aristotelian notion of practical training in habits of virtuous character was largely denigrated within the Arabic philosophical milieu as somewhat trivial and insufficiently intellectual to be transformative of the whole personality and lead one closer to God.

In *Nicomachean Ethics*, X, Aristotle had depicted the spiritual ecstasy of the divine mind's "thinking on thinking," in God's contemplation of his own universal ideas, the causal principles at the heart of all reality, natural and cosmic. Aristotle's God was engaged in theoretical contemplation of the scientific order of the universe, an order of which he was held to be the eternal necessary and ongoing source. To know the underlying order of nature and universe was thought to accomplish (some degree of) the union of human mind with the divine mind, whose "ideas," or causal principles and laws, were embodied in natural and cosmic processes. Hence to study nature and universe was to initiate the human spiritual union with the divine, a spiritual transformation of ecstatic proportion and imbued with mystical joy. In the embrace of the divine knowledge, the philosopher could come to love all nature as God's creation and bring the intellectual dimension of the self toward communion with God. In analogy to the divine mind, the human engagement in theoretical intellectual pursuits—in contrast to the exercise of practical reason—could lead to the fulfillment of what human beings shared with God, theoretical reason. The exercise of this capacity shared by divine and human was the path to human perfection and its accompanying joy. The theoretical life of philosophic exploration and scientific discovery was the answer to the moral question of what kind of life was the best for a human being, fulfilling human desire and perfecting human nature.

The Neoplatonized reading of Aristotle by the *falasifa* developed Aristotle's mythic vision of theoretical intellectual rapture into a doctrine of the union of the human with the divine mind in the act of knowing universal or theoretical principles. Arabic philosophers generally read the latter passages of the *Nicomachean Ethics* along with several from Aristotle's *De Anima* (*On the Soul*), in which the act of thinking was said to unify the subject, the process, and the object of thought or knowledge. The human mind was said to approach union with the divine mind via knowing nature as an expression of God's ideas (natural causal principles) embodied in the world. When a person engaged in philosophy and science and came to know the order of nature, at the same time—because of the Aristotelian principle of the identity of knower and known in the act of theoretical understanding—that amounted to coming to know what was knowable of God's mind. It was only in such knowledge of God's mind, as it was embedded and embodied in the principles of nature, that the human mind could commune and even become, to some extent, one with God—and with all other human minds that were engaged in the same theoretical pursuit.[21]

In the Neoplatonic tradition (and even to some extent in Aristotle himself) the appropriation of these Aristotelian texts of the *Nicomachean Ethics* and the *De Anima* was also suffused with the sublimated eroticism of Plato's *Symposium*, which depicted the cognitive-affective ascent from primitive love-knowledge of singular objects (individual bodies) to the educated desire to know and love the universal, supernal truth, wisdom, and beauty. Hence, the account of the union of the human mind (the subject) with God's mind (the object) in the act of coming to know (the process), when one discovers, by philosophy, God's mind (the laws of nature) embedded in the natural world, was suffused with an erotic longing that led to consummate pleasure (Plato), and the perfection and fulfillment of one's humanity (Aristotle). That theoretical philosophic and scientific endeavor and its accompanying joy (eternally in God and rarely and intermittently in human philosophers) were assimilated to a religious notion of the ecstasy of the mystical union of the human with the divine in the joining of the human mind with (some portion of) the divine mind in coming to know theoretical wisdom (the scientific causal principles of nature). The transcendent spiritual path of human-divine communion was thus open only to philosopher-scientists.

The Aristotelian Neoplatonic story lent support to a rational mysticism that uniquely privileged the spiritual potential of the theoretical philosophic life, reinterpreting moral agency as well in that key. It was held to be a moral path of virtue in two respects: it transformed desire so that all one's passions were directed only toward the knowledge of God, and thereby one's social relations were rendered of only transitory and secondary importance and pleasure; and it fulfilled human nature through perfecting its divine dimension, theoretical reason, in living the truly virtuous human life, that is, a life of intellectual pursuit and engagement. Via a non-zero-sum game of deep and ever broader understanding, competitive striving toward ephemeral ends was transformed into a shared joy of discovery and into a universal empathic embrace of the divine to be found everywhere in the world. Aristotle's God, depicted as engaging in theoretical rational understanding with an accompanying ecstatic joy, became the normative vision of God and the goal of human perfection for the *falasifa*. In this section of the *Nicomachean Ethics*, Aristotle had explicitly rejected the notion that God could ever be engaged in practical instrumental thinking, and hence God himself was explicitly depicted as beyond the ethics of habit and character formation and, even more significant, beyond all practical moral concerns. The God of Aristotle inhabited the nether realms of a paradise of the mind, living a life of philosophy and wisdom, free from the pull of any distracting practical concerns, including those that would require practical moral discernment.

Arabic Greek Intellectual Virtue Versus Latin Moral Free Will

The Aristotelian-Neoplatonic philosophic myth captured the imagination of the *falasifa*, coloring their relative valuation of practical moral versus intellectual concerns. The philosophic myth also stood in the face of the kind of practical reason entailed in direct historical intervention in the world, that is, it contradicted the biblical view of God as the God of history, the God who is a personal actor in the human arena. That notion of God as person served as the inspiration for an account of the divine will, of a will free and unconstrained, originative, omnipotent, and just. In Arabic Aristotelianism, the philosophic vision of God as source of natural processes triumphed over the biblical God of history, the intellectual mystical rapture in the natural world over the human union with the divine will through obedience and faith. Hence the Arabic intellectualist moral attitude was in sharp contrast with the Latin Christian one of Augustine and those within the Augustinian orbit. The latter tradition criticized Aristotle's notion of even practical rational training of moral character as a conception of moral agency that was too intellectualist, naturalist, and determinist, rather than too little so. Averroës, by contrast, for example, argued that it was theoretical intellectual endeavor alone that harbored deep moral and spiritual transformative power, and so he lamented that most people relied on mere moral habituation; as a result, the transformation of desire through the contemplative life was all too rare. In this, Averroës articulated the general disparagement of the practical intellect by the Arabic philosophic tradition as a whole.

Latin Christianity had rejected even the position that morals amount to a matter of social and self-development of a natural practical moral capacity through the social institution of the right habits, as Aristotle's own version of the theory had claimed, as insufficiently grounding moral responsibility in the free will. The danger of the thoroughly naturalist account was seen in the Latin West as the weakening of actions as fully attributable to the individual him- or herself alone as sole originator of action. If morals were determined by biological endowment and early social practices, that is, by nature and context, rather than by the exercise of *freedom from* nature and nurture, it was questionable whether any action or person could be worthy of the full force of the assignment of praise or blame. In the Latin world, Aristotle's program of training in virtuous habits was thus regarded as too naturalist an account of morals, one that did not sufficiently take into account the spiritual capacity of the free will to overcome body and nature and social context. To the extent that it could be assimilated into Latin Christian doctrine, Aristotle's naturalist training of character—and not his doctrine of the taste of eternity adumbrated only in the cultivation of the theoretical philosophic and scientific life for its own sake—was modified by Latin Christian thinkers to include at least some free will.

Within the Arabic philosophical orbit, however, the notion of social and political training in habits of virtuous character was neither controversial nor considered to have gone far enough in the naturalist intellectualist direction. For the commonplace in that philosophical world was that the practical intellect, far from relying too much on the mundane self- and social development of the natural agency of human practical reason—and thereby posing a danger of undermining the moment of "free" personal choice between good and bad as the criterion of moral worth—instead exhibited too little reliance on and development of human reason to be of salvific and

spiritual import in the approach to the divine and hence had too little potential for transforming moral motivation. The Arabic Aristotelian philosophers believed that practical training in good habits was of scarce importance
for the ultimate goal of the rational self-transformation of the head, and
with it the heart, through the deep and broad understanding of nature and
cosmos that was thought to be at the core of the wholesale transformation of
desire that amounted to real moral agency via spiritual communion. Hence,
the Arabic philosophic approach, in contrast to the Latin, was thus to inten-
sify the Aristotelian theoretical intellectualist thrust in religious sensibility
and in moral psychology at the expense of Aristotle's competing emphasis
on the more practical rational deliberative moral outlook.[22]

While the Arabic philosophers intensified Aristotle's moral and spiritual
intellectualism, Latin Christian Aristotelian thinkers emphasized and reinterpreted his account of practical moral deliberation and habit, finding it
more assimilable to accounts of free will. Nevertheless, Aristotle's position
that choice of action followed necessarily in a determinist way from character, desire, and knowledge was not a permissible part of the equation for
Latin Christians. Habit could be allowed in, in an attenuated way, for some
Latin Christian Aristotelians, as helping people make moral choices more
automatically. But that is not what Aristotle actually had held, nor had he
introduced freedom of a voluntarist sort into his account. For Aristotle,
choice was necessarily the result of one's understanding; cognitive grasp of
the good was driven by desires and by the state of one's desires, not by an
independent (that is, free) will. No intervention of a free will (if there were
such a thing) could solve the problem of moral action as Aristotle understood it. Becoming a moral person was a matter of early social practices via
the training of desire through habit, which in turn produced either moral
benightedness or clarity of moral vision. Ultimately, fully perfected moral
agency was an outgrowth of the development of the theoretical intellectual
life, the human virtue par excellence. It was the pursuit of intellectual virtue
that alone could transform and integrate all of a person's desires, directing
them toward the overarching aim of coming to understand the world and
oneself within it. Anyone who achieved that transformation in vision would
display beneficence toward all God's creatures, for love was believed to be
consequent upon knowledge, a theme incipient in Aristotle and developed
by Neoplatonic interpreters upon whom the *falasifa* depended. Nevertheless,
it was a pressing problem for the *falasifa* that lesser mortals than philosophers, namely, the masses, needed to be managed socially and politically.

They were the prime candidates for the coercive institution of the training of moral habit. Intervention in practical moral education and habituation was thus relegated by al-Farabi and the tradition that followed him to the social and political domain. For philosophers, however, there was the heady life of the mind, a life that was held to transform the heart and soul through its reorientation of passion from the mundane and petty to the lofty and enduring.

The Fundamental Moral Insight of *Falsafa*

What all this amounted to was that in the Arabic tradition of *falsafa*, moral agency was understood as a matter of how thinking and motivation, cognition and emotion, were mutually intertwined and interpenetrated. Reason and desire were even considered to be what we would call today *co-constructed*. Action was considered an outgrowth, a necessary outcome, of the emotion-cognition complex. Following Aristotle, it was in the ways that thinking shaped desire and desire shaped thinking that intervention to substantially change moral action was thought to be possible and could be initiated. In other words, in a contemporary idiom, moral transformation was consequent upon a systematic and thoroughgoing transformation in *interpretation*—of the world, of the self, of the self and human person in the world—as it recast meaning and redirected motivation. Habits introduced early in life could, in behaviorist fashion, orient desires in practice so that the understanding of situations was pushed in the right direction. Yet this was a stopgap kind of intervention, neither fully reliable nor self-aware. It took a systematic and wholesale philosophical revolution in self- and world understanding to bring about a transformation in the focus of one's life away from the everyday to the exploration of the cosmic and eternal, reorienting the passions along the way. To grasp the process, think of falling in love. That is Plato's model for the ascent to the intellectual-spiritual life in *The Symposium*. Here was a devotion to classical Greek philosophy—emergent from Plato's *Symposium*, Aristotle's *Nicomachean Ethics*, the *De Anima*, Plotinus's *Enneads*, and all the rest—elevated to high spiritual meaning and thrust upon biblical, post-biblical, and Koranic texts, rituals, and practices. Falling in love with the philosophic-scientific search for truth, equated with communion with God—that was ethics as well as religion. In *The Guide of the Perplexed* Maimonides called it the "true worship."

The Arabic classical rationalist philosophic tradition ended within Islam with the death of Averroës in 1198. Yet it had an afterlife in Judaism, and also in Christianity, as discussed in the last chapter. In Judaism, *falsafa* had its pinnacle in the Arabic language in Maimonides's *Guide of the Perplexed*, but subsequently it continued largely in Hebrew and made a decisive contribution to the Western philosophical orbit—not yet adequately recognized or acknowledged, perhaps—with Spinoza, writing in Latin for a largely Christian philosophical audience. The late great scholar of Jewish and Arabic medieval philosophy Shlomo Pines comments at one point on the irony of Spinoza unknowingly reviving the more radical positions of al-Farabi about the status of religion as mere imagination in his arguments against Maimonides about the Jewish polity.[23] But perhaps there is less irony here than meets the eye, for al-Farabi saw himself as continuing and further developing the approach and work of the Neo-Aristotelian Alexandrian school of philosophy of late antiquity, of which Spinoza, in the seventeenth century, can be considered in some significant respects its last representative.[24]

Maimonides's Belief in the Transformative Power of the Intellectual Life

Within the Arabic tradition of *falsafa*, Maimonides can be considered the philosopher who most ardently embraced the doctrine of the transformation of (base and petty) desires into the joy of theoretical learning, with its accompanying universal love as the only basis for true ethics.[25] He radicalized the doctrine, proposing a version of it that may stand as the purest in the history of philosophy. Maimonides went so far as to claim that the true and the good are in some sense *opposed*.[26] By this he wished to suggest that the theoretical study of nature for its own sake—the true—was the *only* avenue to the spiritual life and the deep transformation of motivation from self-interested to cosmic. Hence the standard training and habituation in moral action, he believed, had no real transformative power at all but could serve merely as a helpful preparation for engaging in philosophy. Nevertheless, it could also contribute to social stability and political utility.

In *The Guide of the Perplexed*—the book that Maimonides wrote for rabbis who had had a philosophic and scientific education as well as a Jewish one—the only route to true ethics was held to be via the transformation of the emotions and desires through theoretical intellectual understanding, that is, by engaging in philosophy and science for their own sake. Maimonides, like al-Farabi, relegated the Aristotelian practical rational training of character in habits of virtue to the coercive practices and suasive rhetorical discourses of political, religious, and social institutions—a position incipient in the *Nicomachean Ethics*. Maimonides regarded the moral rules identified by practical reason and instituted by coercive habitual practices as mere social conventions rather than as real or natural.[27] Human fulfillment, in contrast, was to be accomplished through intellectual pursuits, which, if engaged in with the whole heart as well as the head, were believed to transform desire so that the true knower displayed beneficence toward all. Moral rules, however, were regarded as largely conventional: the rules were socially and practically beneficial for human beings but varied from society to society since they were particularized in different social, legal, cultural, religious, and political systems. What made standard moral virtues, such as courage and moderation, of less importance than intellectual virtues for Maimonides was that the former were relative to human beings, benefiting primarily social interactions. Hence moral norms were held to be not of divine, eternal, infinite value but instead of mere "local" human utility. The natural universe, regarded as a divine creation, in contrast to human society, was held to be a far larger place whose well-being could not be reduced to human interests or values. Moral commandments, for example, were not to be discovered written into nature; instead, the cosmic perspective of nature was to be brought to bear upon human values to render them of spiritual scope and import. In Maimonides's conception, the universe existed only for its own sake and not for the sake of human beings.[28] It was thus up to human beings to bring their thinking about their place in the universe into accord with the infinite scope of the natural cosmos.[29] Thus the universe was not "good" on any human scale or with any human moral meaning.

Maimonides held that the love of God was an outgrowth of coming to know God through the philosophic and scientific investigation of nature. Obedience, in contrast to love, was a product only of the fear of God, and hence of little motivationally transformative power.[30] Maimonides utterly de-anthropomorphized God. God as the source of all natural necessary processes and nothing else—no less than Spinoza's "God or Nature"—could not be obeyed but could only be known and lovingly embraced. Maimonides puts it succinctly. In contemporary terms, one cannot "obey" the laws of motion or the laws of quantum mechanics because one cannot "disobey" them; they are descriptive, not prescriptive. So human choice, too, was at bottom another necessary cause embedded in nature. Maimonides was insistent that God did not issue commandments as a king would to his subjects. That biblical claim was interpreted by Maimonides as a flight of the imagination geared to the most uneducated of the common people or to children. That image of God was not to be taken literally but understood instead as a concession to the ignorant in the interest of suasion and political exigency. It was not to be countenanced by even the semieducated, Maimonides insisted.

A true—that is, intellectual—understanding mandated that God's will and intellect were one and the same. The identity went in the direction of the intellect: God's will ought to be understood as divine intellect, the source of the necessary principles of nature. Maimonides thus believed that natural necessity held sway in the universe—and this was as true of the human person as it was of the objects of physics. Maimonides puts it succinctly: God is "He who arouses a particular volition in the irrational animal and who has *necessitated this particular free choice in the rational animal* and has made the natural things pursue their course—chance being but an excess of what is natural."[31] (Emphasis added.) The human will was as embedded in causal contexts as much as any natural object was or any animal volition—and thus the human will could not be free in the sense of (self-)determination apart from its constitutive natural endowments and contexts. It could not will itself beyond its causes or choose beyond its desires and other motivations.

In the Aristotelian philosophical tradition, mind was believed to be more amenable to scientific natural explanation than matter was. Science was the study of God's mind and matter merely made the rational causal content of God's mind empirically existent. Mind, in the Aristotelian terminology, was "form," and it could also discern form—that is, mind was the source of the intelligibility of things and of their natural causes, the possibility of their classification, and the reason inherent in them that made them what they were, their species and their genus. Matter, body, was unknowable except through its form and formal properties and causes. In contrast to the recalcitrance and only approximate character of the way that physical principles were embodied in particular objects, mind, in the act of knowing, could completely capture, and hence conform to, the divine causal principles it extracted from natural processes by discovering their regularities (according to this reading of *De Anima*). Maimonides's project in the *Guide* was to rethink God along these scientific naturalist lines and then to read that naturalism back into the Bible.

Maimonides argued that the Bible was a mythic rendering of a thorough-going philosophic scientific naturalist perspective, and thus Judaism was a religion of science. Judaism, Maimonides argued, was actually the only truly consummate religion of science because Moses, its philosopher-founder, was a philosopher-king who brought basic principles of philosophy to the masses in a popular idiom so that they could grasp them at a rudimentary level. In addition, the Torah offered a system of law and justice necessary in an un-redeemed world. Implicitly Maimonides cast himself as the new Moses, rendering a naturalist perspective into the idiom of imaginative normative myth and law. It is indeed fascinating that Judaism, the religion of com-mandment, was rethought by Maimonides, its greatest legal scholar since the Talmud, in the Arabic philosophic key as *the* religion of science. That religion of science exalted the ecstatic joy and love of the universe of an Ein-stein in the act of discovery as its model of moral as well as spiritual transfor-mation. By contrast, Aristotle's character education, instituted by the practical intellect, was developed by Maimonides, following al-Farabi, as an "imaginative"—by which Maimonides meant conventional, coercive, and suasive—political program of socialization and law geared to the masses and to society as a body rather than chiefly to the individual.[32]

The Interpenetration of Emotion and Cognition

Maimonides's psychology heightened the importance and role of the inter-twining of desire, emotion, and intellect in moral transformation. The ulti-mate moral life (attainable only by some highly motivated and philosophically educated individuals) thus integrated reason and emotion, passion and thought, affect and cognition, endowing the relation between head and heart with a spiritual meaning and rationale. More than five hundred years later it was up to a modern thinker, Spinoza, to secularize in a quasi-biological and quasi-modern psychology the theological underpinnings and philosophic myth at the heart of the Maimonidean account. Spinoza, a Sephardic Jew from a family of Portuguese Marranos who had recently escaped the Inqui-sition in the Iberian Peninsula and found freedom and security in the Netherlands, followed Maimonides in his radical opposition of the true to the good. Spinoza offered theoretical support from psychology and biology for the moral orientation of the most radical version of *falsafa*, thus bringing Maimonides's Alexandrian Arabic moral psychology into conversation with

a modern European educated audience, and in particular with the Augustinian philosophical psychology of Descartes. Spinoza rethought Maimonides's philosophic route to the perfection of moral agency open to some
individuals with an elite education. He also recast the Maimonidean-Farabian
social psychological and political management of the masses through habit,
social incentives, mythic ideology, and the strong arm of the law into a blueprint for a modern democratic pluralist polity that embraced and enshrined
freedom of thought, speech, and religion.

Maimonides and Spinoza Versus Descartes and Kant

The contrast that I have developed here between Western Latin and Mediterranean Alexandrian paradigms of spirituality, moral psychology, and
moral agency can be given sharper outlines if we juxtapose the doctrine of
human nature embraced by the greatest of modern Western philosophers,
Immanuel Kant, with that of the seminal Jewish Aristotelian, Moses Maimonides. It is a stroke of luck that both wrote lengthy interpretations of the
biblical account of Adam and Eve in the Garden of Eden. Comparing their
understandings of the biblical story is a succinct and engaging way of highlighting their respective beliefs about human nature and the place of human
beings in nature and the cosmos. Both also found in the Genesis story an opportunity to explain how and why human beings came to have a moral sense
and moral values.

Immanuel Kant on the Garden of Eden

> [In late antiquity] the Christian people, and especially their bishops, decided
> which parts of nature were "unearthly" and which were not. [Saint] Ambrose
> led the way.
>
> —PETER BROWN

The German Enlightenment philosopher Immanuel Kant (1724–1804) "is
the central figure in modern philosophy," according to the online *Stanford
Encyclopedia of Philosophy*. Mainstream philosophy even today is practiced
largely in a Kantian key. If all philosophy is in some sense a footnote to
Plato, as has been said, modern and contemporary philosophy is largely a
footnote to Kant. The Kantian "critical philosophy" shapes contemporary

discussions in ethics, political theory, aesthetics, metaphysics, and episte-mology (the theory of knowledge). The basis of all these discussions is Kant's understanding of "human autonomy," which is to say his account of the freedom of the will, an account that we shall see owes a great debt to Des-cartes and the entire Augustinian voluntarist tradition.[33]

Kant's anthropological writings, in contrast with his three *Critiques* (*Cri-tique of Pure Reason*, *Critique of Practical Reason*, and *Critique of the Power of Judgment*), are not central to his corpus. Moreover, narrative, historical, and biblical reflections have a somewhat ambiguous status for him, and such re-flections are largely absent from the *Critiques*. Nevertheless, at the age of sixty-two in 1786, Kant wrote an essay, "Conjectural Beginning of Human History," in which he responded to an interpretation of the Garden of Eden published two years earlier by his former student Johann Gottfried Herder, a noted philosopher. In his essay Kant proposed an alternative interpretation of the Eden account in which he revealed both his own philosophy of his-tory and some deeply held beliefs about human beings.

Both Herder and Kant were inspired by the French philosopher Jean-Jacques Rousseau in what they had to say about Eden, but they assimilated Rousseau in different ways. Herder saw Eden as a picture of human begin-nings in a natural condition of a life of social harmony. As long as they remained true to their spiritual origins, their harmonious existence was unmarred. Herder focused on the biblical verse that stated that Adam and Eve became (in the Hebrew idiom) like *elohim*. *Elohim* can mean "God," or it can mean "gods" or even "rulers" (as Maimonides will inform us). Herder takes it to mean "gods." The two human beings aspired to become like gods with the knowledge of evil, Herder says, and that marked the entry of the possibility of evil into human history in the form of civilization, which would eventually lead to the corruptions of advanced civilization. Adam and Eve distanced themselves from nature and their natural origins, according to Herder, and that eventually led to the building up of state power and the politics of war. War introduced alien elements, disruptive of an innocent, harmonious existence, into the bosom of nature, thus violating the human life that God had intended. Herder writes that it had been "a benevolent thought of providence to give the more easily attained happiness of individ-ual human beings priority over the artificial ends of large societies."[34]

Kant wrote his essay as an "exercise of the imagination," he says, to chal-lenge Herder's view, and he contrasts it with his own more serious history writing. Nevertheless, Kant challenges the reader to "check at every point whether the road which [his] philosophy takes with the help of concepts coincides with the story told in Holy Writ."[35] He makes clear that in retelling the story of the Garden of Eden, his "sole purpose is to consider the development of manners and morals [*des Sittlichen*] in [the human] way of life."[36] The story represents a time before society, when no social conditions had yet arisen. While Adam can talk and think, Kant says, nevertheless he "is guided by instinct alone, that voice of God which is obeyed by all animals."[37] It is at this point, Kant says, that reason "began to stir" and disrupt the beauty, harmony, and pleasure of natural human existence. This is reminiscent of Rousseau. Kant here conceives of reason as disrupting instinct by providing an array of so many possible "foods" that natural human instinctual desire becomes confused. Reason introduces "artificial desires," some of which can be contrary to natural instinct and hence harmful to survival and harmony. So what the story points to, Kant says, is that Adam chose the fruit as an act of reason and against natural instinct.

Kant proposes that reason introduced comparisons between "foods" and awakened the desire for "foodstuffs beyond the bounds of instinctual knowledge."[38] Adam chooses a fruit that his reason discerns as similar to fruits to which instinct directs him through taste. But reason does not keep him from fruits that might harm him, the way natural instinct does. Nevertheless, this move away from instinct marks the human person's "first attempt at becoming conscious of his reason as a power which can extend itself beyond the limits to which all animals are confined."[39] Adam's action is understood by Kant as the first occasion of "reason [doing] violence to the voice of nature (3:1), [and nature's] protest notwithstanding, . . . [this is] the first attempt at free choice." With decisive Augustinian echoes, Kant identifies reason as, and reduces it to, free choice; its exercise is an act (of will) opposed to nature.

What Adam "discovered for himself" in his first use of reason—Kant says that "it sufficed to open man's eyes" (Genesis 3:7)—was "a power of choosing for himself a way of life, of not being bound without alternative to a single way, like the animals." Kant goes on to describe Adam as standing,

> as it were, at the brink of an abyss. Until that moment instinct had directed him toward specific objects of desire. But from these there now opened up an infinity of such objects, and he did not yet know how to *choose* between them. On the other hand, it was impossible for him to return to the state of servitude (i.e., subjection to instinct) from the state of freedom, once he had tasted the latter.[40]

Next Adam discovered that sexual desire can be enhanced by reason and imagination: covering the genitals with the fig leaf (Genesis 3:7), Kant suggests, is an invention of reason to enhance sexual allure. This step, he says, represents "a far greater manifestation of reason than that shown in the earlier stage of development," which was eating the fruit. The fig leaf indicates a beginning of the mastery of reason over impulse in its refusal and manipulation of immediate instinctual desire toward desire's enhancement but also toward desire's spiritualization. "Refusal," Kant writes, "was the feat which brought about the passage from sensual [*empfundenen*] to spiritual [*idealischen*] attractions, from mere animal desire gradually to love, and along with this from the feeling of the merely agreeable to a taste for beauty. . . . The covering of the body is also a first hint at the development of man as a moral creature. This came from the sense of decency [*Sittsamkeit*], which is an inclination to inspire others to respect by proper manners."[41] And it is "epoch making," he says: more important than all the subsequent advances in culture to which it gave rise.[42]

Kant proposes that we can identify in the story four stages in the development of reason. The first stage is choice as the initiation of freedom, and the second is instinctual self-regulation. The third stage, the capacity to envision the future, which is in his view "the most decisive mark of the human's advantage," followed.[43] The capacity to envision future possibility, however, had a downside: anxiety and the fear of death. These were the punishments inherent in being fully human and in having and using reason. So Adam and Eve, in Kant's words, "apparently forswore and decried as a crime the use of reason, which had been the cause of all these ills."[44] That is why in Kant's view the Bible portrays the first use of reason as a rebellion or a sin. Kant remarks later on in the essay that Adam's "abuse of reason [was] in the very first use of reason."[45] The fourth stage of reason's development is also depicted in the biblical story: the mastery of nature. The "fourth and final step which reason took," Kant concludes, "raised man altogether above community with the animals." For "he came to understand, however obscurely, that he is the true end of nature."[46] And Adam realizes from then on that animals were no longer to be seen as fellow creatures, "but as mere means and tools to whatever ends he pleased."[47] With this awareness came the recognition that all fellow human beings share his status and are "equal participant[s] in the gifts of nature." Kant regarded stage four of the development of reason as the true beginning of morals and the basis for the establishment of a civil society: "Thus man had entered into a relation of equality with all rational beings, whatever their ranks (3:22), with respect to the claim of being an end in himself, respected as such by everyone, a being which no one might treat as a mere means to ulterior ends." Kant concludes, "Reason makes [man] an end in himself." Hence, "this last step of reason is . . . man's release from the womb of nature."[48]

The post-Eden life, brought on by Adam's act of choice against nature, although fraught with fear, envy, and social conflict, is regarded by Kant as "nothing less than progress toward perfection."[49] For the exercise of reason, while it initiates social conflicts that make life difficult, is nevertheless the source of human progress. Adam's Fall sets into motion a conflict between human natural animality (the instinctual choice of a fruit naturally good for the human species) and human rationality (the "free" choice of a fruit that seems desirable because of rational speculation). Human beings have two dispositions, Kant writes: "man as animal" and "man as a moral species." So Kant regarded the Fall as both the beginning of morality and at the same time its violation: "the history of nature therefore begins with good, for it is the work of God, while the history of *freedom* begins with wickedness, for it is the work of man."[50] (This statement of Kant's could have come straight out of the mouth of Augustine.) Human nature is bifurcated between the natural and the free: the natural is bodily and necessary, while the free—which by implication refers to the mind as beyond nature and undetermined or underdetermined by it—is choice as the distinguishing characteristic of reason. Rather than being characterized by a search for knowledge, reason is portrayed by Kant as essentially rising above natural endowment and tendency and freely choosing—a doctrine that Augustine more or less invented and which is prevalent in his own account of Adam's Fall. Hence the capacity for free choice, in keeping with the entire Augustinian tradition from Augustine himself through Descartes, is identified as the essence of reason (mind) and thus the essence of man. Kant portrays body and mind, natural necessity and free choice, as in conflict, and that internal war initiates and shapes the course of human history.

To his "Conjectural Beginning of Human History," Kant adds a final section, "The End of History." Human history, he speculates, is a tale of increasing conflict and growing inequality. Such conflict, however, sets progress in motion, because as time moves on, "the human species is irresistibly turned away from the task assigned to it by nature, [which is] the progressive cultivation of its disposition to goodness. . . . [For] its destiny . . . is not to live in brutish pleasure or slavish servitude, but to rule over the earth."[51] The first part of this quotation rejects outright the pursuit of human natural goodness as appropriate to human ideals. The natural in the human personality—even and especially our natural goodness—is to be repudiated and disavowed. Kant concludes that "war is an indispensable means to the still further development of human culture." A golden age of "universal contentment with the mere satisfaction of natural needs" and of "universal human equality and perpetual peace"—a "Robinson Crusoe" world, Kant calls it—would be neither possible nor desirable. We live in a post-Edenic world, fraught with the troublesome aftermath of Adam's choice, which is also our own free choice. Here we hear the Augustinian echoes in Kant's interpretation of the tale: the reduction of reason to (voluntarist) choice, the opposition of the natural, equated with the body (even the urge to the natural good and social harmony), to the mind as free from nature; the attribution to Adam of full responsibility for sin in his act of free will as the beginning of the human moral capacity; the relegation of the natural to the body alone, whereas mind, as choice, is beyond natural causal determination and hence free.

Maimonides on the Garden of Eden, Human Nature, and Moral Agency

> Philosophy, insofar as it celebrates the truly divine principles of the visible cosmos, is prayer.
> —CHRISTINA D'ANCONA ON THE GREEK NEOPLATONIC PHILOSOPHY

Maimonides's interpretation of the Garden of Eden focuses on human reason as much as Kant's does. Nevertheless, Maimonides's interpretation differs from Kant's and even stands opposed to it in important ways. Both philosophers present the Garden of Eden as a story about the endowment of human beings with reason as their most precious and unique possession. For both, reason is at the heart of human nature and sets the human ideal as well. Yet there is a marked contrast in Maimonides's and Kant's understandings of what reason is, its role in human life, and its place in nature. What each means by reason is quite different. For Kant, it represents and initiates the repudiation and transcendence of nature, with its concomitant freedom of the will, which rules over nature and body. For Maimonides, it points to the life of scientific discovery leading to mental union or psychological identification with nature and the divine that is within it, and its concomitant love of and riveting rapture in God's cosmos. Kant's conception of human potential and ideal and Maimonides's crystallize and encapsulate quite different and to some extent clashing religio-cultural values and worldviews.

The interpretation of the Garden of Eden is central to *The Guide of the Perplexed* and is introduced by Maimonides close to the beginning of it, setting the stage for much of what follows. Two chapters in their entirety are devoted to the interpretation of Adam and Eve.[52] In the first of these the story is brought up in the context of a discussion of creation. Here Maimonides writes that the biblical account of the six days of creation alludes to a scientific account of cosmology. The Garden of Eden story, by contrast, is written in an entirely different literary genre from the account of creation, for it is a mere fictional tale and is neither chronological nor historical. In this respect Maimonides and Kant are in agreement: the Garden of Eden story is conjectural rather than historical. Kant, however, does include a stab at a philosophy of history as well, emergent from his reading of Eden and its aftermath. Adam was not a historical figure, Maimonides says, and so the story should be understood as an allegory.[53] The second treatment of Adam and Eve, in *Guide* II, 30, emphasizes the fictional character of the biblical Garden. It begins with an allusion to Plato's etiological fantasy from the *Symposium* (a comic absurdity that Plato puts in the mouth of Aristophanes) of the initial androgyny of the human species, and its embodiment in three types of bouncing balls (three sexes) with two sets of genitals, one on each side, male-male, female-male, and female-female. Perhaps Maimonides is hinting that the Eden story is as much a tall tale as is the Platonic Aristophanes's account of the original hermaphrodites.[54]

Maimonides says that the setting for the Garden of Eden story is the sixth day of creation, so neither cosmic nature nor human nature, established earlier within the cycle of creation, appears to be changed for the worse by Adam's disobedience. (We saw earlier that Augustine had to develop a complicated theory explaining how a decline in nature as a whole and in human nature could be reconciled with the biblical text.) Maimonides is implicitly challenging the Christian doctrine of Adam's disobedience as an original sin that affects and infects all human beings and nature itself from then on. The Torah offers in the Garden of Eden tale not a chronology of historical personages, and certainly not a Fall, but instead, Maimonides proposes, a symbolic narrative to convey an abstract analysis of human nature both in its inception and for all time. Its purpose is to convey to the reader a picture of the ideal human life and how that ideal can be approached given the exigencies of bodily existence and societal constraints.

In the first part of the *Guide*, Maimonides presents his interpretation of human nature and the ideal life that he believes the Garden of Eden story is pointing to. He introduces his understanding of the story as a response to an anonymous learned man who years ago, he says, put up a challenge to the biblical text. The learned man is said to have argued that the story of the Garden of Eden portrays Adam as being given knowledge (by eating fruit from the tree of knowledge) as punishment for his sin. The anonymous challenger concludes that the Bible maintains that the ideal human being, Adam, was better off in his original condition, in which he was "like the beasts, devoid of intellect." All humanity after Adam must henceforth endure the sorry condition of Adam's punishment, namely, knowledge, he says. So the story's moral, the challenger contends, is that Adam represents the human ideal as naive and without knowledge, and knowledge is the price he (and we) must pay for his sin.

Kant might have agreed with Maimonides's anonymous challenger, since he argued that reason was a sort of punishment, in that Adam's first exercise of reason in free choice wrested him from an idyllic harmony with nature, making his life difficult and filled with anxiety and conflict. Yet Maimonides will have none of it. Maimonides provides his interpretation as a defense of the ideal of the intellectual life as being spent in philosophic and scientific exploration, against the anonymous challenger's objection. He sets out to demonstrate that in this story the Bible, contrary to first impressions, is pointing to human theoretical reason as Adam's greatest natural and spiritual endowment rather than as his punishment.

Maimonides's Version of the Garden of Eden Story

For Maimonides the Garden of Eden is a kind of just-so story, something like Rudyard Kipling's "How the Leopard Got His Spots." It is not that there were ever any unspotted leopards, but rather Kipling imagines a fanciful explanation for the spotted nature of leopards. Maimonides takes the Garden of Eden story as offering a fanciful answer to the question of why human beings are so dominated by imagination rather than able to easily follow their reason. A close parallel might be Plato's philosophical just-so story in the *Phaedrus* of the soul falling to earth and coming into a body. In falling to earth, Plato imagines, the soul loses much of the knowledge it had at its origin in the heavenly realm of true ideas. Through practicing philosophy, however, the fallen soul can regain (or, as he says, "recollect") some of its knowledge; in doing so, it sprouts wings, Plato's myth goes, and can then fly high and return to its supernal home in the realm of divine knowledge. This myth tells us something about Plato's view of the human soul as rational, quintessentially a rational knower, albeit a flawed one, for Plato fancifully portrays the soul as harboring truths it can recall only in fragments once it has fallen to earth. It is the practice of philosophy that Plato suggests can reawaken the soul's memories and initiate its ascent, returning it to its home in the divine realm of truth. For Maimonides, the Garden of Eden, like Plato's myth, gives us a vision of what our human ideals should be by telling us what our origins were and hence who we truly are and what we ought to aspire to. It also gives us a fanciful and metaphorical answer to the question of why we often fail to develop our rational potential and what obstacles stand in our way.

Maimonides argues that the point of the story is that engaging the intellect in the theoretical life of devotion to the pursuit of knowledge and the practice of scientific exploration for their own sake is the highest human ideal and that their partial loss is Adam's punishment and tragedy. In the Garden of Eden story, initially Adam was not only rational—and certainly not "without intellect, as beasts are," as the anonymous challenger proposes—but, in Maimonides's estimation, the perfect philosopher. For Maimonides says of Adam, "Through the intellect one distinguishes between truth and falsehood, and that was found in [Adam] in its perfection and integrity."[55] So not only did Adam initially have a rational capacity, but he also had a perfected (rational) intellect engaged in theoretical scientific and philosophical pursuits. It was instead his acquiescence to imaginative bodily impulse, according to Maimonides, that led Adam to turn away from the intellectual joys of Eden. Because he pursued bodily desires and pleasures rather than intellectual ones, his punishment was the refocusing of his attention upon the body via the imagination, a faculty that both perceives bodily pleasures but also can be engaged to regulate the passions. Thus Maimonides begins his interpretation of the meaning of the Garden of Eden story by conceding to the anonymous challenger his point that Adam acquired a kind of knowledge and cognitive capacity for his disobedience.

In reply to the challenger, however, Maimonides argues that the knowledge Adam acquired and the capacity he now became endowed with are not *rational*. They are instead imaginative. What Adam gained (or enhanced) was an inferior kind of bodily cognition that in his original perfection he had not needed. What Adam acquired in the story, Maimonides says, was a full-fledged imagination, a capacity and exercise that will be necessary for life in the real world outside paradise but also distracting. What the story tells us, Maimonides proposes, is that Adam receives an enhanced imaginative power to help him control the raging desires that were unleashed by his impulsive act.[56] So Adam's disobedience, Maimonides says, resulted in his becoming endowed with a faculty equipped to "apprehend conventional knowledge and categories." This was a capacity that Adam neither had nor needed prior to his disobedience.

From the beginning Adam had had bodily desires and the capacity to envision possible desires and pleasures, for his impulsive act of eating the fruit depends on these. Yet he did not initially perceive, nor was he able to construct, moral conventions, or what we call moral values. So the capacity that Adam gained, in Maimonides's retelling of the story, is actually what is designated in the Aristotelian tradition as the "practical intellect" or practical reason. Maimonides in the *Guide*, however, does away with practical reason and assimilates its capacities to the imagination. In the *Guide*, the term *reason* refers strictly to (Aristotle's) theoretical reason, the pursuit of philosophy and science for their own sake. Practical intellect, on the other hand, as a separate capacity of reason, has fallen out of the picture and become a feature of the imagination. Hence Maimonides says of Adam that by "becoming endowed with the faculty of apprehending generally accepted things, he became absorbed in judging things to be bad or fine."[57] Such conventional moral values are introduced by the enhanced (and educated and perfected) imagination to regulate and constrain the passions. Maimonides regards Adam as having the sorry fate of needing moral restrictions and constrictions—moral values—to regulate his life, a life that is now to be carried out in a social and political world.[58]

If it had not been for the bodily distractions of the imagination in the first place, moral habits and regulations, moral values, would not have been called for. Maimonides elaborates on the conventional nature and lowly status of moral values, contrasting them with the intellectual virtue that governs and fills to the brim the philosophic life dedicated to the rational pursuit of knowledge for its own sake:

> Fine and bad . . . belong to things generally accepted as known, not to those cognized by the intellect. For one does not say: it is fine that heaven is spherical, and it is bad that the earth is flat; rather one says true and false with regard

to these assertions . . . Now man in virtue of his intellect knows truth from falsehood. . . . [W]ith regard to what is of necessity, there is no good and evil at all.[59]

Adam, Maimonides says, "was punished by being deprived of intellectual apprehension."[60] This kind of apprehension included no practical thinking or moral values but was focused on knowledge for its own sake. Adam's divine punishment consists in no longer being able to live a life fully devoted to science and theoretical investigation, which is to say a life devoted to reason. He is now in a situation that forces him to turn to the imagination, educating it and perfecting it so that it can take care of all the sorry practical concerns of normal life. In Maimonides's reading of the story as an allegory for the human need to take care of business, so to speak, rather than spend all one's time on higher intellectual pursuits, it is only Adam's turn away from the pleasures of the mind to those of the body that necessitates and brings into the world the invention of moral norms.

Shlomo Pines argues that Maimonides radically opposes the true to the good, casting the turn toward the good and away from the true as the fall of man.[61] For Maimonides asserts of Adam in Eden that "when man was in his most perfect and excellent state, in accordance with his inborn disposition and possessed of intellectual cognition, he had no faculty that was engaged in any way in the consideration of generally accepted things and he did not apprehend them." There was no need for practical intellect in Eden. Instead, Adam could focus on theoretical contemplation alone. Hence morals and the moral capacity are compromises necessary for living in a post-Eden world, a world of bodily needs and social necessities.

In his idyllic Eden, neither bodily concerns nor social conflicts pressed upon Adam. That original paradise remains for him—and for us, for all humanity—the ideal, even amid changed circumstances. In Maimonides's interpretation of the Eden story, it is Adam's circumstances that primarily undergo transformation, rather than his nature. Adam's nature undergoes an *enhancement* to help him address his new, more difficult life, for the pressing needs of the body (and the social body) must now be satisfied through labor and social organization. Adam must now form relationships with others who have the same pressing needs, and develop social-political institutions to regulate their harmonious and measured satisfaction.

It is Adam's sorry fate that he must now become a *moral being*, living by moral norms formed by habit and socialization and the coercive arm of authority and law and social institutions. In paradise, Adam could have lived by the joys and desires of the intellect alone, pleasures that directed his passions toward the universal and the eternal—that is, nature and science. Now, however, Adam is Everyman, living in an unredeemed world where social and political exigencies necessitate that he focus on justice and morals, for he no longer has the luxury of an Eden that allowed him an uninterrupted intellectual life of the transcendent pleasures of the mind, pleasures that resulted in the love of God, and the universal love of all God's creations. The rational contemplative life of the scientist or philosopher is now to be available to him only in spurts rather than continuously.

We see here that in *falsafa*, the contemplative life remained wedded to the rigorous rational enterprises of philosophy rather than taking the direction of mystical contemplation of a monastic turn separated and walled off from the life of scientific discovery. Nor, for Maimonides, was philosophic contemplation to be isolated from the practical social and political life of the community—at least, not in this world. In the next world, in Eden, God could (figuratively in the midrash) be occupied wholly with the study of Torah, much as Aristotle's God was occupied with contemplating the principles of natural science, of which he was the source. But in this world, the philosopher-king, rather than the philosopher-scientist, was the supreme model—for Maimonides, that was Moses the prophet. The distinction that Maimonides drew between philosophy and prophecy was that prophets were Platonic philosopher-kings, the Arabic philosophic tradition having embraced Plato's *Republic* in political philosophy. Prophets were committed to society, instituting conventional morals and systems of justice as well as offering to society at large tastes of rational truths and pointing the way to their further discovery. Prophets came back into Plato's Cave to instruct and to better the life of the community and of the polis through education and governance. Pure philosophers might all too readily remain basking in the blinding light of truth.

Maimonides on the Contemplative Life: *Torah Lishma* Meets Aristotle

Maimonides is squarely in the tradition of Greek classical moral theory. He is also squarely in the rabbinic tradition of *torah lishma*, the study of the Bible as an end in itself, as the aim of the spiritual life and the path of human fulfillment. Maimonides understood moral success as a kind of knowledge (even for the non-philosophic masses, whose moral values he relegates to the imaginative cognitive capacity) and moral failure as due to cognitive error or ignorance. He was heir to the intensification of that tradition in the Arabic philosophical appropriation of it, for he, too, made an Aristotelian distinction between theoretical intellect and its supreme virtue versus practical intellect and its necessary but lesser virtue. He, like Aristotle and in keeping with his fellow *falasifa*, gave primacy to the theoretical intellect as *the* path to human fulfillment and perfection and spiritual transformation and ecstasy. Maimonides's innovation was to radicalize the clear intellectualist tendency of Arabic Aristotelianism by proposing an *opposition* between the true and the good. Nevertheless, he remained committed to the social and political necessity of instituting, institutionalizing, and training society at large in moral values.[62]

For Maimonides, Eden becomes the memory—and governing human ideal—of an intellectual paradise. It is the post-Eden Adam, however, the Adam who is fully human and living in a real world peopled by actual flesh-and-blood human beings in social groupings, who is the inventor of moral values. He does so not because he has acquired reason, as the Kantian mythic Adam had, but rather because he has lost it—or lost its dominance. Body and mind have come into conflict in both Maimonides's and Kant's reading of the story. For Maimonides, however, that conflict does not produce reason, as Kant proposes, but is instead a signal of the difficulty of exercising reason and living a life devoted to reason (theoretical intellectual engagement) in a world under the sway of body and community. Survival needs and social commitments tear a person away from the joys of following the intellect wherever curiosity, rigor, and wide-ranging interests lead.

Maimonides's disobedient Adam stands for both human limitations and the human condition and also for the non-philosophic masses in need of external controls to curb their appetites. Adam in Eden stood for the ideal philosopher living in an ivory tower paradise, his intellectual joy and love freely turning him away from the body and away from selfish and irrational motives toward other people. Post-Eden Adam has an enhanced practical imagination that can be perfected so that it can exercise its important moral and political role in a non-Edenic world. Post-Eden Adam must turn to the regulation of his newly aroused passions, moderating them so that they conform to socially beneficial ends—for he is now in society.

Maimonides writes in his chapters on prophecy and politics in the *Guide* that all societies, because they regulate populations of potentially disobedient Adams, must depend on the political and social use of the imagination to implement and maintain a moral life.[63] Both the development and the introduction of moral values depend upon an imaginative politics, according to Maimonides, who here is following his great mentor al-Farabi. The identification and institution of virtuous habits, the shaping of systems of justice, and the suasive power of rhetorical moral urging are all products of the imagination in its perfected form. None of these suasive, institutional, and coercive tactics would be necessary in Eden, in which intellectual joy and love toward all are the basic human motives. The very existence of moral values and norms bears witness to the unredeemed state of the world and are a concession to it. The transformation of motives brought about by the refocusing of desire in the act of the joyous pursuit of understanding the underlying scientific order is open only to a philosophic and scientific elite.

Nevertheless, Maimonides retained a small opening to spiritual beatitude and its concomitant moral transformation for the Jewish community as a whole, for Adam's Eden gave the Jewish community a glimpse of philosophic heaven, a hidden yet discoverable path to the life of philosophic and scientific (and hence spiritual) devotion, the true path to the transformation of head and heart toward spiritual perfection. What was true of the literary purpose of the Eden story was generally true of the Bible as a whole in Maimonides's view. He proposed that the Bible was an imaginative invention of the prophets to instruct the semieducated masses, via stories and tall tales, in some basic philosophic principles. He argued that the Bible had a dual aim: to introduce Jewish society at large to a basic scientific naturalist outlook on the world, on one hand, and, on the other, to institute and convey authoritatively social rules and regulations and a political system of justice to foster conventional moral life. So the Adam story was the Torah's imaginative rendering of a philosophic insight about the nature of human virtue, the intellectual passionate desire driving the path to its attainment, and the ideal of the transformation of head and heart as one.[64]

Morals, however, were the stuff of politics rather than of the intellectual-spiritual life. They were about socialization, institutional incentives and constraints, and rhetorical suasion. They resulted from the perfected functioning of the imagination—as did the Bible itself, which Maimonides said was a work of the human (prophetic) imagination in the service of both science and justice. Maimonides proposed that the Torah was precisely the work of Moses's imagination, and in it Moses translated a basic rational naturalist causal outlook into mythic and historical idioms and also created a politics that instituted a moral vision of social justice. Moral virtue concerned the harmonious life of the group, whereas intellectual virtue concerned the spiritual life of the individual, in Maimonides's estimation. Only the latter and not the former could lead toward the perfection of the divine within the human spirit.

Maimonides's Psychology of Moral Agency

There is no room for free will or even free choice in Maimonides's conception of human nature and of the origins of morals. Our capacity for morals is cognitive yet replete with the affective fullness that all cognition involves. Theoretical thinking has as its concomitant a love of the object of knowledge. Practical thinking (for Maimonides a component of imagination rather than reason) has as its motivational component fear or awe. Fear and awe motivate obedience, whereas the pursuit of understanding motivates love and empathic identification, Maimonides clearly says at the end of the *Guide*. The imaginative capacity to pinpoint moral values and institute them in society, according to Maimonides, is certainly not the essence of what makes us human, nor does it mediate between the human person and God. It is not a spiritual capacity. Moral values arise only when the necessity of obedience enters the picture—but that is in the context of human social institutions and political arrangements. Obedience is a feature of human society alone and not of the relationship to God, which is infused with love alone. It is a feature of strictly human arrangements. Theoretical intellect is the common thread between human and divine, between the mind that created nature, the universe that manifests it, and that exalted creation, the human mind, which alone can come to know (and hence love) the divine code. And it is rapture that takes over the mind in the scientific quest to understand creation, a rapture that connects the divine code in all creation in loving embrace.

The Bible, on the other hand, is principally a political text, developed by its political founder, the philosopher-king Moses, and it is human, all too human. It is, in short, the blueprint for how to live in a post-Edenic world: it offers ethics, a political constitution, and a little basic philosophy for the masses so that they gain a quasi-scientific worldview and give up their primitive anthropomorphic view of God as a bearded man in the sky issuing commandments for life on earth. Maimonides views Moses, in composing the Torah, as rendering a thoroughly rational naturalistic quasi-scientific outlook into imaginative religious symbols, metaphors, and language so as to convey them to a society as a whole; and he sees Moses as also imaginatively putting into effect, via the Torah, moral practices instituted within a political system of justice backed by suasive and authoritative narratives as well as the force of law. The Mosaic constitution institutes practical virtue in that it is well ordered, promoting mutual accord and justice, which Maimonides defines in Aristotelian fashion as the mean between excess and defect (*Guide* II, 39), and it aims at theoretical virtue, a return to Paradise (to the extent that that is possible), in that it promotes in individuals a rudimentary philosophic and scientific wisdom (and worldview). In this way it awakens the desire for theoretical understanding, the true human ideal and the only approach to the divine. There is hope that obedience will turn into love—but it is the rare human person whose thirst for knowledge is allowed enough sway to turn her heart to a love of the divine in creation, awakening true moral generosity. The moral life is instead made secure for all, in deed but not in heart, by the obedience awakened by fear and awe.

Maimonides's vision is profoundly non-Kantian and very Aristotelian. In fact, it perhaps out-Aristotles Aristotle on the transformative power of the intellectual life. Maimonides is on firm Aristotelian ground in distinguishing sharply between the theoretical and practical intellects insofar as Aristotle repeatedly tells the reader that "practical wisdom is not scientific knowledge."[65] Aristotle distinguishes between the theoretical and the practical in content as well as in mental operations. Theoretical knowledge is beyond mere human concern insofar as man is not the best thing in the world. Practical wisdom, in contrast, is relative to (mere) human concerns and ends.[66] Theoretical thinking is the highest human virtue and also divine; it is an activity engaged in for its own sake, and it results in human fulfillment and joy.[67] Yet Maimonides removes from the province of reason the practical thinking that produces a moral and political life. It is still cognitive but no longer rational; ethics and justice are instead matters merely of the imagination. They have no spiritual transformative force; they lack the power to transform the emotions in all but superficial and external ways. They leave the inner personality, the rational soul, untouched. They manage but do not heal.

In the *Guide*, reason always points to the philosophic life and never to practical moral deliberation. Reason means the passionate engagement in theoretical learning for its own sake, which is intellectual love. Maimonides designates it the true divine worship, bringing about human fulfillment and unmixed joy. Indirectly it also heals the soul, by calming and minimizing the unruly passions that are the cause of so much pain and conflict. So theoretical wisdom, in the end, is the ideal source of moral transformation, while training in the virtues is a mere second-class approximation and behaviorist imitation of what intellectual virtue brings about without even trying. The paradigm, for Maimonides, is the biblical Moses. Maimonides says of him that

> because of the greatness of his apprehension [of God's ways, the divine governance, which is God's attributes of action or physics] and his renouncing everything that is other than God . . . [a]nd because of his great joy in that which he apprehended, he did neither eat bread nor drink water. For his intellect attained such strength that all the gross faculties in the body ceased to function.[68]

For, according to Maimonides:

> The philosophers have already explained that the bodily faculties impede in youth the attainment of most of the moral virtues and all the more that of pure thought, which is achieved through the perfection of the intelligibles that lead to passionate love of Him, may He be exalted.[69]

Furthermore:

> [True] worship ought only to be engaged in after intellectual conception has been achieved. If . . . you have apprehended God and His acts [i.e., physics] in accordance with what is required by the intellect, you should afterwards engage in totally devoting yourself to Him, endeavor to come closer to Him, and strengthen the bond between you and Him—that is, the intellect. . . . The Torah has made it clear that this last worship to which we have drawn attention . . . can only be engaged in after apprehension is achieved; it says: To love the Lord your God, and to serve Him with all your heart and with all your soul. Now we have made it clear several times that love is proportionate to apprehension [i.e., theoretical understanding].[70]

This passage is Spinoza's source for his famous doctrine of the intellectual love of God, a Maimonidean account of the life of the mind that Spinoza now explicitly understands and designates as ethics.

Maimonides Versus Kant Redux

The contrast between Maimonides and Kant on the Garden of Eden could not be more stark. Nature for Maimonides represents the opening of the path to a system of rational explanation that can unlock the divine secrets of the universe and fulfill the human soul in the only way possible, through intellectual discovery. Following the tracks of nature is to follow the divine code, the only route to God's mind, the only path to union with all things (in God) through love, and the experience of the rapture of eternity. The ideal human posture is the loving embrace of nature and the discovery of the God-filled universe as one's true home. It is a contemplative and empathic stance, a noninvasive, conceptual embrace. For Kant, in contrast, nature was confined to body; reason, a uniquely human capacity, was the proof that human beings were above nature and could impose their will upon the inchoateness of matter, both in the self and in the world. Reason, in Kant's view, supplied Adam with four new capacities: (1) freedom, the power to choose for oneself a way of life; (2) a degree of control over sensual impulse, over the natural bodily self; (3) the expectation of a future; and (4) the uniquely human status of being an end, whereas all other creatures and things in nature were merely means to human ends. This last capacity is what Kant designated as the human "release from the womb of nature." (We can perhaps sense here a misogyny: woman as nature and man as reason who controls the womb and also repudiates it.) So Adam's disobedience is interpreted as the first act of freedom from nature.

That ethical life originates in and is initiated by an act of wresting and differentiating the human from the natural world. The escape from Eden (if we can call it that) sets the trajectory of history as a progressive distancing of the human from nature, internal and external, and the progressive harnessing of nature for human ends. The unique dignity of the human species is held to consist in this God-like distancing and controlling posture, power over all: it is that power, the power of the will, over body and world that is designated *freedom*. For Kant the advent of the human marks the repudiation of the human beings as natural and within the natural world. It is the hope of completely willed mastery over natural impulses and natural processes. Adam learns that he can choose his actions and invent himself and control both himself and the world for ends he invents. Ethics is confining oneself to acting upon this dignity, which asserts the human superiority to all natural being. Acting according to one's human dignity should be humanity's sole motive. It means subduing the natural self and seeing all other human beings as having equal status above nature. (Although women and "uncivilized peoples" have less reason, Kant says, they still are of equal moral status as rational and above the natural.)

We can discern the Augustinian underpinnings of the assimilation of reason to will in the Kantian moral vision. Nevertheless, we note that the overt Augustinian theology of direct divine intervention has been eliminated from the Kantian picture. A theology that distributed free will between human person and a personal God, thereby preserving some human limit and humility, was eliminated in a secularized modernity. The human person has now absorbed much of the divine side of free will—a position adumbrated in Descartes's philosophy. Kantian obedience is self-legislation, not submission to divine legislation. Yet *individual* free will it remains.[71]

For Maimonides, in contrast with Kant, intellectual engagement could never amount to an overriding of nature. Human flourishing is the natural end of a rational animal and does not consist in the subduing of nature, thought of as external to the human, toward human ends. Instead, knowing is both the fulfillment of human nature and also the embrace of the natural world through understanding its underlying rational scientific basis, to the extent possible, as an expression of the divine, as the divine necessity, or (in Maimonides's naturalizing theological language) as the divine "action." So for Maimonides, it is only through creation—of which we, body and mind, are a part—that we approach God. Nature is ripe with that divine possibility, and while no doubt we have unique access to some small region of the divine universe, the dignity of nature is not uniquely our own.

Hence for Maimonides, human beings are deeply embedded within nature as expressions of nature and, as such, potentially privy to (some of) the workings of nature via the theoretical path of discovery and understanding. The human mind is as natural as the body, and theoretical reason enables it to approach the underlying principles of nature synoptically. Standard moral virtues are not a human perfection in the Maimonidean conception but a sorry necessity of the difficult existential conditions of human life. Nevertheless, the social organizations and conventional virtues that further human practical well-being also ought to support the engagement of all those who are capable of pursuing the true human natural goal: knowledge for its own sake.

Enter Spinoza

Now we have made it clear several times that love is proportionate to appre-hension . . . Intellectual thought in constantly loving Him should be aimed at.

—MAIMONIDES

The mind's intellectual love towards God is part of the infinite love wherewith God loves himself.

—SPINOZA

Our mind, insofar as it understands, is an eternal mode of thinking which is determined by another eternal mode of thinking . . . and so on ad infinitum, with the result that they all together constitute the eternal and infinite intel-lect of God.

—SPINOZA

It was this beatific vision of the philosopher-scientist fully engaged in the search for knowledge—with its accompanying intense joy in and expansive love toward the divine expression discoverable in the natural universe—that inspired not only Moses Maimonides but also Baruch Spinoza, who thought through the Maimonidean vision to try to explain in secular, proto-scientific terms the moral psychology at its heart. Spinoza set himself to making a modern, rational, and philosophically compelling case for Maimonides's understanding of the intellectual aim of human life and the transformation of moral motivation that it offered.

From Maimonides's hints about the intellectual love of God Spinoza derived the outlines of a psychological theory of the ways in which emotion and cognition are bound together and hence must be transformed together in(to) moral agency.[72] And from Maimonides's Alexandrian doctrine of the identity of knower and known—that is, the identity of the laws of nature in God's mind, as embodied in natural processes, and discoverable by and in the human mind—in pursuing knowledge for its own sake, Spinoza derived a theory of progressive personal identification with "God or nature" (as Spinoza coined the phrase) through rigorous philosophic engagement. A path of knowledge and self-knowledge, Spinoza proposed, could lead to the dis-covery of how one fit within, and was oneself a product of, the ubiquitous and eternal divine laws of nature. Such knowledge, in discovering God within oneself, and oneself within the divine universe, was replete with pure love and rapture, inducing a moral transformation of the personality toward joy, universal compassion, and freedom. "We take pleasure in whatever we understand . . . accompanied by the idea of God [that is, nature] as cause," Spinoza wrote.[73] This is Spinoza's intellectual love of God.[74]

An Aristotelian Jewish philosophical education was still alive and well within the "Portuguese" (Jewish Sephardic) community in Amsterdam. Central to that education was Maimonides's *Guide of the Perplexed*, yet other Jewish philosophical works within the (once) Jewish Arabic Aristotelian curriculum also engaged Spinoza, including those of the twelfth century's Abraham Ibn Ezra, the thirteenth century's Levi ben Gershom or Gersonides, and the fourteenth century's Hasdai Crescas.[75] Spinoza was particularly inspired by distinctive positions and innovations that Maimonides had introduced in the *Guide* that extended and purified and even radicalized the *falasifa*'s naturalism and intellectualism in moral psychology.

Spinoza developed a theory of individual moral psychology based on Maimonides's reliance on the interdependence of reason and emotion to initiate spiritual and moral transformation. He also adopted the latter's determinist naturalism—that is, his denial of free will in his embrace of a thoroughgoing natural causal universe.[76] There was an emotional ecstasy and expansive communion of human mind with the divine to be found in the discovery that one's own reason could uncover the rational workings of nature, for the reason in mind and the reason in world expressed an underlying unity, one that could be made real through the practice of philosophy and science. The human mind was potentially as divine and as wide as the scope of the nature that could be discovered and explained by it.

Spinoza thought through and appropriated a number of Maimonidean doctrines and approaches, including (1) the reduction of the good to the true; (2) the reduction of the will to the intellect in both God and the human person; (3) the assimilation of the Aristotelian practical intellect to the imagination; (4) the conventionality of moral values and their instrumental social utility, in contrast with intellectual virtue, which was the true human aim; (5) the natural causal determinism of the universe and all within it, including the human mind, human choice, and action; and (6) the accomplishment of human perfection and moral transformation through the intellectual love of and identification with the presence of the divine reason in nature.[77] Spinoza recast these Maimonidean Aristotelian doctrines into a modern psychological idiom, recruiting them into new theories of his own, especially into ethics, in an innovative account of moral development and agency. He thus refashioned and streamlined an understanding of human nature, of the goal of human life, and of the origin of moral values, largely drawing on the intellectualist sources and philosophic traditions of Alexandria, in contrast with his older contemporary Descartes, who could be said to have fashioned a modern account of human nature and moral psychology out of the voluntarism of the Christianity of late antique Rome.[78]

Spinoza's Aim of a Scientific Psychology

> I have made a ceaseless effort not to ridicule, not to bewail, not to scorn human actions, but to understand them; and to this end I have looked upon passions, such as love, hatred, anger, envy, ambition, pity, and the other perturbations of the mind, not in the light of vices of human nature, but as properties just as pertinent to it as are heat, cold, storm, thunder, and the like to the nature of the atmosphere.
>
> —SPINOZA

> He who clearly and distinctly understands himself and his emotions loves God, and the more so the more he understands himself and his emotions.
>
> —SPINOZA

Spinoza fashioned a psychology of moral transformation for the individual from Maimonides's Einsteinian rapturous intellectual identification of the reason inherent in the mind with the rational order of the universe discovered through philosophy and science. He adopted and adapted Maimonides's account of how theoretical understanding could heal the soul, creating the truly morally virtuous person from the inside out, via a spiritual transformation in which, through the exercise of reason, one came to understand and even experience the underlying unity of all things and of oneself within that unity.

Spinoza recruited the theoretical rational devotion of the Maimonidean scientist to a search to know oneself. The understanding of oneself would engage and bring together all the sciences. His aim was a scientific explanation not merely of general causal laws that obtain in the various arenas of nature—and the divinity inherent in a mind that can come to know nature in this way—but specifically of a particular individual, namely, oneself. The aim was to discover how one's self could have arisen in all its complexity and individuality. What were the underlying causes, physical and mental, biological and cultural and social, that came together to give rise to this individual?

Spinoza placed the study of the emotions as central to the philosophic and scientific study of the individual. Spinoza's path of self-discovery was aimed at explaining one's unique emotional character and history with reference to the general laws of a scientific psychology. A large portion of Spinoza's *Ethics* presents the principles of a general psychological theory. He outlined a threefold path of self-discovery by which individuals could come to understand themselves, their own particular emotional causal development, in terms of the principles of his general psychology. That scientific psychological explanation could then be set within ever broader and multiple theoretical scientific disciplines, ultimately setting the psychological self within the universal scientific explanatory understanding of the entire universe.

Spinoza introduced two innovations into the Maimonidean account that refocused it decisively on moral psychology. The first was the view that reason could be brought to bear to explain every person as the unique and particular individual that he or she is. Reason, he held, can explain not only what members of a species share with others of their kind or the universal principles of natural processes, as Maimonides (as an Aristotelian) had held, but reason can also explain the individual. The second innovation was the nearly earthshaking claim of the identity of mind and body. A person, Spinoza argued, is a psychophysical whole, rather than two natures tied together. The mind and body, Spinoza argued against Descartes, are one thing conceived and explainable in two ways, rather than two things of different kinds and origins barely held together, as Descartes maintained and as religious tradition took as a commonplace. A person can be explained via two types of necessary causal explanation, the mental and the physical, but is one thing. And the natural universe is one thing, too, physical and also mental in the sense of the information, the natural laws, that its regular causal processes embody and express.[79]

Spinoza, in embracing the mechanical causality of the seventeenth century, jettisoned the Aristotelian formal causes and species essences and conceived reason as capable of explaining each thing, including oneself, as an individual in terms of the causes of its coming into being and of its unique nature.[80] Moreover, in claiming that body and mind were in identity,

Spinoza regarded each person's unique mind, as much as each body, as capable of being explained in the mechanical terms of its particular causal makeup. In this, he retained his rootedness in the Aristotelian Maimonidean naturalism of the mind as well as the body. It was the mind that was quintessentially transparent and amenable to rational scientific explanation in the Aristotelian conception. But he added to this naturalism of the mind and of its general underlying presence in nature a new amenability to rational causal explanation of individual instances and manifestations of material processes.

Maimonides's conception of the transformation of the mind when it came to discover the rational principles (for example, biological classification and physical laws) embedded in nature came to be recast by Spinoza as a science of the individual embodied mind as a product of nature—namely, psychology. Spinoza indicated that he hoped to write a work on medicine that would do for the "enminded body" what the *Ethics* had done for the embodied mind, but he died before he could turn to write it. If Spinoza had lived longer, or had lived in the present, perhaps his book on psychosomatic medicine might have been something like a neurological account of the emotions and would have made a contribution to affective neuroscience and perhaps even to psychopharmacology. In the *Ethics* Spinoza proposed a path by which a revolution in self-understanding and self-feeling could be brought about by tracing the broadest possible grasp of oneself as the particular and unique product of one's own constitutive psychological causes, that is, thoughts and feelings.

Spinoza's overall argument in the *Ethics* was that the theoretical scientific rational explanation of the causes of one's emotions could lead to the consummate transformation of motivation from painful, fanatical, narrowly self-serving, automatic, and unreflective to a joyous, self-aware, and compassionate embrace of nature and the world. Spinoza thus drew on and developed the basic Greek classical insight, most radically presented in Maimonides's intellectualism, that cognition and emotion are transformed together. Moral development is the task of reason when reason is educated and put to use according to a specific philosophic understanding and praxis. Moral motivation is transformed by the knowledge of the union (or what we would call today the psychological identification) of oneself with the universe—an identification accomplished when one traces the causes of one's own individuality, of one's own emotional and motivational makeup, back to their causes in the natural and social universe.

Spinoza's Moral Psychology: The Education of Desire

God does not act from freedom of will.

—SPINOZA

Will and intellect are one and the same thing.

—SPINOZA

That thing is said to be free [*liber*] which exists solely from the necessity of its own nature.

—SPINOZA

Blessedness is not the reward of virtue, but virtue itself. We do not enjoy blessedness because we keep our lusts in check. On the contrary, it is because we enjoy blessedness that we are able to keep our lusts in check.

—SPINOZA

We can pick out five ways that Spinoza revolutionized the study of human nature and the determination of the proper aim of a human life (that is, ethics and moral psychology). First, it was to be a study of mind and body as one entity, one thing, one nature (not two natures tied together). Second, that single nature was both an embodied mind and an enminded body. Third, while human nature, like all nature, had infinite expressions, the essence underlying those expressions was the fundamental desire to live and survive, what Spinoza termed the *conatus*. Fourth, the study was to include ways to explain the particular individual—not only the general features and principles that all shared, but also how each person came to be the individual he or she was. And fifth, the study of human beings was analogous to the study of any natural object insofar as human beings were like everything else in nature, even if, in them, the ubiquitous universal forces and laws that explained everything were nuanced in some systematically regular ways. Hence the study of human nature—to discover what human well-being, human flourishing, and the transformation of motivation toward universal love (that is, ethics) consisted in—was a case study within the study of nature writ large. It was to be carried out with reference to natural principles and causes that Spinoza insisted were everywhere the same.

That Spinoza put a basic global psychophysical desire, a striving, at the core of the person redefined human nature in a significant way. He claimed that "desire is the very essence of man," thus dislodging reason as the central feature of what makes us human.[81] Spinoza proposed that at the core of each person there is a basic desire to survive and maintain oneself. This desire expressed itself as a striving to protect oneself and to further one's own internal organization and self-coherence. Spinoza believed it was inherent in all organic individuals, from plants to humans, and he speculated that even inorganic objects such as rocks had a rudimentary form of it. He recruited the seventeenth-century term *conatus* to capture this striving.[82] Spinoza in a sense had demoted reason to a feature and kind of desire, one expression of the conatus among others. In this move Spinoza not only redefined the human person as more than rational but even rethought the nature of the mind, which he conceived as embodied and affective.

The mind, as an expression of desire, was no longer to be fully explainable by the study of knowledge (epistemology)—how we come to know and what is knowable. Instead, the study of the human mind would now become the study of human desire, which is to say a biologically informed psychology.[83] In this one move, Spinoza recast the science of human nature. Epistemology, the science of knowledge, was thus placed subordinate to psychology, and cognitive transformation through rational self-understanding became a means to the human end of the development and transformation of the desire at the heart and core of the person. Spinoza had taken Maimonides seriously about the emotionally rich character of thinking and the consequent inherently motivationally transformative potential of rational understanding, and he tried to explain in secular scientific terms Maimonides's more mythic grasp.

The emotional, desire-related character of all thinking, of the mind (as well as of the body), emerged from what we might identify today as Spinoza's prescient understanding that human beings are fundamentally organic beings and must be understood in terms of biology. The biological account of the mind that Spinoza developed was not, however, materially reductive but tried to explain how desire drives thinking and belief.[84] Insofar as Spinoza held that the body and mind are one entity, one thing described in two different ways, he conceived the conatus, the basic striving, the desire of the person to survive and maintain and further oneself, as underlying and driving both bodily and mental processes. Since all mental processes expressed the conatus, perceiving and thinking were necessarily always full of desire and emotion. What that meant was that the conatus of the mind, in its desire to maintain and protect one's own mental states (including beliefs), was as strong as the desire to maintain one's own bodily survival and integrity. The conatus expressing itself in the mind was Spinoza's explanation for why, for example, an attack on a belief one holds or on one's character is felt just as harshly and personally as a punch in the stomach.

The conatus was the engine of all the mind's perceiving, thinking, and feeling, as much as it was of the body's maintenance of life through its organic systems. Spinoza regarded mental operations, including thinking, as being carried out with the same causal necessity as bodily processes such as respiration. Thus the mind in its awareness, in its thinking and self-reflection, was expressing and enacting its striving, its desire to maintain itself, to further itself. The mind, as much as the body, was filled with desire, and that desire in its perceiving and thinking, just as in the body, was directed toward survival, organic stability, and self-maintenance. Spinoza also held that the mind was identical with its processes. It was not a thing, a container of ideas, as Descartes believed, but instead simply the mental processes themselves being carried out. That meant that the boundaries of the mind were not those of the skull and the skin. Hence the mind could never be for Spinoza a separate and discrete capacity, a cognitive apparatus divorced from the biological purposes of the embodied organism as a whole or divorced from its environments and causes. The desiring of the mind was open to and engaged in the world.

Spinoza defined the mind as the awareness of the body, the body made conscious, and further self-reflection on that process. As a consequence, the mind was not independent of the body but articulated bodily experience. Specifically, the mind was the awareness of one's interactions with the environment from the perspective of the body, of the embodied person, and it was driven by the desire to keep oneself alive and whole. Because desire drove the mind toward maintaining its own thoughts and feelings, it could not be expected to provide objective views of the world from a detached perspective—a perspective that one could then freely choose to intervene in. Instead, because the mind's aim was to survive and to survive as itself, it provided skewed, self-serving, and self-related views of the environment.

Spinoza held that every perception, every interaction, every thought, and even the slightest awareness delivered a jolt of pleasure or pain, for every experience either enhanced or lessened the striving of the conatus toward survival. The drive to survive and survive as oneself was not always constant, Spinoza held, but underwent constant enhancement and detraction, empowerment and weakening, in response to every experience.[85] The weakness of the conatus Spinoza called passivity, and its strength he called activity. The human aim was to enhance one's "activity"—that was Spinoza's answer to Descartes's free will voluntarism. To be active rather than passive meant to strengthen one's capacity to act; it was to become free, to become a true moral agent—but not via free will.

Spinoza's Moral Agency Without Free Will

Men are deceived in thinking themselves free, a belief that consists only in this, that they are conscious of their actions and ignorant of the causes by which they are determined. Therefore the idea of their freedom is simply the ignorance of the causes of their actions.

—SPINOZA

The more perfection a thing has, the more active and the less passive it is. Conversely, the more active it is, the more perfect it is.

—SPINOZA

Spinoza followed Descartes in claiming that emotions could be either passive or active and that the burden of moral psychology was to render passive emotions into active ones.[86] Descartes claimed that free will could exercise control over its own thoughts and feelings, and from them over the body, and via the body over the world. It was the exercise of free will, Descartes insisted, that could bring the emotions under control and render motivations active expressions of the mind, imposing the mind's activity upon the body and upon the world. The only alternative, he said, was a passive surrender of the mind to the body. Thus body and mind were in constant internal struggle for control and mastery. Free agency was taken by Descartes to be the triumph of the mind over the body and world via the active choice of the will.

Spinoza's notion of activity skirted any recourse to free will, a notion he thought incoherent and magical. As Spinoza put it (echoing Maimonides), "Will cannot be called a free cause, but only a necessary cause." For Spinoza, in contrast to Descartes, body and mind were active or passive as the relative condition of a single thing in relation to its environment, rather than as a composite entity in a never-ending internal and external power struggle. Spinoza recruited the new seventeenth-century notion of internal coherence or equilibrium to his concept of the conatus, dubbing it the "ratio of motion and rest" and its "idea." He reinterpreted the term *activity* in terms of internal self-organization (mental as well as physical) and its enhancement, and he got rid of its Cartesian meaning of having control over body and world. Spinoza used the term *activity* to mean the urge of an organism to maintain a dynamically stable internal organization and equilibrium in the face of the inputs from the environment and its outputs to the environment. Rather than a struggle between mind and body (which was impossible, according to Spinoza) and between the human person and the environment, Spinoza's notion of activity pointed instead to an urge toward harmonious and progressive mutual integration of individual and natural environments (including the human) as the source of enhanced agency.[87] Spinoza argued that activity, or agency, the ability to act from oneself, consisted in the increased empowerment of the conatus, of its striving for survival—an enhancement of one's self-organizing dynamic stability and hence of one's internal coherence in response to the environment.

Spinoza argued that the conatus was passive or active in relation to the environment. The aim of a life well lived was to become active. Passivity indicated a relation between self and world in which the immediate environment was taken in uncritically and one's actions were unreflectively expressive of that environment. (For example, someone who was a member of an organization, let's say of the Nazi SS, and who was at the cognitive affective primitive level of psychological passivity would simply take on the SS identity in context and act accordingly. Most did.) Spinoza regarded the taking in of the immediate environment as the playing out of passive desires, or desire's passivity. Passive agency was about the passive way that the conatus, the desire at the core of the self, related to its environment and contexts. It was driven largely, he said, by hope and fear, the emotional motivations that prompt obedience rather than love. To render the conatus active was to gain broader perspectives on the immediate environment so that one did not automatically and unreflectively enact its incentives and absorb uncritically its proffered roles. Yet it did not mean to control or have power over the environment. It meant instead to develop a dynamic interplay of self and environment from which both grew and were transformed. To render the conatus active was to have one's actions motivated only by love and joy, pleasure and desire—but not by all pleasures and desires, and never by fear or even hope, the last being an emotion that harbored anxiety as well as optimism.[88]

To wrest oneself from the immediate environment not only necessitated envisioning larger and multiple perspectives from which to look back upon one's context but also entailed expanding the scope of the self, the conatus, that was doing the viewing. The person had to come to identify his or her own conatus as the conatus of more than the role he or she had been assigned in narrow context. One's conatus had to be detached from hopes and fears, which coercively kept one passively bound to the narrow situation. Spinoza envisioned the broadening of the conatus—which we can now understand as the broadening of identity, of the scope of the self—as a developmental process, taking place in three stages. The initial stage was the condition of most people prior to following the path toward moral development outlined in the *Ethics*. It was a state of passive, uncritical, and obedient swallowing of the local environment and the submission to its situational incentives. The two subsequent stages involved the tracing of the causes of the particular configuration of one's conatus, of the conatus one is and not just has—that is, one's emotional history and makeup. That progressive inclusion and integration of one's causes within (self-)awareness would then enable one to act to maintain and further a more broadly encompassing self, a conatus that sought to maintain more than its definition in and by the immediate environment. The conatus had now become aware of itself as the conatus of its (progressively internalized) constitutive causes rather than merely of its present moment and situation. The person now identified him- or herself with a larger world and that larger world's survival and betterment. When the mind comes to know, and to know itself as knowing, it takes pleasure in its own enhanced power of understanding, which is also its enhanced capacity to act and contribute to the broader environment (activity).[89]

In the path of moral development, one was to come to understand that the conatus was in fact a desire to maintain and preserve its constitutive causes, those larger contexts of which one was the product.[90] The constitutive causes came together in oneself as the desire to preserve that configuration. But that very understanding was to include within the conatus the broader causal contexts—ad infinitum, which Spinoza called "God or nature." The desire for self, one's self-organizing striving, which Spinoza identified as love and joy, was asymptotically expansive of the boundaries of the conatus/oneself, as one understood oneself as all the causes, ad infinitum, that had come together to produce that self. It was also the result of an internal search that one actively initiated and carried out, an understanding of one's own self in its environments and in the larger scheme of things that was truly one's own. Rather than being the result of an external, and hence coercive, set of incentives inherent in the immediate context to which one passively responded by accepting its interpretations of oneself and one's immediate world, a Spinozist education developed the independence of one's own mind as the progressive grasp of the self as the product of how nature and nurture had come together to produce this unique point, or what we might call a niche. Understanding oneself was an active process, one that one owned. Passivity depressed the conatus, whereas activity empowered it. Obedience and obligation were always, for Spinoza as they had been for Maimonides, forms of self-diminishment in response to fear and hence not to be embraced as the true moral posture. They were political expedients necessary in an unredeemed, post-Eden world. True morals, by contrast, stemmed from love and joy in an expansive sense of engagement and empowerment and belonging in harmony with nature and universe.

Spinoza's notion of activity was that the conatus was a self-organizing energy that integrated larger and broader causal environmental understandings of oneself within its "ratio," or homeodynamic stability. As a result of the dynamic harmonizing of self within the world and the world within the self, one was no longer so buffeted by waves of immediate circumstance—by situational fears and hopes, Spinoza said—but instead was rooted in a stabler and broader constitution of self.[91] That broader constitution of self, a constitution capable of extending to infinity—that is, of harmoniously integrating the world into itself as itself and of integrating itself into the world as world—was Spinoza's working out of a practice of theoretical self-understanding that would lead to a desire to maintain and further a self now understood to be integral to its world. It was to love the world as oneself, and to realize that self and world survived together or not at all. It was to rid oneself of the false notion of a free will that could (and ought to) exercise power over the body and the world. That Cartesian notion of activity was to be replaced with one that envisioned activity as acting as a broader and broader portion of the environment as it was progressively understood and felt to be constitutive of self. A harmonious relation between self and world, a sense of belonging in nature and being part of a nature working itself out in the self, was the moral posture that replaced that of an ongoing power struggle between body and mind and between human person and nature.[92]

I call Spinoza's path of transformation in moral motivation (exposing its kinship with the Arabic *falsafa*) the "education of desire." Spinoza called his aim of the intellectual love of God or nature "independence of mind," and he warned that it could be only a "private virtue." By that he meant the path he outlined in the *Ethics* could help one gain freedom from the seductive and coercive incentives of immediate situations and also from the turns that fate metes out through understanding one's interests and identity much more broadly. Nevertheless Spinoza warned that few people could achieve such moral heights and hence society as a whole could not depend on its citizens' individual moral transformation of motivation toward beneficence toward all and toward the universe. A politics that instituted and institutionalized conventional moral values was necessary for general justice and peace.

Spinoza's Maimonidean Politics of Morals

The supreme mystery of despotism, its prop and stay, is to keep men in a state of deception, and with the specious title of religion to cloak the fear by which they must be held in check, so that they will fight for their servitude as if for salvation.

—SPINOZA

That the mind was an expression of the conatus and not just the body, Spinoza believed, turned out to be a double-edged sword. He regarded the desires and pleasures of the mind as the source of the moral problem as well as of its resolution, for more often than not those desires and pleasures led to the corruption of judgment. The pursuit of the pleasures of self-serving beliefs and the avoidance of painful ones, especially about ourselves, was a ubiquitous reality. Wishful thinking left one vulnerable to (passive in the face of) social and political manipulation as well as unrealistic assessments of the world. Instead of the pursuit of true beliefs—realistic beliefs about oneself and the world, about what is actually beneficial and what is harmful— the conatus of the mind was driven by what Freud later referred to as the pleasure principle. It led all too easily to the disastrous pleasures of self-deception and mythic beliefs about self and world rather than to true ones. It is to this moral problem, stemming from his theoretical biological psychology, that Spinoza offered an updated, secularized, and biologized Alexandrian-Maimonidean resolution for the individual capable of deep transformation through rigorous, honest self-examination combined with broad psychological and scientific understanding.

The lucky person trained in Spinoza's path of philosophical-psychological self-examination could gain a measure of enhanced agency through the realistic grasp of his or her own motives and capacities, as well as contexts and environments. Along with that honest grasp of self in world and world in self, Spinoza envisioned a transforming love, a progressive sense of identification, an embrace of the cultural and social worlds and natural processes that intersected to make up the self. The conatus was to be globalized to include all its constituent causes and effects, that is, all of nature, human and universal. "That thing is said to be free [*liber*]," Spinoza wrote, "which exists solely from the necessity of its own nature."[93] Its "own nature," however, inclusive of everything and everyone who had contributed to it from time immemorial, spanned the universe and all eternity. Spinoza wrote: "In the mind there is no absolute, or free, will. The mind is determined to this or that volition by a cause, which is likewise determined by another cause, and this again by another, and so ad infinitum."[94] To feel in and as oneself the desire of a self embedded within and constituted by the infinite universe was to be free, without any external limit. To love and promote the well-being of the universe and all within it was to love all others as oneself, to find oneself in them and they within one's own conatus.

For the society as a whole, however, there would have to be a reliance on social incentives, political management, and coercion, induced and enforced conventional moral values that mimicked in the polity, as best as possible, what a Spinozist transformation in moral motivation could create from the inside out, from heart to act. Spinoza turned to Maimonides again for a political theory, updating Maimonides's social psychology and somewhat authoritarian and elitist political constitution to reflect his own modern liberal commitments to democracy, freedom of speech, and the multireligious polity.

A Final Note on Where We Are Today: History Repeats Itself—and Then Repeats Itself

In his 1905 book, *Five Types of Ethical Theory*, the British philosopher C.D. Broad, following up on a suggestion by fellow British moral philosopher G.E. Moore, presented Spinoza's ethics as outside the limits of all legitimate theoretical accounts of ethics. Spinoza's determinism and naturalism had led him to commit what Moore regarded as the "naturalistic fallacy." To commit the naturalistic fallacy made ethics impossible, according to Broad, because if all things were determined and nature could not be intervened in, then all moral judgments were impossible. So Spinoza's ethics was taken to vitiate ethics rather than account for it and explain it.[95] The entire edifice of modern ethical theory stands upon Spinoza being cast out. Because of his denial of free will and embrace of a determinist and naturalist account of the human mind and human psychology, Spinoza still occupies the outside, the dangerous and prohibited beyond, whose boundary must not be approached, let alone breached. Most of the secondary literature on Spinoza's ethical theory by his philosophical partisans tries to defend Spinoza against the charge of having committed the naturalistic fallacy. Spinoza didn't really reject free will but is a sort of compatibilist, these apologists claim. Hence most of his defenders argue that Spinoza's modern excommunication from philosophy is unfounded and should be rescinded because Broad got it wrong and Spinoza's ethics is within the boundaries set by Broad. But these defenders of Spinoza are barking up the wrong tree. For what Moore and Broad did, ironically, was to reinstate the Catholic condemnations in 1277 of Averroism—moral naturalism and determinism and intellectualism— and reassert free will voluntarism as the only true (Christian) doctrine.

What we have here is a cultural battle and a religious one. We can even discern the outlines of the hidden and disguised religious character of the argument, for the move that Broad makes is not so much to debate Spinoza but to put him beyond the pale of acceptable, legitimate philosophical opinion. Broad's intent is, in effect, to excommunicate Spinoza from the philosophical canon. The battle over ethics seems to have reached its most extreme and oppositional formulations in modernity, and that story is still playing itself out; it is a story we now recognize as a clash of civilizations, not just of philosophers and desiccated theories. We are still beholden to a hidden and disguised and disavowed theology: Is the human relationship to God via wisdom or via will? Via science or via obedience? Does free will seem so intuitive a feature of human nature that it must be saved at all costs? Whichever set of intuitions about human nature seems obvious, natural, and unavoidable to each of us reflects long and detailed contrasting histories of which we are mostly unaware.

Now that the cultural battle lines have been drawn, it is time to reframe the question. Rather than asking if you are "with us or against us" when it comes to free will versus determinist naturalism, or Descartes and Kant versus Maimonides and Spinoza, instead let's turn to a more detached and global inquiry into what discoveries in the new brain sciences are exposing about how the mind actually works. The remainder of this book will be devoted to what we can learn about moral agency from a review of some pertinent discoveries in neuroscience and related fields of inquiry. Once we have gained a sense of where current research is leading in the understanding of the relations among cognition, emotion, and action, we can then return to the embattled philosophical traditions and see if any philosophers of the past can help us think through in creative ways the implications of the new thinking about the brain for moral agency—why we are moral, why we fail to be moral, and how to get people to be more moral.

6

Surveying the Field: How the New Brain Sciences Are Exploring How and Why We Are (and Are Not) Ethical

Here are two of the biggest questions in moral psychology: (1) Where do moral beliefs and motivations come from? (2) How does moral judgment work? All questions are easy . . . once you have clear answers to these questions.

—JONATHAN HAIDT AND FREDERIK BJORKLUND

Introduction: Ethical Naturalism with a Nod to Free Choice

Philosophers and scientists have begun to collaborate to design experiments to investigate moral psychology. They've begun to think together about how all the new information about how the brain works changes how we have to think about why and when people act morally and why and when they don't. A three-volume compendium, *Moral Psychology*, came out in 2008 with essays by many of the foremost scientists and philosophers working in the field.[1] Collected here are articles describing cutting-edge work; each article also has appended to it the responses of others in the field, who raise the current debates. The first volume is *The Evolution of Morality: Adaptations and Innateness*, the second is *The Cognitive Science of Morality: Intuition and Diversity*, and the third is *The Neuroscience of Morality: Emotion, Brain Disorders, and Development*. It is safe to say that all the philosophers and scientists represented in the three volumes—more or less the major players in the field of rethinking philosophical ethics in the light of the new brain sciences—advocate some version of ethical naturalism. This is the view that what we can learn from the sciences has some bearing (though there is a

240

range of opinion on what that bearing might be) on the origin and content of morals and how and why we enact them or fail to enact them. Many other philosophers, however, hold the standard view that only by investigating our moral concepts by philosophical or conceptual analysis can we come to understand our human moral nature and engagements. So science can explain only the "wetware" underlying our concepts; it cannot contribute to any understanding of what ethics is. Other material mechanisms could underlie those concepts and it would make no substantive difference in human moral life. That view is challenged in these volumes. In their introductory essay for the three volumes, "Naturalizing Ethics," Owen Flanagan, Hagop Sarkissian, and David Wong set out what they mean by the phrase "naturalizing ethics." Central to that endeavor is getting rid of any appeal to God or the divine—or to a faculty of free will:

> Let us call an individual a scientific naturalist if she does not permit the invocation of supernatural forces in understanding, explaining, and accounting for what happens in this world. . . .
>
> A naturalist cannot accept . . . the notion found in Kant that humans have metaphysical freedom of the will. . . . The twentieth-century philosopher Roderick Chisholm (1966) puts the point . . . this way: "If we are responsible . . . then we have a prerogative some would attribute only to God: each of us when we act, is a prime mover unmoved. In doing what we do, we cause certain things to happen, and nothing—or no one—causes us to cause those events to happen." . . . This sort of free will violates the basic laws of science, so the naturalist must offer a different analysis.[2]

Flanagan, Sarkissian, and Wong make absolutely clear in this essay that they hold that "there is no such thing as 'will' and there is no such thing as 'free will.' . . . There is no faculty of will in the human mind/brain."[3] While they grant that there is no such thing as a free will in the Cartesian-Kantian sense and that scientifically no such organ or natural capacity as a "will" can be found by empirical investigation into the components and operations of the brain, they argue, nevertheless, that we still choose and make choices in a way that *originates* our actions:

> Persons make choices. Typically they do so with live options before them. If new reasons present themselves, they can change course. . . . Persons experience themselves choosing, intending, and willing. Ethics sees persons as choosing

and thus . . . look[s] for . . . voluntary action that involves reasoning, deliberation, and choice.[4]

These reasons, resulting from conscious thinking and deliberation, have
causal power or efficacy, the authors say; they are reasons that can be causes
of our actions, in fact, *originating* causes of our actions.[5] And so the authors
call themselves "neocompatibilists," which is philosophical jargon for saying
they believe that human free choice—that is, originating choice—is compatible with naturalist determinism, the view that all actions are situated
within and emerge from layers and systems of causes, both internal and environmental. Standard philosophical compatibilism serves to carve out interior consciousness as a realm apart, one immune both to scientific explanation
and to multicausal contextual assignment of responsibility. The authors
maintain, however, that we can and ought to hold both that we choose and
act freely via rational thinking, *originating reasons*, and at the same time that
we are fully determined from the standpoint of natural lines and webs of
causation, both material and psychosocial.

Flanagan, Sarkissian, and Wong reject the common sop that somehow
the indeterminism of quantum physics helps us out here. First, there is no
evidence that the neurons of the brain are subject to indeterminacy in the
way, say, the firing of electrons is (and in fact there is much evidence against
it); even if that were the case, however, they point out that the indeterminacy
of some outcomes in the brain would not help with establishing personal
causal origination of actions. For randomness in fact would make us *more*
rather than *less* subject to unexpected turns of fate, as the Epicureans were
well aware with their theory of the occasional random swerve of atoms.
Only an open or closet theist would benefit from such hoped-for indeterminacy in the brain, because space would then be opened for a divine hand to
intervene—but not a human one. Moreover, human free choice would not
be made possible by neuronal randomness in any case (and all the evidence
so far seems to be against it) because no conscious human choice could ever
operate to refashion neural networks directly at the neuronal level. Neural
networks change through experience, not through will. One can't just say, "I
think I'll connect my love of chocolate with my fear of heights and see if
I can get myself to fear chocolate." We do not have direct access to neurons
and their patterns of firing any more than we have the capacity for direct
internal intervention into the functioning of our liver, even if the liver sometimes were to function randomly. And, as a colleague once proposed to me

in defending free will, where would the "I" come from that is other than that of the neurons, the "I" of the gaps between them, if such indeterminacy were the case? But that's magical thinking and the authors of this essay reject it. Only divine intervention could work like that—the hand of the biblical God reaching down miraculously and intervening in nature, as in the rescue of the Israelites from the Egyptians at the Sea of Reeds. But divine intervention is, of course, exactly what all forms of naturalism preclude.

Flanagan, Sarkissian, and Wong opt for a kind of two-truth theory or perspective. The human perspective of experiencing ourselves as choosers and deciders, originators of our actions, and hence alone and individually morally responsible for them, is valid because it is phenomenologically true or real—it is a true (and universal) account of how we experience ourselves as human beings, they propose. At the same time, they hold that the naturalist perspective of our actions and choices as within and resultant from the complex interactions of various biological, psychological, social, cultural, historical, quantum, and cosmological causal systems is also true. Nevertheless, the authors claim that by giving up the idea that there is a faculty in our minds that is a will, free or otherwise (as the philosopher Hume did in the seventeenth century), science can be left to its kind of explanation—causal determinism—and at the same time, a second description from the insider human subjective perspective of experience can also be true. The insider subjective description of ourselves as originating our actions from ourselves alone is compatible with the scientific causal explanation, they say, because it makes no claims about a faculty in the natural brain that can be located and proven to exist or proven not to exist. One explanation is about natural causes, and the other is about presumably universal human subjective experience. The human and the natural are separate self-contained systems of explanation. The authors have saved as much of the Cartesian-Kantian philosophical tradition of a subjective turn inward as they possibly can while giving some opening to an account of human beings as products of nature.[6] But do the facts bear out the authors' saving operation? Can any facts challenge this viewpoint or is it beyond any empirical challenge? If the latter, is this in the end a metaphysical question and thus a kind of article of faith?

Some of the discoveries made by the new brain sciences point toward an answer to this question—a plausible answer if not a definitive one, for how could one ever definitively prove or disprove divine intervention or an article of faith? The validity of compatibilism (neo- or otherwise) and hence of the unassailable truth of the internal perspective comes into question because there is quite a bit of evidence suggesting that we harbor an illusion that we originate our actions. The evidence—I'll describe the relevant experiments in some detail in subsequent chapters—suggests that we do not have conscious access to the internal psychological causes of our actions. Much evidence indicates that we become consciously aware of decisions or choices only after we make them and begin enacting them. This surprising discovery has been confirmed in experiment after experiment and the evidence is quite robust.[7] So the reasons we impute to why we perform an action are, or tend to be, ex post facto reconstructions ("confabulations," psychologists call them)—that is, illusions. (At least, this is generally true, although it is perhaps possible for some few people to be trained to some extent in rigorous introspection. Later I'll say a lot more about how the psychology of decision making works.) Philosophers would certainly hold that having an illusion does not in and of itself prove that it is a false belief—being paranoid does not preclude the possibility that there are in fact people out to get you. Nevertheless, the explanation of special access and hence the unassailability of our internal perspective no longer seems to be the most plausible and parsimonious explanation if it can be shown that we tend toward such an illusion.

Another source of evidence against the universality of the presupposition of free choice comes from cultural anthropology. For example, one can point to the particular Buddhism of the Khmer, whose belief in karma has a profound inhibiting effect on their conception of the capacity to take action. They hold that a person's current character as well as situation and fate are entirely predetermined and merited by actions in a prior life and hence outside a person's control. So parents generally wait to see what a child's character and fate seem to be before urging him or her, however slightly, toward positive actions and engagements. This Cambodian cultural myth generally engenders a kind of quietism, to the point that some Cambodians feel that they merited the genocide of the Khmer Rouge.[8] This view is in stark contrast to the Western one, which would seem to encourage a radical kind of self- and world-making, as if no cultural background, biological conditions, psychological inheritance, or situational context could in the end be thought too significant to overcome. That this philosophic move has as marked a religio-cultural flavor and provincialism as the Khmer belief does in its cultural context can be seen in the particular Augustinian cast to the Western conception and implications of theological predestination/predeterminism in comparison with the Khmer notion. Calvinist predeterminism assigns to the divine will all power, in analogy to the Khmer doctrine's assignment of all power to the operation of a karmic moral rule of fate. Moreover, the conceptual underpinnings of divine predestination are not naturalistic despite their determinism (as I pointed out in relation to Augustine), for the reduction of all nature to the intervention of divine fiat is exactly the opposite of assigning all causes to the working out of independently functioning and necessary laws. Yet in stark contrast with Khmer quietism, the Calvinist myth functions to encourage a wild activism, in order to prove to the world that one is in fact saved rather than damned.

So Calvinism, paradoxically, functions pretty much the way the myth of free choice and free will does to elicit concerted and even frenzied action in pursuit of the afterlife. Such is the power of provincial cultural presuppositions to dominate thought across generations and to do so beneath awareness. The Khmer doctrine of karma, while superficially similar to Calvinist predestinarianism, ends up driving an entirely different theological anthropology and ethics. The brain science discovery of motivational amnesia suggests that such cultural myths, the Western Augustinian as well as the Khmer, come about because they fill a gap left by our actual ignorance of the causes of our actions—a finding that Spinoza anticipated. Our still dominant and pervasive Latin Christian culture fills the gap with a divine-like human freedom, while the Khmer culture fills the gap with a divine reward system that is retrospective rather than prospective. The latter myth justifies and rationalizes one's self and the present situation and sets them in stone, so to speak, whereas the former makes the present and oneself completely open to personal (divine-like) control, hence our notion of "freedom." We are enacting our own provincialism and calling it "universal." Our unconscious subjectivity, at least in the case of some of our deepest presuppositions, turns out to be in fact cultural and hence not universal at all. And the cultural filling in of aspects of the deepest self would appear to take place just at those junctures where self-transparency fails us. So the universal feature that we share as human beings is not the freedom of our interior consciousness from natural causal determination (as we are brought up to believe) but instead the general inaccessibility to our conscious minds of the causes of our actions. So the architecture of our brains is actually driving our tendency to cultural myth making here, both theirs and ours. We are most unfree when we think ourselves most free—and there's neuroscientific evidence for that, too.

It is hardly surprising, then, to find that while Flanagan, Sarkissian, and Wong end their essay with a plea for ethics as a kind of evolutionary way of seeing human beings as within and part of their natural and social worlds, they also return to their vision of the freedom available to us to *choose* our values and lives. On one hand, they write, "If ethics is like any science or is part of any science, it is part of *human ecology*, concerned with saying what contributes to the well-being of humans, human groups, and human individuals in particular natural and social environments."[9] And they remark on the local and contingent nature of many moral values as relative to flexible notions of healthy community and social and individual flourishing. They call themselves "pluralistic relativists" and "pragmatic human ecologists." Yet they conclude their essay with a plea for voluntarism: they insist that we must express our freedom to choose our lives, our identities, and our actions. They call upon each of us to choose to "deploy our critical capacities in judging the quality and worth of alternative ways of being . . . [by] deploying our agentic capacities to modify ourselves by engaging in identity experimentation and meaning locations within the vast space of possibilities that have been and are being tried by our fellows."[10] We have landed very close to home: right back in the Augustinian tradition, dressed up a bit in contemporary and scientific language and filled in with a larger and more accurate range of how human beings live and what they value.

William Casebeer, in his response to Flanagan, Sarkissian, and Wong, reads the brain science data as allowing "executive functions" to take the place of free choice in ethical decision making. Casebeer aims to revive and revise Aristotle's biological conception of the human person in terms of natural and inherent aims and the dynamic operation of organic systematicity. Flourishing can be understood from a biological standpoint as "proper functioning," yet still incorporate a range of flexibility and various human possibilities and capacities, he says. Casebeer proposes that these proper human functions are "natural facts" that are normative because they involve an internal standard in the way physical health does. You know when your heart function is optimal, and you certainly know when it isn't. So, too, for the body as a whole and the person as a whole, Casebeer argues. He proposes that moral statements can be reduced to statements about functions, that human beings have multiple functions, and that these functions are relative to environments and have co-evolved with those environments. So the state of various types of functioning can be determined and evaluated. Casebeer supports a version of compatibilism, he says, whereby the controlling or executive functions of the brain can be assessed and responsibility assigned in terms of the proper operation of the directing cognitive systems.

He bases this claim on a distinction between "well-ordered and disordered cognitive systems" as a basis for "maintain[ing] attributions of responsibility."[11] He remarks that he follows the argument of Patricia Churchland that the distinction between well-functioning and poorly functioning brain systems of executive control "might allow us to salvage understandings of moral responsibility which are generally compatible with those required by traditional moral theory." I take this to mean that we are responsible for our decisions as originating actions in some sense (even though our cognitive capacities are embedded in webs of causes) because they are those of a decision-making capacity producing action. So despite the nod to Aristotle's biological systems theory, we are still working with a definition of moral agency as conscious originating (free) choice—with cognitive systems doing the choosing. That Casebeer offers his theory as a version of compatibilism suggests he is arguing that we derive an internal feeling of the origination of our actions because their proximate cause (that is, the last cause in the series) is within the executive areas of the brain. Yet he is not claiming that executive functions and decisions are themselves uncaused or that their relative state of good functioning (flourishing) is self-originating. (I will return to the discussion of executive capacities in subsequent chapters to determine whether they can be recruited to offer a scientific warrant for conscious originative choice or even a consistent feeling of originative choice.) So Casebeer, too, is making a valiant effort to salvage a version of the Augustinian tradition of free choice as underlying and necessary for moral agency and responsibility—by assigning it to a natural brain mechanism.

The Scientific Search for the Sources of Moral Agency: An Overview and a Sampling

The first volume, *The Evolution of Morality*, includes essays on whether human morality is innate and the direct result of evolutionary processes. What could have led to moral sentiments, beliefs, and rules? they ask. Are these distinct, hardwired capacities in the brain? Moreover, what do claims of an evolutionary history mean for the prescriptivity, the authority, of morals? Several essays propose that specific moral rules (in one form or another) are innate, and the authors speculate on evolutionary scenarios that could have given rise to these hardwired moral injunctions. A good example is an essay by Debra Lieberman of the University of Hawaii about the incest taboo.

Lieberman argues that a negative moral sentiment has come to trigger the avoidance of sexual relations among near genetic relatives as the evolutionary result of a history of the detrimental consequences of inbreeding.[12] Geoffrey Miller of the University of New Mexico argues in another chapter that sexual selection works to favor mates with moral virtues and hence their predominance in the gene pool. Other chapters argue for and against particular sets of innate moral rules and principles. One recurring argument is that moral rules are analogous to Noam Chomsky's claim of an innate natural universal grammar, which every particular language instantiates in its own cultural idiom. (Chomsky's claim of the innateness of grammar is no longer universally accepted, however; more on this later.) Most of the authors argue that if morality is a direct result of evolution, it has to be innate and content specific—in other words, a particular evolutionary history resulted in a specific innate moral sentiment or rule. By contrast, Chandra Sekhar Sripada argues in the first volume that the innateness of a moral structure of some kind in the brain would not have to prescribe particular moral norms but instead could produce tendencies or innate (moral) biases toward certain moral norms. Sripada's theory of innate biases is intended to help account for cultural differences and changes in moral sentiments and values over time. The theory of a combination of innate moral modules plus a sociocultural overlay occurs in a number of different versions in the three volumes.

A number of the cognitive scientists who contributed to volume 2, *The Cognitive Science of Morality*, approach the investigation of the moral capacity by observing patterns in the cognitive processes involved in forming moral judgments, emotions, and actions, and they also invoke a Chomskian model of an innate and specific human language ability as the relevant analogy.[13] Marc D. Hauser, Liane Young, and Fiery Cushman, in "Reviving Rawls's Linguistic Analogy: Operative Principles and the Causal Structure of Moral Actions," argue that all human beings have a moral *faculty* and not just a moral capacity. The authors make their case by citing Chomsky's theory of a discrete universally human and unique linguistic faculty. Hauser, Young, and Cushman point out that they regard many "domain-general" cognitive systems as providing inputs constitutive of moral judgment. Nevertheless, they are "committed to the existence of some cognitive mechanisms that are specific to the domain of morality." These constitute, they say, a "moral faculty":

These [dedicated moral] systems are not responsible for generating representations of actions, intentions, causes, and outcomes; rather they are responsible for combining these representations in a productive fashion, ultimately generating a moral judgment. Our thesis is that the moral faculty applies general principles to specific examples, implementing an appropriate set of representations. We refer to these principles as an individual's "knowledge of morality" and, by analogy to language, posit that these principles are both unconsciously operative and inaccessible.[14]

A moral faculty, according to this approach, is an innate universal specialized system of moral judgment causally responsible for moral appraisal; it is discrete and hence independent of (other) reasoning and also of emotion. The authors propose a research program to test whether several identifiable and universal moral principles are operative in moral action and justification, in the same way grammatical principles are held to be universally operant in human linguistic performance. The arguments both for and against such a moral faculty turn on the available evidence for such a specialized discrete and localized unitary mechanism. The team has begun to develop a battery of paired dilemmas to pinpoint principles, to determine whether these principles guide moral judgment, and to explore whether subjects refer to these principles in their moral justifications. So far they have been engaged in exploring three moral principles: (1) harm by commission is worse than harm by omission; (2) harm intended as the means to a goal is worse than unintended harm in pursuit of a goal; and (3) harm involving physical contact is worse than the same harm without physical contact.

Jonathan Haidt and Frederik Bjorklund, in "Social Intuitionists Answer Six Questions About Moral Psychology," also present a Chomskian case for moral judgment as produced by a discrete hardwired faculty, yet they allow for a social process that nuances it.[15] According to them, the moral faculty produces quick, automatic intuitions below the level of conscious thought, reasoning, and decision making.[16] They argue that moral intuitions are products of human evolution. Haidt and Bjorklund count themselves as Humeans on the origin of morals: they say that morals arise as immediate sentiments (of right and wrong) that are universal to the human species. They make the case for morality as originating in what they call "a small set of innately prepared, affectively valenced moral intuitions." They also put the point in more colloquial terms: these are evolved "quick intuitions, gut feelings, and moral emotions."[17] They argue for a version of modularity: "Most psychologists," they write, "accept Fodor's (1983) claim that many aspects of perceptual and linguistic processing are the output of modules which are informationally encapsulated special purpose processing mechanisms. Informational processing means that the module works on its own proprietary inputs. Knowledge elsewhere in the mind will not affect the output of the module." Fodor himself dismisses modularity when it comes to "higher cognition"—a position that, if true, would invalidate not only Haidt and Bjorklund's model but that of Hauser and colleagues as well as many others. Nevertheless, Haidt and Bjorklund maintain that for their own theory, "all we need to say is that higher cognitive processes are modularized 'to some interesting degree'" and that "there can be bits of mental processing that are to some degree module-like."[18]

Haidt and Bjorklund also cite evidence for what they call the "inescapably affective mind," by which they mean that the mind is always judging everything we encounter on a scale of good and bad. They call this our ever-present "like-ometer." They argue that research has shown the brain is composed of two systems of ongoing valuation: an evolutionarily ancient, automatic, very fast, affective system and a "phylogenetically newer," slower, motivationally weaker, cognitive one. They cite studies by Gazzaniga of split-brain patients that expose the left side of the brain as the "interpreter" offering a post hoc running commentary on behavior. Psychologists call this kind of rationalization the mind engaging in "confabulation." Conscious verbal reasoning, Gazzaniga says, is thus not at all the command center of action but instead "more like a press secretary whose job it is to offer convincing explanations for whatever the person happens to do."[19] And other research has given support to this contention, demonstrating that our everyday thinking largely serves to bolster our already favored opinions.[20] (I'll have a lot to say about the self-servingness of our beliefs in a later chapter.)

So a set of hardwired intuitions makes morals possible and also sets constraints for the social nuancing of them, Haidt and Bjorklund argue. They regard the social contribution to morals as extremely important, and they hypothesize that it works in the following way: moral intuitions are immediate, evolutionarily set beliefs and motivations, but ex post facto reasoning comes in to justify, explain, specify, and extend those insights. That reasoning is socially influenced and transmitted, so when someone has a (hardwired) moral intuition, he or she interprets it via the prism of the social reasoning and understanding that he or she has heard from others. Reasoning is important but occurs and is influential almost entirely in social transmission rather than in an internal personal and individual reasoning process that gives rise to a moral judgment. While moral reasoning is rarely ever the source of an individual's moral judgments, they maintain, "moral reasons passed between people have a causal force" nevertheless.[21] So one's own moral reasoning is fundamentally an exercise in rationalizing one's intuitions. This is true across the board, they propose, except in highly specialized contexts such as the philosophy classroom. They raise this question:

> Can anyone ever engage in open-minded, non–post hoc reasoning [i.e., moral reasoning that does not consist in rationalizations after the fact] in private? Yes . . . [but that is] hypothesized to occur somewhat rarely outside of highly specialized subcultures such as that of philosophy, which provides years of training in unnatural modes of human thought.[22]

Moral intuitions are set by a process of evolution plus social exchanges, they say, hence their term for it: a social intuitionist model (or SIM). It is a species-wide model offering moral insights that evolved *for* and *in* the species, so it offers what they regard as true or real morals for all human beings while also allowing for cultural nuances.

Haidt and Bjorklund see the unconscious and conscious systems as generally in competition, but the edge is almost always given to the unconscious quick one. "Modern social cognition research," Haidt and Bjorklund remark, "is largely about the disconnect between automatic processes, which are fast and effortless, and controlled processes, which are slow, conscious, and heavily dependent on verbal thinking."[23] Their hypothesis about how SIM works is that in the great majority of morally "eliciting situation[s]," a person's unconscious and primitive "like-ometer" settles on a moral judgment via the quick and automatic route. Then either more recently evolved conscious cognitive processes (reason) rationalize the already made moral judgment or social persuasion intrudes to modify the quick automatic judgment toward that of social partners. This social persuasion also generally operates, they maintain, via automatic, unconscious influences rather than via conscious reasoning and rational discourse.[24] In rare cases, two other routes of moral judgment are possible: in the first, consciously reasoning together with others, such as in a philosophy class, results in a moral judgment; in the other, private conscious rational reflection results in such valuation. But these cases, they say, are unusual.

Thus moral judgment, in Haidt and Bjorklund's estimation, turns out to be generally, and in almost all cases other than highly specialized ones, the result of the primitive automatic affective cognitive system. Higher cognition comes in largely as ex post facto rationalization, on one hand, and as social persuasion that merely nuances the primitive, on the other. As the authors suggest, this is a contemporary version of Hume's notion of reason as slave to the passions, and hence of how (primitive) emotions drive morals.[25] Moral judgment can be stopped in its tracks, however—for example, by other intuitions or by socially modified and mediated valuations. They offer an example of a prejudiced immediate feeling toward someone based on race or sex or something of that kind. A person can block that fast judgment in the light of other valuations; in their view, most of these other valuations would be obtained at least initially via social persuasion. Nevertheless, they maintain that "moral reasoning is . . . usually engaged in after a moral judgment is made, in which a person searches for arguments that will support an already-made judgment."[26] Haidt and Bjorklund conclude by identifying a set of innate moral modules that they believe are species-wide and develop in children on a set schedule if all goes well. These are innate intuitions that become externalized, rather than external social values that become internalized on society's schedule, they say. Haidt and Bjorklund do not take a hard-and-fast position on the degree of innate modularity they are arguing for. They say it is somewhere on a continuum between simple preparedness and a discrete modularity according to which capacities are hardwired by evolution to meet the long-enduring structures of the environment in which human beings evolved and live.[27]

Nevertheless, Haidt and Bjorklund express a particular affinity for a version of modularity in which each capacity is unconnected to the others: "Because cognitive modules are each the result of a different phylogenetic history, there is no reason to expect them all to be built on the same general pattern and elegantly interconnected."[28] Whatever the precise degree of innateness and discrete modularity turns out to be, they believe that five innate human moral modules can be identified, and children come to express them in stages: (1) sensitivity to harm and expressions of caring, (2) fairness and reciprocity, (3) recognition of hierarchy and respect for authority, (4) concern for purity and sanctity, and (5) recognition of in- and out-groups and the boundaries between them. Society helps that externalization develop by, for example, teaching certain games that help kids accomplish a given moral developmental stage. The authors cite the example of the game of "cooties," which they say is played all over the United States: it is based on the purity foundation and involves recognition of status and in- and out-groups. Eight-to-ten-year-olds play it, they say, because that is the time when the fourth developmental stage is coming online.[29] Society nuances the expressions of these basic modules, since the modular capacities under-determine the precise morals and virtues a society develops along each of the trajectories. Virtues are thus "constrained social constructions."[30] Cultures and societies both nuance the basic moral insights and selectively emphasize some over others.[31] Finally, Haidt and Bjorklund add to their innate moral modules approach the explanation that people differ in their moral perfor-mance because of differences in innate temperament.[32] Yet they make clear that temperament does not affect any global moral traits, since global moral tendencies, in contrast to situational virtues, have been ruled out by contem-porary research in social psychology. So in the end they characterize their view of morals or virtues as "a set of skills needed to navigate the social world" that are "finely tuned automatic reactions to complex social situa-tions."[33] It would seem that what is proposed here are hardwired global responses that are perhaps nuanced to specific situations via the social transmissions that are taken in as unconscious skills.

The responses to Haidt and Bjorklund raise the question of whether the innate and social aspects of the theory can be held together and exactly in what ways different aspects of ethics are assigned to each dimension. One of the critical responses to this essay suggests that the social persuasion dimen-sion of the SIM is problematic because it introduces into intuitionism a social reasoning process driven by bias (rationalizing). The basic virtue of moral intuitions seems to be that, as purported products of human evolu-tion, they are presumed to be universally hardwired action prompts that are of objective prosocial benefit to the species. The social enculturation dimen-sion fits uneasily with this innate intuitionism. The social suasion link seems in fact to vitiate the purported objectivity and reliability of the intuitions, since it is based on shared rationalizations and on inducing an overriding social conformity.[34] It appears that at the end of the essay, where the authors are trying to address objections that have been raised about the SIM in the past, they introduce both far more work for the social nuancing dimension of the model and also some new theoretical dimensions that don't seem to fit with it comfortably.

Haidt and Bjorklund's commitment to hardwired innate moral modules seems to be driven by their presumption that if moral intuitions evolved over the existence of the species to contribute to its perpetuation, they would be "real moral truths," if only species-wide. It is wrong to murder or commit incest (as another version of this argument goes) because we have a species-wide innate taboo against murder or incest. But we have all kinds of capacities, including both prosocial ones and aggressive competitive ones. Why would or should the prosocial ones carry more moral authority and justification for us? Are innate tendencies that hoodwink our moral intentions rather than serve them, such as self-deception, any less real, less human, or less the products of evolution? Another essay commenting on this paper makes the point that a theory aimed at describing how morals come about has somehow segued into a theory of moral justification, an account of why these intuitions are true and authoritative for humans.[35] Walter Sinnott-Armstrong in his essay later in volume 2 also challenges the widespread version of moral intuitionism that holds that the immediacy of moral belief not only is descriptive of how moral beliefs arise psychologically but also serves as justification for such beliefs.[36] I would extend this critique to raise the question of what justifies presuming that prosocial human tendencies are a natural category, "ethics," that can be identified as different from other innate human psychological tendencies and therefore has some special normative status, whereas other innate tendencies (such as some cognitive framing, for example) are considered to interfere with ethics rather than contribute to it.

What Does It Mean to "Naturalize" Ethics?

More important to my purpose is the problematic character of the implicit assumption that the naturalizing project is the search for hardwired, discrete morals (whether rules, beliefs, or sentiments). The authors' presumption that ethics is "natural" thus entails that it is broadly cross-cultural, always the same beneath the cultural overlays. Naturalizing ethics in this way implies that ethics is *hardwired* and hence to be contrasted with the conventional and cultural, which is either left out of the picture or brought in as an overlay that adds a small nuance. This view of what counts as nature conceives the natural as marking out an arena that excludes anything that is a product of nurture. But the very contrast between nature and nurture itself is a cultural artifact, for it is a claim with a specific cultural history and provincialism, namely Augustine's two cosmic principles, as I showed in the chapter on the Augustinian legacy. So we are still in an Augustinian cultural world, one that drives the presuppositions of what we are looking for when we search for ethics in nature. The approach Haidt and Bjorklund take, which involves searching for the hardwired module that culture nuances, reappears in many versions in the essays in the three volumes and beyond. I'll give a few examples. In volume 1, Peter Tse complicates and nuances how moral innateness could possibly work by proposing that morals need not be directly written into the brain, so to speak, but instead could have arisen as a consequence of the evolution of symbolic thought. Symbolic thought is generally held by brain scientists to result from the "conceptual binding" of many modules, enabling the formation of a single representation ranging over many domains. Binding also makes it possible for one thing to stand for something else—which is the basis of abstraction (including categorical abstractions), symbolization, and metaphorical thinking, among other cognitive capacities. Tse proposes that moral categories arise from this kind of process. Conceptual binding is influenced by experience, but what is bound together are hardwired modules resulting from evolutionary history. Tse proposes that moral categories might have arisen in this way, and he speculates that a category of evil might have arisen from binding together tokenism, sadism, and rejection of the body.

A salient and influential example of the explanatory strategy of searching to identify some form of nurture overlying hardwired moral modules that are the product of evolution is Ursula Goodenough and Terrence W. Deacon's "From Biology to Consciousness to Morality." The authors propose a speculative evolutionary scenario that has some similarities to both Tse's and Haidt and Bjorklund's approaches.[37] Like Tse, they credit wide-ranging connective representational blending and the sign capacity, by which one thing can stand for and point to another, for the way that morals emerge from the evolutionary inheritance. Yet they also introduce into the mix what they regard as a uniquely human capacity for self-reflection. They propose that uniquely human morals arise from self-reflection upon inherited primate hardwired modular (that is, discrete) social abilities inherited from humans' primate ancestors. This human moral capacity was made possible by language, they argue, since they regard self-reflection as a product of symbolic language. It is implied that language and the flexibility it offers introduce a chasm between lower primates and humans. The human moral capacity is thus bootstrapped onto a capacity Goodenough and Deacon regard as uniquely human: language. (The claim that self-reflection is a product of language is controversial and not universally accepted, however.) Goodenough and Deacon propose that each individual engages in self-reflection upon four evolutionary inherited hardwired primate prosocial capacities: empathy, strategic reciprocity, nurture, and hierarchy. These primate proto-morals result in a set of universally human, self-discovered virtues that bubble up in each of us as subjective moral intuitions. They believe that each of us discovers versions of these four virtues within our consciousness as more or less unanalyzable (yet culturally nuanced) basic moral experiences, and thereby we each lay claim to our universal human moral nature as a feeling of inner compulsion.[38] Goodenough and Deacon, like Tse, admit that this evolutionary approach to the nature and origin of morals can be and must always remain speculative: "The [evolutionary] scenario is by definition a speculation (what actually happened may never be known), but we find the scenario heuristic, helping us to focus on what is distinctive about human mentality."[39]

Yet the speculative character of the "evolution of morality" scenarios makes them far too open to being blank slates upon which we project cultural or other prejudices and then attribute those projections to the very makeup of the brain. What are these other than just-so stories to justify whichever selves we wish to justify? Male over female? Western over non-Western? Human over animal? Should we now embrace hierarchy as a true value for us rather than strive for equality, because hierarchy has an evolutionary history? What about fairness, which also often appears in the catalogue of traits inherited from primate kin? It is hardly a surprise that prosocial behaviors in primates and other animals had survival value, as Darwin pointed out. A Hobbesian view of relentless competitive struggle was not Darwin's assessment of evolution but a social Darwinist misreading of both nature and Darwin. Morality is not a mere veneer of sociality over brutish, ruthless nature, nor is it merely Freud's social restraints on aggressive and sexual drives.[40] Goodenough and Deacon have divided up the human brain and assigned certain traits (empathy, hierarchy, etc.) to its animality, and others to a unique humanity (self-reflection borne of another allegedly uniquely human capacity, language). So we get a human version of prosocial emotional traits shared across species with a cognitive ability supplied by a unique human evolutionary inheritance.

Too Much Philosophy—and Too Little

Goodenough and Deacon's model is a kind of hybrid of a Kantian criterion for autonomous moral judgment as self-reflection (that is, being able to reflect upon one's own motives and behavior and articulate the reasons for them) and a Humean notion of innate moral sentiments that are natural virtues, functioning beneath the level of thought and spanning humans and our close animal kin. So what we seem to have here is an implicit and pre-supposed Humean-Kantian hybrid model of morals read into—or out of—the evolutionary evidence. The implication is that self-reflection (via language) offers an evolutionary basis for self-origination, that is, free choice influenced by moral sentiments of evolutionary origin, but not, apparently, compelled by them. What this amounts to is that we are free human beings strapped onto (or sitting atop) evolutionarily determined bodies harboring set patterns of emotional response. There is a veiled appeal to unarticulated assumptions—some version of the Augustinian free choice tradition—about how moral agency works, so that the philosophic search is directed at iden-tifying underlying brain mechanisms that can help explain the presupposed way that moral agency is held to operate (namely, some version of rational free choice). This is surely reading our provincial religio-cultural legacy into nature, rather than vice versa. Those who fail to toe the line are often subject to withering criticism. A case in point is the primatologist Frans de Waal and his investigation of sympathy and other prosocial behaviors in primates. De Waal's philosophical critics largely presuppose a Kantian morality in which free, conscious, rational decision making in conformity to universal impartial moral principles is the sine qua non of true ethics, and they roundly fault de Waal for not holding such Kantian presuppositions about moral agency and for not judging nonhuman primates by that standard.[41]

So an overarching problem with the various studies is that the authors generally fail to raise the issue of how to frame or define the problem of moral psychology. As a result, there is something rather haphazard as well as culturally narrow about the projects. The haphazardness can be detected in the isolated character of some of the focuses, for example, of incest as a po-tentially hardwired moral taboo. The question of what relationship a pur-ported innate feeling of the moral repulsiveness of incest might have to the larger moral domain, if there is such a domain, is not raised. Is the human moral capacity a collection of such discrete hardwired feelings of universal moral repulsiveness or moral approbation? Goodenough and Deacon's moral experience argument presupposes that morals are both innate and an evolutionary inheritance, and hence also individual and a subjective human category of experience that we impose upon the world. Should we just accept this Kantian presupposition? Why should we think with Flanagan and others that to have ethics at all we have to freely choose (whatever that means) our identities and life paths? If sexual selection favors reproductive partners who are morally virtuous, what version of virtuousness would that be? Is this a reproductive preference for cooperation over competition? For fair play over deceit? Moreover, what would such a claim, if true, actually explain? Why is this a claim of the evolutionary transmission of moral values rather than of merely prosocial psychological traits (if traits are global in that way, which other research has shown they in fact are not)? What is systematically lacking is a sense of a consciously articulated and rationally and empirically defended common domain or set of phenomena as the subject of exploration rather than a hodgepodge of presupposed ethical examples whose status is ambiguous.

Unexamined Culturally Narrow Presuppositions and the Hope of a Remedy

If unexamined and unconscious cultural presuppositions are driving the search for ethics—and hence at least in part the way the evidence is selected and interpreted—what would happen if different cultural assumptions guided the search into the discoveries in the new brain sciences? Why not search instead among our evolutionary kin for protoexamples of the classical Greek notion of ethics as the cultivation of beauty and the beautiful life, as harmony and balance in self and relationships? Why take for granted that ethics consists in moral principles that are innate ideas that come to us in flashes of insight, as Descartes did, or that they are innate and universal sentiments of sympathy beneath the level of thought, as Hume did, rather than habits developed through repetition and experience, as the standard Western reading of Aristotle had it (and as Casebeer and the character education advocates purport to embrace while in the end appealing to some form of free choice)? Or why not presuppose that ethics consists in intellectual honesty and clarity of moral vision, while moral evil is a form of ignorance and intellectual benightedness, as the ancient Greek dramatists, Plato, and the Arabic Aristotle did? Why not presuppose that ethics is a type of health and general well-being and good functioning, while immorality is a diseased state of pain and impaired functioning that requires remedy, as the Stoics did? (Casebeer starts off in this direction but then superimposes upon it a free will compatibilist theory.) These are all metaphors we live by, as George Lakoff puts it. We are too prone to reading our deep and unconscious cultural metaphors back into nature, discovering within it, lo and behold, our own assumptions and impositions. It would be better to have on hand multiple possible models of what ethics is about and to hold them lightly so as not to narrow the search from the get-go and presuppose its findings or stuff them into a ready-to-hand bell jar. The following chapters will try to do just that: I will broaden the search for the human moral capacity by extending the cultural bounds of what we are looking for to include (but not be confined to) Aristotelian-Spinozist understandings.

Nevertheless, Goodenough and Deacon, Haidt and Bjorklund, and Tse have the advantage of a more nuanced insight into what they are looking for over the claims of simple inheritance of evolutionary hardwired moral rules or virtues (for example, of Lieberman and Miller). Claiming, for example, that an incest taboo is simply a hardwired modular (isolated) emotional response to a triggering situation has the disadvantage of emptying it of moral content. If morals consist of automatic and isolated responses—like our knee reflexes—the one thing they can't be is chosen (an Augustinian model) or contextually meaningful and shaped (an Aristotelian habituation or an Aristotelian-Spinozist intellectualist model). In guiding what to look for, the implicit hardwired modular model jettisons nurture and reduces nature to material causes alone, thereby rendering immoral actions and differences in the behavior of the same people in different situations largely inexplicable. Nevertheless, we are back at the Augustinian reduction of nature to the material—yet another unexamined Western cultural presupposition implicitly framing the search for a science of ethics. There is too little self-awareness of cultural location and history driving these studies. The search for ethics in human and other primate minds is still too beholden to narrow unconscious cultural presuppositions and a very specific philosophical history, and the theological tradition from which it emerged, both about what ethics looks like and what the science can contribute.

I hoped to avoid the problem of what counts as ethics by preceding this chapter and the next with three chapters on what ethics actually looks like on the ground, beginning with an initial chapter on our American view of ethics. Then in the following two chapters I attempted to show why our standard American way of thinking about ethics got it mostly wrong because it can't explain the examples of ethics from the Holocaust and the social psychology studies of ethics that I introduced in Chapters 2 and 3. There is just too much presumed confidence that we all know what we are referring to when we talk about ethics—that we at least know what the problem is, if not the solution. But I am trying to argue from a number of different standpoints that that is not the case. Instead we harbor unwarranted assumptions, some of which are culturally provincial to Americans and Europeans. Moreover, not only the presuppositions about ethics but also the science we have so far seen brought in as explanatory often has a piecemeal flavor to it, rather than being a more systematic and holistic look at what evolutionary biology generally now thinks can be said about the brain and what that might inform us about human agency generally—and not just about the narrower domain of moral agency. Perhaps we ought to be focusing our attention on a very general human tendency for the emergence of norms and for the demand or desire for normative performance of self and others, instead of on a presupposed narrower domain of the purportedly legitimately "ethical." Where does the striving for normativity come from both in evolution and in the brain, and how does it operate as a mechanism? What is the range of such normative demands—what do they look like in all kinds of practices, thought, institutions, social contexts, relationships, and cultures?

In the next chapter I will present an overview of some relevant current research on neuroplasticity, the vast flexibility and changeability of the brain/mind through experience. Scientists are just beginning to explore and grasp the extent of the brain/mind's plasticity. If we take that plasticity seriously, there are some discoveries about the brain from even further afield that I believe need to be taken into account when rethinking ethics. My tack will be to further complicate the analysis of ethics by introducing some other discoveries about the mind/brain and hence to try to situate ethics within an even larger range of human engagements. By going further afield I hope to contextualize ethics more broadly in culture, in social and psychoanalytic and neuropsychoanalytic psychology, in political psychology, in neurochemistry, and even in open adaptive systems theory and the like. I shall propose, among other things, that we need to think of ethics in terms of the overall nature and life path of a "self." I will not, of course, try to resurrect the Cartesian substantive self, the self as "thinking thing." Instead I will point to the growing evidence for a notion of the self as a neurological and unconscious process of self-mapping—where our limbs are, how we are feeling today, what hurts, how pollution is affecting us, et cetera—and also as emerging from basic emotional systems that are homologous in all mammals. Both neurological mapping and our basic emotions give us a feeling of self in ongoing and remembered responses to the environment, and especially the social environment.

Breaking Out of the Box

At present some of those engaged in research in the new moral psychologies, in contrast to Hauser and colleagues, deny the likelihood of a moral faculty that is a specific innate brain system dedicated to morality.[42] Jesse Prinz argues that all of the data cited by Hauser and colleagues can be explained with reference to general-purpose emotion systems and socially transmitted rules. In my next chapter I will begin a discussion of neuroplasticity that introduces a seminal essay by the affective neuroscientists Panksepp and Panksepp in which they argue strongly against the claim that higher functions are either hardwired or encapsulated (modular).[43] I think we should leave it to the neuroscientists who study the actual mechanisms of moral cognition and affect to settle the question, or at least weigh in.

In another essay Jesse Prinz challenges claims of moral innateness: he questions not only whether there is an innate moral faculty but even whether there is any innate moral content (rules, principles, or sentiments). His argument poses a challenge not only to strong innateness arguments such as Lieberman's about an inherited hardwired incest taboo but also to softer innateness arguments such as Sripada's innate moral biases, Tse's category of moral evils, and Goodenough and Deacon's universal moral intuitions based on self-reflection upon inherited primate prosociality. Prinz argues that if morals were innate (and hence hardwired or the result of hardwiring), the moral capacity would be a localized and fixed brain area or system rather than a broadly distributed and plastic set of connections resulting from how experience variously shapes the brain's neural networks. Prinz argues against there being such an encapsulated moral faculty. Innateness, he says, generally presupposes both functional and anatomical modularity. Modularity means that a capacity is localized in the brain and processes information specific to that domain without access to other such modules. "To be anatomically modular is to be located within proprietary circuits of the brain," he writes.[44] Rather, he provides evidence that this is not the case when it comes to morals. He says evidence supports the idea that "moral stimuli re-cruit domain-general emotion regions and regions associated with all man-ner of social reasoning." Hence "there is no strong evidence for a functional or anatomical module in the moral domain."[45]As a result, without modular-ity, it is hard to argue for the innateness of morality. Morals are more likely a result of the recruitment of other capacities, especially of broader social cognition and emotions, for new or different uses than the purposes for which they evolved. Prinz concludes that if neuroplasticity is dominant in human behavior, then morality cannot be directly innate but must instead result indirectly from evolutionary processes of environmental fit that affect other mechanisms that come to have effects upon, or can be targeted for use in, moral life as well. He argues that the evidence points toward the second alternative, that morality is a "by-product" of faculties that evolved for other purposes.

An important bone of contention in the debate between those who favor the innateness and modularity of the moral capacity and those who reject it involves the distinction between the violation of conventional rules and the violation of moral norms. Many of those who argue for innateness hold the view that the distinction between convention and morality supports the hardwired modularity of a universal moral core of sentiments and beliefs, which can be contrasted with merely social conventions that differ across cultural context. Morals, they say, are therefore natural, unlike social and cultural conventions. Prinz shows that evidence from recent experiments designed to test just that hypothesis contests the distinction. A crucial piece of evidence comes from studies of psychopaths, who some scientists main-tain cannot distinguish between moral and conventional rules. Neverthe-less, evidence going back decades and confirmed again and again suggests that psychopaths do not have a specific, "selective" moral deficit but instead have a diminished capacity to feel negative emotions or recognize them in others. As a result, they do not respond normally to fear conditioning. They have diminished startle responses, little depression, high pain thresholds, and difficulty in recognizing facial expressions of sadness, pain, and disgust. He writes, "Without negative emotions, psychopaths cannot undergo the kind of conditioning process that allows us to build up moral rules from basic emotions. Psychopathy is not a moral deficit but an emotional deficit with moral consequences."[46] Brain imaging studies of healthy people view-ing pictures that portrayed morally fraught scenes involving physical assault, abandoned children, and the like confirmed the evidence regarding psychopathy, Prinz says, because they showed that these scenes (in contrast to scenes that were merely disgusting or disturbing but did not raise moral issues) activated areas of the brain devoted to social cognition. Hence the evidence for a specific moral brain module and capacity in either anatomy or function is lacking, he contends.[47] Moral sentiment and thinking seem to draw on social cognition and emotion.

Prinz's view of the moral capacity as a human construction rather than an innate inheritance, while a minority view among the contributors to the three volumes, nevertheless sets the stage for a broader and integrative view of the nature and origins of the moral capacity. Still, adequate evidence has not yet been brought in support of morals being an outgrowth of other capacities rather than a unique and modular capacity in itself. Prinz makes just this point in his response to the critique of his essay. He admits that he "need[s] to show that the data are rich enough to allow the acquisition of moral competence without a domain-specific learning mechanism." He proposes that "we should be open to the possibility that moral competence, like religion, tool use, and the arts is a byproduct of more general psychological resources."[48] The jury still seems to be out, the evidence yet too thin on both sides. Prinz's own view is of morals as sentimental norms of moral praise and blame.[49] The debate between those who argue that ethics is fundamentally an emotional capacity with cognitive effects and those who think it is essentially cognitive with emotional consequences is also visible in many of the papers.

Muddying the Waters I: Cognitive Heuristics

Gerd Gigerenzer begins his essay in volume 2 with a description taken from Christopher Browning's book about Reserve Police Battalion 101 (which I discussed at length in Chapter 2).[50] Gigerenzer analyzes the moral failure here in terms of heuristics, rapid cognitive processes that take place below the level of consciousness. These heuristics are sorts of rules of thumb that we employ in making decisions, but they are beneath awareness and happen automatically. The unconscious rule that drove the men in Police Battalion 101, according to Gigerenzer, is "Don't break ranks." That is a "social heuristic" that drives behavior, he claims. The next example of a social heuristic that Gigerenzer introduces comes from a comparison of the number of Americans willing to be organ donors versus the number willing in various

European nations. Twenty-eight percent of Americans are willing to be organ donors, compared to 17 percent of the British and more than 99 percent of the French and Hungarians. What do these statistics show—that the Americans and British are selfish and the French and Hungarians are of a different moral stripe? Not so, according to the author. Rather, there is a simple heuristic rule of thumb that accounts for these startling discrepancies. In the United States and Britain, a citizen has to opt in to an organ donor program, whereas in France and Hungary one has to opt out. Such a small and seemingly insignificant variable, Gigerenzer says, makes all the difference, for doing nothing (not opting out) is far easier than doing something (opting in). This is called the "default rule." The available evidence, Gigerenzer contends, indicates that it is not preference or moral principle that is the deciding factor here but instead a seemingly morally irrelevant factor: the way the choice is constructed rather than its substance. Nevertheless, in the United States more people violate the default rule by opting in than violate it in France by opting out.

Having used the notion of unconscious cognitive heuristics to throw a monkey wrench into the possibility, and certainly the reliability, of morally intentional action, Gigerenzer then proposes to reconstruct ethics on a sounder basis. He characterizes heuristics as both fast and, because they involve decisions based upon little information, "frugal." From the perspective of fast and frugal heuristics, morals cannot be relied upon to function at the level of the individual but instead must be induced through situational, social, and institutional nonmoral mechanisms. So heuristics represent a psychological phenomenon that poses a challenge to the existence of hardwired moral emotions, which are claimed to steer all of us away automatically and unconsciously from incest (Lieberman), for example, or toward empathy (Goodenough and Deacon). Heuristics have to be taken into account, Gigerenzer maintains, and the situation and the environment must be constructed carefully so that they shape action toward positive ends. They cannot be ignored or discounted without moral danger, as the example of Reserve Police Battalion 101 illustrates. Gigerenzer contends that it is the structure of the *context* that must be manipulated to produce morally desirable action from individuals, instead of relying on individual moral choice or moral training. Neither individual moral emotional motivation nor cognitive moral judgment can be relied upon to override the operation of seemingly trivial cognitive heuristics, Gigerenzer concludes. Cognitive heuristics are morally neutral but can be recruited toward morally better or worse ends—and therein lie both the danger and the hope.

The science of heuristics focuses on these three questions, Gigerenzer says: (1) What is in the adaptive toolbox, that is, what are the various heuristics? (2) Which environments can be structured to take advantage of these heuristics? (3) How can the heuristics be manipulated to solve specific human problems and how can environments be structured to take advantage of the heuristics? Gigerenzer's approach transforms ethics into a range of ecological problems to be addressed at the systems level rather than at the level of the individual mind and individual choices and actions. Gigerenzer calls the moral thinking involved "ecological rationality" and describes the solution as cognitive environmental engineering—the design of environments to fit the human mind. The heuristic analysis of Reserve Police Battalion 101, for example, focuses on the context and situation rather than on personality traits (such as the authoritarian personality) or culture (anti-Semitic prejudice, for example). This approach is sensitive to context rather than being focused on the individual. "Heuristics," Gigerenzer concludes, "provide explanations of actual behavior; they are not normative ideals. Their existence, however, poses normative questions."[51] Social heuristics pose a challenge to all the varieties of Western moral theories:

> If moral action is based on fast and frugal heuristics, it may conflict with traditional standards of morality and justice. Heuristics seem to have little in common with consequentialist views that assume that people (should) make an exhaustive analysis of the consequences of each action, nor with the striving for purity of heart that Kant considered to be an absolute obligation of humans. And they do not easily fit a neo-Aristotelian theory of virtue or Kohlberg's sophisticated postconventional moral reasoning.[52]

The advantages of heuristics are clear: they actually explain (some) behavior in terms of its causes, and that explanation can serve as the basis for interventions that produce morally desirable results. Knowledge of heuristics makes changes in behavior possible and controllable. The cost is in identifiable moral action for its own sake, out of moral motives. Gigerenzer also challenges the assumption of so many brain scientists that there is a discrete moral domain or capacity that can be discovered within the mind's architecture or functioning. Rather, the notion of heuristics reduces the moral domain to a larger social domain and puts the moral focus on the group and the context rather than on the individual. As Gigerenzer puts it, "Heuristics that underlie moral actions are largely the same as those underlying behavior that is not morally tinged."[53] Moreover, the causal explanation of moral action motivated by heuristics bypasses the debate about whether moral actions are motivated by emotions or cognition. Instead the relevant distinction is whether an action is motivated by unconscious or conscious reasons. As I have previously noted, the evidence that conscious reasons actually cause behavior is in dispute, and in fact there is growing evidence that what we think are our reasons for our actions largely cash out as ex post facto rationalizations for our decisions and actions. The absence of conscious awareness of the cognitive heuristics that drive many of our actions makes them like other decision making rather than unlike it, Gigerenzer points out. Decision making both within and outside the moral domain is largely unconscious. (More about our sources and levels of awareness or consiousness will be introduced later on.) But heuristics, unlike many other kinds of unconscious causes of decisions, are easily made accessible to consciousness and can be taught.

Heuristics are not just products of our individual brains but are often present as the socially embedded unconscious rules of institutions, rules that often conflict with the purported public principles of those institutions, Gigerenzer says. He offers as an example the British legal system, in which magistrates are supposed to follow due process in bringing a defendant to justice but in fact instead follow a heuristic to protect their own institution from being held liable if a defendant out on bail commits another crime. Hence they err on the side of jailing rather than allowing bail. So the facts on the ground show that the institution rewards institutional self-protection over justice for defendants and for the public. It can do so because there is no feedback mechanism that provides evidence of how well it is actually protecting both defendants and the public. He calls this and similar cases "split-brain institutions," likening it to what happens in split-brain patients, where the left side of the brain serves only to rationalize and confabulate what the right side is doing. Gigerenzer remarks that medical institutions are particularly prone to the heuristic of protecting the employees and the institution instead of the patients, those whom the institution is ostensibly designed to benefit. Again, this is so because there is no feedback mechanism to provide evidence of how well the institution's aim is being carried out, but there is plenty of blame for the employees if a miss occurs. So the employees act to avoid the possibility of blame (and potential lawsuits) rather than in pursuit of the ostensible purpose of the institution. Part of the heuristic is to keep out of consciousness what is really happening and to give it a public name and face that ascribes to itself an ideal that is in fact avoided in the interest of the institution's self-protection. As a consequence, overtreating and overmedicating become the standard practice. Individual self-deception and institutional collusion keep the whole thing going. Gigerenzer remarks that although there is a general consensus among many in the field that "heuristics are always second-best solutions, which describe what people do but do not qualify as guidelines for moral action," he thinks heuristics in some cases, such as ecological ones, can be prescriptive as well as descriptive.[54]

The evidence from heuristics points to several hypotheses about ethics. First, the locus of ethics here is the interaction of mind and environment. It is in this in-between area that ethics should be sought, not in the mind alone, and especially not in the individual mind alone. Second, moral decision making is to be sought within unconscious cognitive processes, not just, or perhaps not even primarily, in conscious thinking. I will return in later chapters to elaborate on the evidence for these two points.

Muddying the Waters II: Cognitive Framing

Does ethics consist in moral intuitions, as those who advocate hardwired moral modules in one form or another contend? Do we have strong moral beliefs that arise immediately from an objective grasp of a situation? In his essay "Framing Moral Intuitions," Walter Sinnott-Armstrong challenges the claim that moral beliefs arise as innate, basic, and discrete human responses to a set of clearly discernible universal and repeated situations. Those who advocate innate moral intuitions hold that there are intuitive moral responses that are not derived from other beliefs, or justified in terms of other beliefs, but are instead singular responses that feel right or wrong just in themselves. Sinnott-Armstrong provides evidence that we have no such basic moral intuitions. His data come from studies of cognition that expose how seemingly morally irrelevant factors in the framing of moral questions affect moral judgments in ways that they would not if intuitionism were true. For example, modifying word choice and context, without changing the basic situation described or its meaning, can systematically affect moral intuitions.[55] (These findings are consistent with those of the heuristics experiments discussed previously.) Sinnott-Armstrong first cites a famous experiment of Tversky and Kahneman, who were the first to study framing effects. Subjects were asked either of two versions of a hypothetical story. Both versions amounted to the same thing but were framed differently. This was the first version:

> Imagine that the U.S. is preparing for an outbreak of an unusual Asian disease which is expected to kill 600 people. Two alternative programs to fight the disease, A and B, have been proposed. Assume that the exact scientific estimates of the consequences of the programs are as follows: If program A is adopted, 200 people will be saved. If program B is adopted, there is a ⅓ probability that 600 people will be saved, and a ⅔ probability that no people will be saved. Which of the two programs would you favor?

The second version presented the same story but worded the choice between the alternatives in the following way:

> If program C is adopted, 400 people will die. If program D is adopted there is a ⅓ probability that nobody will die and a ⅔ probability that 600 people will die.

Clearly, program A is the same as C and program B is the same as D. The two versions differ merely in their wording, with the first version referring to how many people will be saved and the second version to how many people will die. The results, however, did not reflect this equivalence. Instead, while 72 percent of the subjects chose program A over B, only 22 percent chose program C (the equivalent of A) over D (the equivalent of B). It appears that the language of saving versus of dying made all the difference. But perhaps the issue here is more substantive, Sinnott-Armstrong suggests, and that what we are seeing is not just a wording difference but also a real difference in connotation between a saving intervention and its lack. Saving sounds worth doing no matter what.

In another experiment two groups were given hypothetical scenarios of a classic moral story about a trolley car that has lost its brakes and is hurtling down the tracks toward five innocent people who will lose their lives if the trolley continues straight ahead, but only one person will lose his life if the trolley is diverted onto a side track by activating a switch. Imagine you are a bystander who can flip the switch. Most people agree that it is not morally wrong to flip the switch and cause the death of one person to save five. Experimenters tested students to see if the same moral conclusion held across different wordings. Half of the questionnaires used the word *kill*, and the students were asked to make a moral choice between throwing the switch, which would result in the death of one innocent person, or doing nothing, which would result in the deaths of five innocent people. The second group received questionnaires that used the word *save*, and they had a choice between throwing the switch, which would result in five innocent people being saved, and doing nothing, which would result in one innocent person being saved. The responses slightly favored action when the question was asked in the *save* version and slightly disagreed with taking action when it was posed in the *kill* version. The experimenters found that the wording effect accounted for about a quarter of the total variance. Nothing had been changed—not consequences, intentions, or facts—except the wording.

In another experiment the change in wording was even less substantive since it introduced only a change in the order of the scenarios presented. In the first version of this experiment the change in order had no effect, but in the second version it did. In the second version a seemingly irrelevant and negligible change had a significant effect on the moral judgments people gave. In the first version of this experiment 180 students were asked how strongly they disagreed or agreed, on a scale of +5 to -5, with each of several versions of moral dilemmas set out on a form. There were three pairs of forms. Form 1 posed three moral problems. The first was the trolley problem just mentioned. The second gave a hypothetical case of saving five dying people by scanning the brain of a healthy person, but the healthy person would die as a result. In the third problem, the only way to save five dying people was to transplant the organs of one healthy person who would as a result die. All of the scenarios were described in the language of saving. An alternative version of this form had exactly the same scenarios but in reverse order. In this experiment, no difference was noted—that is, no framing effect was found.

A second experiment, however, did show framing effects. In this experiment the trolley problem was set out first. Then a variant of the trolley problem was introduced as the second moral dilemma: according to this new scenario, a *button* could be pushed that would cause the train to jump the tracks, saving the five but causing the death of one. The third moral problem was also a trolley scenario: in this one, only pushing a very large person onto the track in front of the trolley would stop it from killing the five. In the case of this set of forms, the order in which the three problems was presented had a significant effect on the moral judgments. There were two findings: people approved of a moral action far more when it appeared first in the sequence rather than last. Second, when the button scenario followed the original trolley problem, there was far more approval of it than when it followed the pushing-the-person dilemma. The experimenters concluded that when dilemmas are more homogeneous (as was the case in the second experiment), the context has an effect. Maintaining consistency with the initial judgment seemed to play a role in the subsequent moral evaluations. Other (and more realistic) cognitive framing experiments of moral evaluation exhibited a similar ordering effect and also a general tendency for later ratings to be more severe than earlier ones. Subjects showed a general tendency to increasing blame over time. In all the cases neither the facts, nor the consequences, knowledge, or intention, changed in the different hypothetical situations presented. Yet the valuations of the same situations did change, exhibiting an ordering effect.[56] Sinnott-Armstrong interprets the data in terms of his presupposition that morals consist in unchanging real truths and that cognitive framing interferes with these truths. In his view, since moral truth is always the same, the cognitive framing effect interferes with accurate moral "perception." He concludes that the evidence of a framing effect shows that discrete and ubiquitous innate moral intuitions of situations are not adequately reliable, so moral intuitions are not justified or justifiable without further inferences. "Framing effects," he says, "signal unreliability."[57] He uses these examples to defeat the case for moral intuitionism as an accurate and universally human innate ability in the way language is thought to be, since the "studies show that moral intuitions are subject to framing effects in many circumstances."[58]

The framing effect poses a challenge to the many versions of the claim that there are hardwired discrete moral mechanisms. Haidt and Bjorklund's assumption that there are fixed ethical situations that are readily apparent and have clear moral meanings that trigger set moral responses is a case in point. They offer a diagram of the SIM with a single circle containing the eliciting situation, and an arrow points to everything else that follows. Sinnott-Armstrong's cognitive framing effects make such an initial clarity and univocality of meaning of the ethical situation highly unlikely, for even small changes in wording or in context seem to affect the meaning of any given situation, moral or otherwise. In the next chapter I will cite an array of research on how large-scale cognitive framing is ubiquitous and includes the many different cultural, social, and historical as well as personal psychological meanings we attribute to our worlds and to the array of situations within them. The possible meanings and interpretations of situations, it will become apparent, are almost infinite. Hence Sinnott-Armstrong's conclusion of the mere error-proneness of our "perceptions" of situations is not the only way to read the data he presents.

But why does Sinnott-Armstrong presuppose that cognitive framing is merely a source of error in moral reasoning and decision making rather than evidence for how human moral thinking actually operates? Cass Sunstein, in contrast to Sinnott-Armstrong, argues for a larger role for cognitive framing in ethics. He proposes that the discussion of cognitive shaping of moral decisions needs to go beyond unconscious fast heuristics and focus more generally on cognitive frames, particularly on the cognitive framing of moral situations. Cognitive frames are largely unconscious, yet they are not fast heuristic rules of thumb. They consist in implicit gestalt (global) interpretations of situations.

In the next chapter I will elaborate on the importance and ubiquity of unconscious sociocultural cognitive framing driving the interpretation of situations. Cognitive moral framing implies that ethics is not merely about action, choosing and adhering to principles or having moral sentiments, but instead is embedded in the interpretations given to situations. So rather than merely playing the role of interference, cognitive framing has a large positive role to play in ethics. What Sinnott-Armstrong has identified as errors in ethical valuation may be instead just what ethics as practiced looks like.

Muddying the Waters III: Conscious and Unconscious Choices and Decisions

Let us grant for the moment Haidt and Bjorklund's presupposition that innate morals, if they could be found, are moral truths for all human beings. Yet even the intuitionist's "ethical facts" version of the project to naturalize ethics runs aground. For Haidt and Bjorklund base their claim of innate moral intuitions on their prior claim that research has shown a strong line of demarcation between unconscious, automatic, rapid mental processes and conscious, highly verbal, slow ones. The assumption of that division in the mind presupposes that, in principle, there could be evolved moral intuitions that in some respects remain segmented and isolated (modular) from other areas of mental functioning like culture and language; they bubble up and then are nuanced with contextual factors. But in fact there is increasing evidence that is erasing the distinction between unconscious and conscious processes that the authors depend upon—which, in effect, is a line they are attempting to draw between nature and nurture. While I agree with the authors about the central importance for rethinking ethics of the discovery that "consciousness is at the level of a choice that has already been made," there is now a growing body of evidence that unconscious processes encompass and govern most *executive*, that is to say highly cognized, decision-making processes as well as more primitive automatic ones.[59] This new research indicates that not just primitive intuitions are unconscious, but so is much of the culturally rich and complex thinking involved in judgments and decisions. Haidt and Bjorklund should perhaps welcome this new direction, for giving up the presupposition of innate, modular moral intuitions would make it possible for them to reconcile their claims of social skills and situationism with their claims to automaticity and the unconsciousness of moral judgment and action. So a smoother and more complex interaction of the innate with the social and cultural (what's learned), even at the most unconscious level, appears likely in the light of research on neuroplasticity and on cognitive framing. Moreover, the presumption that "animal" equals "hardwired" and "human" equals "softwired," and hence that our animal evolutionary inheritance and human capacities can be neatly divided and assigned discrete levels of function, is also being challenged.

Darcia Narvaez makes the point that it is likely that moral perceptions of situations—really, interpretations of them (as well as responses to them)—are part of, and beholden to, large-scale complex systems of meaning derived from experience and "softwired" into the brain. She calls for a "biopsychosocial" approach of "ecological contextualism" to be brought to bear upon understanding ethics in the light of the increasing scientific acceptance of the neuroplasticity of the neocortex. Narvaez suggests that the unconscious automaticity of moral judgment comes not from its hardwired evolutionary origins (as Haidt and others presume) but instead from the transition of a consciously learned skill to a developed unconscious expertise. This is the way we all experience the learning process—for example, driving a car or riding a bicycle. The novice-to-expert paradigm, she says, is ubiquitous in cognitive science research. And she calls attention to the social contexts in which we develop cognitive schemas; we are not mere passive perceivers.[60] Narvaez introduces a model of "cognitive-affective-action schemas," which are products of both human construction and cultural inheritance, as a possible model for rethinking ethics in a deeply and broadly contextualized way.[61] Yet she is careful not to reduce morals to mere social conformity, and accuses Haidt and Bjorklund of doing just that. Narvaez suggests that we can avoid that trap when we pay attention to our role in the development of our morals. Citing the dangers of Nazi Germany or the Cambodians under Pol Pot, she remarks, "It is shocking to read Haidt and Bjorklund assert that 'a fully enculturated person is a virtuous person.'"[62] Is Narvaez's reliance on our "active" role in creating and adhering to morals a way of falling back on some version of free will yet again? We can choose to opt out of the immoral, she says, a view that once again seems to make moral action and responsibility depend upon the mystery of human freedom of choice. I think there are better, less magical ways of getting out of this mess. Narvaez nevertheless has both aptly captured the need for contextual explanation and also turned our attention to the moral dangers of social context. She has brought us back to real moral life and away from the contrived, decontextualized hypothetical moral problems of so much philosophical and psychological research, typified by the trolley problem.

Narvaez's cognitive-affective-action schemas (and the work of the cognitive linguist George Lakoff on cognitive framing in morals and politics, to which I will turn in the next chapter) suggest that we must go beyond the SIM's rationalization/reason model of ethics to bring in culture and narrative, self and society and situation. Not only does the initial moral eliciting situation appear to be embedded within cultural traditions and narratives, but so does the ethical action. I will argue that looking at cognitive framing can expose how cultural and other contextual interpretations and narratives in fact drive a foregone moral conclusion. (This supports the classical Greek and Spinozist positions that ethics is about how intellectual understanding shapes desire and directs motivation, and highlights how moral virtue lies in honesty, completeness, and ultimately independence from immediate social meanings and pressures.) Brain research by Michael Gazzaniga and others indicates that human beings have a tendency to persistence in belief over truth, tradition over innovation, and the status quo over revolution. The resistance to conceptual rethinking documented in Thomas Kuhn's famous

Structure of Scientific Revolutions is now explicable by neuroscience. So all kinds of theories that presuppose discrete and localized capacities, including Haidt and Bjorklund's SIM, may have to be questioned and rethought in far more broadly contextualized and flexible ways if the evidence eventually pushes us definitively toward the nonmodular, nonlocalized neuroplastic explanations of thinking and action. A naturalistic account of ethics could then be based on a more integrative and plastic account of the brain and mind, one in which nature includes nurture rather than is defined in contrast with it or as an overlay upon it.

Muddying the Waters IV: Anthropological and Sociological Evidence for the Cultural Variability of Morals

John M. Doris and Alexandra Plakias further deepen our sense of the importance of taking social context into account in coming to understand how ethics operates. They draw upon a growing body of evidence from anthropology and other social science research to introduce salient examples of substantive cultural differences in morals. These, they argue, make it unlikely that we can identify a set of universal moral principles or intuitions that are hardwired products of evolution beneath all the cultural differences.[63] The intractability of moral disagreements also contributes to the unlikelihood that there are real "moral facts" (although, of course, not all agree that such disagreements are substantially irresolvable). In the late 1970s the philosopher J.L. Mackie argued against the possibility of moral agreement, a position that Doris and Plakias argue is now supported by empirical research. Moral judgments, Mackie suggested, can better be explained as reflecting ways of life rather than as perceptions of moral facts, the way that, say, perceptions of objects reproduce them mentally.[64] There are no moral facts, according to this argument, to make moral judgments either true or false in the way that there are physical objects whose descriptions can be compared to the things described.

Doris and Plakias call to our attention the work of anthropologist Michelle Moody-Adams in which she examines the ethnographic literature to see if a universal ethic can be found beneath the cultural diversity. While Moody-Adams is wary of drawing conclusions from an anthropological literature that has had little sensitivity to philosophical nuance, she nevertheless embraces a strong view of the situational meaning of behavior and language. According to Doris and Plakias, she holds that "all institutions, utterances, and behaviors have meanings that are peculiar to their cultural milieu, so that we cannot be certain that participants in cross-cultural disagreements are talking about the same thing."[65] Moody-Adams questions whether cross-cultural differences can even be identified, let alone confirmed. She also doubts that cultures have any kind of systematic internal consistency in morals that would lend itself to cross-cultural comparison. An early (1954) ethnographic study, *Hopi Ethics*, by philosopher Richard Brandt found moral diversity between Hopis and white Americans. The salient evidence for substantive moral disagreement between the two groups was in attitudes about children inflicting pain on pet animals. The Hopis had no moral repulsion to it, whereas American whites pretty much did across the board. Yet given both the paucity of reliable evidence and also the disagreements among philosopher-anthropologists, Doris and Plakias regard the anthropological literature as leaving the question open, and instead they turn to evidence from a large body of cultural psychology research on similarities and differences in emotional and cognitive responses.

Doris and Plakias discuss at some length a study by Richard Nisbett and his colleagues of regional patterns of violence in the American non-Hispanic South (the old Confederate states) versus the American North. Nisbett and colleagues applied the principles and tools of cognitive social psychology to the problem.[66] The American South, they concluded, exhibits persistent patterns common to what anthropologists call "cultures of honor": a focus on the importance of male reputations for strength and of quick retaliation for insult or other violation. The culture of honor, they suggest, explains a disparity in violence that persists between North and South. The South has a greater percentage of homicides resulting from arguments, but not of homicides committed in the course of robbery or other felonies; white southerners are more likely to view violence as justified in response to insult and as a sign of "manhood" than do northerners; legal statistics show that southern states permit more violence in defense of property and person than do northern states. The Nisbett group not only looked at statistics but conducted an experiment that involved sending job application letters to employers all over the country from a fictitious applicant who revealed in the letter that he had a single blemish on his record—he had been convicted of manslaughter for accidentally killing a rival in a barroom brawl after the rival had boasted of sleeping with his fiancée. In another letter, a fictitious applicant admits that he had been convicted of stealing a car when his family was short of money. No regional differences were found in the second scenario, but the southern employers were more sympathetic to the first fictitious applicant's violent action (to maintain his honor) than were northern employers.[67]

In another experiment by Nisbett and colleagues, northern and southern male subjects were unsuspectingly tested for testosterone levels after receiving seemingly inadvertent insults. The subjects were told they were having their blood sugar measured as they performed certain tasks. After a control sample was taken, each subject walked down a narrow corridor where he was bumped into by an experiment confederate who also called him an "asshole." Saliva samples taken immediately afterward were analyzed for levels of cortisol and testosterone, the first indicating stress, anxiety, and arousal and the second indicating aggression and dominance. Southern subjects had great increases in these hormones, while northern subjects showed small increases. Nisbett and colleagues concluded that southerners exhibit a stronger response to insult than northerners.[68] Doris and Plakias propose that this case shows a substantive difference in moral values down to the bone—to the hormonal level. And hence the cultural difference is fundamental and results in a substantive and irresolvable moral disagreement on the permissibility of violence between individuals. They also say that this is a case in which ideal conditions cannot be imagined in which the disagreement would disappear. So Doris and Plakias say that Nisbett's group has come up with evidence of an identifiable and solid case of fundamental and irresolvable disagreement in morals—a case that poses a challenge to the likelihood of hardwired moral intuitions and basic moral facts.

Brian Leiter, in commenting on Doris and Plakias's essay, points out that the partiality we all feel toward our own family, nation, and other groups amounts to familiar and ubiquitous evidence against the possibility of universal solutions to all moral problems. That impartiality should be the norm, he quips, may be exactly the moral value on which no agreement can be found.[69] Another commentator finds the ubiquity of disagreement unproblematic. He proposes turning to a medical model of ethics: real values are to be found on the analogy of health rather than as discrete hardwired principles in our brains. He specifies that he means a model of physical health rather than of psychological health, which he regards as too fraught to be useful.[70] With this, he leaves us hanging.

Muddying the Waters V: Moral Responsibility Does Not Depend on Free Will or Free Decision

Julia Driver presents evidence from empirical studies that free causal agency in performing an action is not in fact the basis people use for holding others morally responsible for their actions, as we in the West presume. Interestingly, even in our own cultural world, in which a strong linkage of causal origination and responsibility is presumed and institutionalized in law and elsewhere, the way people actually assign responsibility in many cases does not follow the cultural rule.[71]

Driver first points out that we often call something the "cause" of an event by picking it out of a large array of other factors that we then label as "background conditions" or "causal factors." In fact, what we say is the cause often means simply the assignment of responsibility. So Driver brings up a classic case of a man who carefully stored his wood in his basement. A pyromaniac entered the cellar and burned the house down. While we would call the home owner's storing of the wood a background condition, we assign full causal responsibility to the pyromaniac. Both are causal factors, yet it is the unusual and reprehensible action that gets called the cause of the fire. In a contrasting case, a home owner who leaves his home unlocked and has his spoons stolen is assigned some causation in the stealing of the spoons. So the concept of "cause" here is a simple stand-in for the assignment of responsibility. From a strictly causal perspective, all kinds of factors contributed to what happened in each case, including in the first case the presence of oxygen in the atmosphere, which enabled the wood to burn. So it's not really "cause" in the strict sense that's being talked about or assigned but in fact it is responsibility that is snuck in as if it were a neutral description of simply what happened. Driver says other cases reveal that we assign responsibility even in the "complete and total absence of causal connection."[72]

From these thought experiments Driver now turns to empirical studies of how people assign moral responsibility and blame in order to determine how the attribution of cause enters in. Generally people are reluctant to say that someone is a cause of something bad if that person is not blameworthy, she says. Driver describes a study by M.D. Alicke from 1992 in which subjects were asked to assign "primary causation" in various scenarios. The experiment went as follows:

John was driving over the speed limit (about 40 mph in a 30 mph zone) in order to get home in time to . . .

Socially desirable motive

. . . hide an anniversary present for his parents that he had left out in the open before they could see it.

Socially undesirable motive

. . . hide a vial of cocaine he had left out in the open before his parents could see it.

Other cause

Oil spill   As John came to an intersection he applied his brakes, but was unable to stop as quickly as usual because of some oil that had spilled on the road. As a result, John hit a car that was coming from the other direction.

Tree branch   As John came to an intersection, he failed to see a stop sign that was covered by a large tree branch. As a result, John hit the car that was coming from the other direction.

Other car   As John came into an intersection, he applied his brakes, but was unable to avoid a car that ran through a stop sign without making any attempt to slow down. As a result, John hit the car that was coming from the other direction.

Consequences of accident

John hit the driver on the driver's side, causing him multiple lacerations, a broken collarbone, and a fractured arm. John was uninjured in the accident. Complete the following sentence: The primary cause of the accident was . . .[73]

Alicke found that subjects who were given the scenario with John's motive as the socially undesirable one considered John the "primary cause" of the accident far more frequently than did subjects who were given the scenario in which John was assigned the socially desirable motive. Joshua Knobe of the University of North Carolina points out that causal attributions do not work in the neutral descriptive way that social scientists believe they do. Instead, moral or other normative valuations slip into the descriptions. He suggests that people most often follow a particular sequence: first they judge that someone has acted wrongly, then they attribute causal primacy to that person, and finally they assign blame to that person.[74] So the assignment of primary cause is the *effect* of already having made a moral judgment; it is not that the cause is assigned independently of moral valuation and then made, but the assignment of cause instead is actually dependent upon the moral judgment already in effect and is an expression of it. Driver concludes that "normative considerations influence causal attributions."[75]

In addition, causal primacy is often attributed to an act that, in contrast to others, is out of the norm or the ordinary course of things. Consequently it's not necessarily the action closest to the outcome that's considered the primary cause. Driver cites Joel Feinberg's assessment of the 1962 Cuban missile crisis as "caused" by the Soviet construction of missile bases on Cuba, since the introduction of those bases was a radical departure from previous standard practice. So it is the unusualness that is the salient feature that draws to it the attribution of primary cause. But does the unusualness criterion cover the moral cases? If so it would seem to bring us back to more morally neutral descriptions. Driver says the jury is still out on this. Yet she thinks that the evidence of Alicke's empirical study and the thought experiments of Knobe and others provide solid support for the view that the assignment of cause is consequent upon a prior moral judgment and against the standard view that the moral judgment of responsibility *results* from an independent and objective analysis of causes. Joshua Knobe and Ben Fraser, in their comments on Driver's essay, provide further evidence that bears out the suspicion that it is not, in fact, the atypicality of an action that suggests to people its causal primacy but its *moral value*.[76] For example, in an experiment Knobe and Fraser conducted, subjects were given a story about a philosophy department. The receptionist in the department would keep her desk stocked with pens, and while administrative assistants were allowed to take pens from there professors were supposed to buy their own. Yet invariably professors, not just administrative assistants, took the pens. One Monday an administrative assistant and Professor Smith walked by the receptionist's desk and both took pens. Later that day the receptionist had to take an important phone message and there were no pens left. Subjects were asked whether they agreed or disagreed with the following two statements: "Professor Smith caused the problem" and "The administrative assistant caused the problem." Subjects overwhelmingly agreed that the professor had caused the problem. This was the case even though both behaviors were typical. Yet only the professor's action was morally wrong. Because here the prior moral assignment of responsibility seems to have driven the assignment of cause, atypicality cannot be a factor, they concluded. These results and those of other relevant experiments led Knobe and Fraser to conclude that "moral considerations . . . play a fundamental role in the way people think about causation."[77] We could extrapolate that the theological, philosophical, and cultural claim that moral responsibility depends upon the capacity to (causally) originate one's actions rests on a conceptual and linguistic confusion, as well as upon its political and social utility as a myth.

The second commentary on Driver's essay addresses legal responsibility. For someone to be criminally responsible for an action, he or she must have acted from a guilty mind, John Deigh points out. This is different from the assignment of cause, for someone can be the immediate cause of someone's death but not be blameworthy, as a postman who delivers a bomb may be the immediate cause of a death but certainly is not responsible for it. Deigh further raises the objection to the causal account of responsibility by citing the principle of complicity. Someone who joins a criminal enterprise is responsible for the enterprise's harmful consequences whether or not those consequences are caused by the person's own actions or by those of his or her accomplices. Hence being criminally responsible is detached from having one's actions actually cause the event. Both commentaries go further than Driver in uncoupling moral responsibility and cause. Deigh ends his essay by citing Hart and Honoré's observation in 1959 that causation in the law either refers to an outcome that wouldn't have occurred had it not been for the defendant or simply means that the defendant is morally responsible. So the very notion of being *the* cause collapses into that of moral responsibility, both in the law and in common parlance.[78]

We have come full circle, back to the problem of free will in ethics. Freedom of the will or of decision presupposes that to be responsible a person has to have fully or at least mostly originated an action. But it turns out that that is not the way we actually assign moral responsibility. According to the research just presented, we determine the moral responsibility and *then* assign the cause. So the causal claim, the free will claim, is actually mere shorthand for saying "You're morally responsible for having done X." So why do we think we need to be "free" to be morally responsible? What does that freedom mean and entail? Why do Westerners hold on for dear life to the notion of free will or decision as if our civilization would collapse without it, whereas ancient Greeks, Cambodians, and so many others do not have such a notion yet have morals and assign moral responsibility? And the facts on the ground even in our own Western practice seem to highlight that problem. We have just seen that we don't need to assume a person has free will or choice to hold that person morally responsible. Instead we seem to assign free will, the causal origination of a (culpable) action, to just that person we have already decided is morally responsible. So we don't really mean exactly what we say—we're misdescribing, misunderstanding our own words and actions. And that misunderstanding has all kinds of cultural, social, political, legal, and personal repercussions, not least of which is an overwhelming tendency to blame the individual (it's his free will, after all) and to let the group and the institution off the hook.

7

Beginning Again: The Blessing and Curse of Neuroplasticity: Interpretation (Almost) All the Way Down

Western philosophical accounts of morality are outdated in important respects, for example in ascribing too much volition and intentionality to moral behavior.

—MARC BEKOFF AND JESSICA PIERCE

Where there's flexibility and plasticity in behavior . . . there's agency. This is the very reason we do not include insects among our moral animals, because as far as we know their behavioral patterns are rigid. . . . And that is why we set threshold requirements for our moral animals: flexibility, plasticity, emotional complexity, and a particular set of cognitive skills.

—MARC BEKOFF AND JESSICA PIERCE

I have striven not to laugh at human actions, not to weep at them, nor to hate them, but to understand them.

—BARUCH SPINOZA

Introduction

It is worth reminding ourselves that the nature/nurture dichotomy is generally considered dead.

—MARC BEKOFF AND JESSICA PIERCE

Experiments on rats and rhesus monkeys show that, unlike human beings, they pass the Milgram test—they exhibit a fellow feeling for suffering members of their species that roughly two-thirds of human beings, cross-culturally, either lack or suppress. Scientists, if not the rest of us, have known this since the 1960s: the experiment on rats was published in 1959 and the one on rhesus monkeys in 1964. And the results have held up to the test of time. A recent controlled study with rats at the University of Chicago offers stronger evidence than has been available before that rats will act to aid another rat in distress and will do so even in preference to reward. The findings were reported in the journal *Science*.[1] The experiment involved pairs of rats that normally share a cage. One of the rats was set free to roam, and the other was trapped in a closed see-through plastic tube that could be opened only from the outside. The free rat was able to see and hear its former cagemate. That rat attempted by trial and error to free its cagemate and, once it learned how, it did so right away. Experiments with placing a toy rat in the tube instead of a real rat or with an empty tube did not elicit the freeing behavior. Nor did separating the two rats change the saving behavior, so scientists concluded that an expected reward of social contact between the two could also be ruled out. In addition, putting chocolate in a second closed tube did not distract the rats: the tube of the trapped animal was just as likely to be opened first as was the one with the chocolate stash. Peggy Mason, one of the experimenters, remarked: "That was very compelling. It said to us that essentially helping their cagemate is on a par with chocolate. He can hog the entire chocolate stash if he wanted to, and he does not. We were shocked." Jean Decety, the study leader, concluded, "This is the first [reliable] evidence of helping behaviour triggered by empathy in rats."[2]

Affective neuroscientist Jaak Panksepp has identified seven basic homologous hardwired emotional systems that all mammals have, including human beings. Panksepp, in commenting on this study of empathy in rats in the same issue of *Science*, writes, "Although we currently look to mirror-neuron zones of the neocortex for evidence of the highest mind functions such as compassion, empathic tendencies are surely also promoted by the more ancient primary-process emotional networks that are essential foundations for mental life."[3]

Human beings may be capable of a more finely tuned empathic understanding of others, and their suffering and their fates, yet when push comes to shove it's monkeys and rats that are statistically far more likely to enact their emotional fellow feeling.[4] What are we to make of these remarkable findings? Are they just weird and anomalous, and so we should simply shrug and carry on? I don't think so. In fact, I think that both the discoveries themselves and our standard ways of ignoring them are both significant. For both the discoveries and their marginalization reveal that we have not yet fully become aware of how to interpret and understand the evolutionary continuity of ourselves within the mammal kingdom. Our long-standing beliefs and assumptions about our exalted and unique human moral capacity persist more or less unshaken despite at least some evidence to the contrary and despite well-established Darwinian theory that calls them into question. Our human self-congratulation that we are above and beyond all (other) animals as well as our human reluctance to relinquish beliefs in the face of mounting evidence to the contrary both tell us something about ourselves.

That nonhuman social mammals and primates may have a reliable prosocial and perhaps even moral sense and capacity is the contention of biologist Marc Bekoff and philosopher Jessica Pierce.[5] Marc Bekoff has spent a lifetime as an animal ethologist, and both his observations and the research of others have led him and Pierce to the remarkable conclusion that the evolutionary record must be read not only up but also down—our capacities are also those of other social mammals, to a degree.[6] Bekoff and Pierce put the evolutionary cut not between ourselves and other primates, ourselves and other mammals, but instead between the social mammals (including human beings, other primates, cetaceans such as whales and dolphins, rats and mice, elephants, and social predators such as wolves and dogs) and, on the other side, the social insects (ants and bees, for example). When it comes to the social insects, they say, we are indeed looking at hardwired prosocial instincts rather than at morality. For insect sociality surely lacks flexibility and thought; *it is more programmed than interpretive.* In social mammals, however, including human beings, the neural architecture that produces empathy (and the other prosocial capacities) is the same across species.[7] Bekoff and Pierce take seriously Darwin's insight that the relationship of human beings to other mammals, and even more so to other primates, is a matter of degree rather than of kind.

Bekoff and Pierce propose that there is even a moral continuity between the animal and the human resulting from a shared neuroflexibility, and not only shared hardwired prosocial behaviors, as most scientists have presumed. This means that social mammals (including humans) generally filter their more primitive prosocial capacities via some thought, learned experience, and somewhat complex emotions, so there is some flexibility and also variation in their behavior. This view is in sharp contrast to prevailing arguments, which have emphasized that human beings are genetically hardwired for automatic moral feelings or behavioral prompts that still drive their behavior either completely or to a large degree—such as an aversion to incest or a sexual attraction to do-gooders. If Bekoff and Pierce are right, then what human beings, our close primate kin, and the other social mammals share are basic prosocial capacities that are mediated by various degrees of flexible thought and emotions, resulting in various degrees of moral behavior, both prosocial and normative. Researchers have documented cooperation, empathy, and even justice (that is, normative rules that promote social harmony when maintained and elicit social punishment and ostracism when broken) in the behavior of a range of social mammals and particularly primates. Cooperation, for example, is evident in all kinds of behaviors: grooming, group hunting, care for babies, alliance formation, and even play.[8]

Within the array of social mammals, Bekoff and Pierce describe what they call "nested levels of increasing complexity" as organisms exhibit a higher degree of social complexity and of intelligence—the difference between rats and primates, for example.[9] Bekoff and Pierce's model may help us understand where we stand within nature, especially within the arena of social mammals. We no longer need presume the direct or only slightly indirect operation of hardwired instincts in us in order to preserve and respect evolutionary continuity. Bekoff and Pierce have cut us free from the kind of reductionism to hardwired mechanisms that until now we have thought were entailed by shared homologous evolutionary inheritances.

Why Do Human Beings So Often Fail to Act Morally?

"Everything in the brain is interconnected—and that of course is the problem."
—JAAK PANKSEPP, December 13, 2012, lecturer in Puerto Vallarta mexico.

Philosophers, scientists, and regular people in the West have long been asking, in one way or another, why human beings uniquely have ethics, a moral capacity that is one of the glories of what it means to be human rather than animal. But I think based on the evidence about other social mammals and the growing evidence from the neurosciences, we need to ask a different question: why do human beings fail so often and so disastrously at moral behavior, both on the personal level and more egregiously on the societal and political levels? When I tell people I casually meet (not philosophers or neuroscientists) that I am writing a book on ethics, almost invariably they say something like, "Boy, we sure need that. Something really has gone wrong. *Are* there any ethics anymore? Why do people, the society around us, the corporations and the banks, and the government more often seem to neglect ethics rather than adhere to ethics?" So intuitively, from lived experience rather than traditions of scholarship and science, people are raising what I think is the right question: What has gone wrong? The evidence for empathy and prosocial actions in some species of nonhuman mammals may offer us a clue to what has gone wrong in human beings, despite growing evidence from the neurosciences that human beings may have an even greater basic sociality than other social mammals.

Zoologist and conservationist William Temple Hornaday, director of the New York Zoological Park and a founder of the National Zoo, commented on the superiority of animal "morality" when he wrote in 1922: "It is quite possible that there are . . . a number of intelligent men and women who are not yet aware of the fact that wild animals have moral codes, and that on average they live up to them better than men do theirs."[10] It turns out that our human moral problem may be a matter of too much thinking, too much interpretation. Because human actions, the meanings we ascribe to them, and the beliefs we have about them can take such a vast number of different avenues, we can lose sight of the good.

Openness confuses our prosocial capacities. The evidence from the neurosciences lends support to the view that human beings are off the charts when it comes to the openness and flexibility of our behavior, of our *interpretations* of behavior. From the standpoint of both evolutionary biology and neurobiology, our human openness in behavior and belief is on a scale and range so vast that it may turn out to be, in terms of morality, a double-edged sword. This is because the prosocial or moral capacities (like the empathic feelings of rats and capuchin monkeys for others in pain, or reciprocity in all kinds of species) that are evolutionary inheritances from the social mammals are mediated by thinking, that is, by language, culture, social group—which is to say, by interpretation. In other species, behavioral possibilities may be narrower because the meanings of behavior are not subject to such variable interpretations.

Recent science leads us toward a second insight: that cooperation is an evolutionary norm, and that humans have as extraordinary a capacity for cooperation as we do for behavioral openness. Bekoff and Pierce comment that "humans have achieved a level of social complexity unparalleled in other species."[11] And tendencies toward cooperation, shared experience, emotional contagion, and social needs may be even more dominant in humans than in other primates and mammals. While we tend to think that evolution is all about individuals competing within a species for survival and reproductive success, Martin Nowak, director of the Program for Evolutionary Dynamics at Harvard, has singled out cooperation as the third basic principle of evolution, along with mutation and selection. The overwhelming character of social interactions across living beings is cooperative rather than antagonistic. The very fact of the evolution of multicellular living things bespeaks cooperation, Nowak remarks. Yet this vast human cooperative capacity can be directed toward evil ends as well as good ones.

Third, new research is revealing interconnections between our prosocial capacities and group feelings and the urge to further and protect the self. Our prosocial tendencies are also enmeshed with self-serving urges. So our third moral problem is a double-edged sword as well: our seemingly moral actions are colored by self-serving and self-furthering motivations and aims. But there is some good news as well, for the reverse is also the case: our selfishness is not always in the service of the narrowest interests within our skin or pocketbook. For the self, it turns out, is not "atomic," not isolated and contained just within our skin. Rather, there is a self beyond itself, beyond the physical body. There is emerging evidence that the self incorporates into itself relationships, family, the social and cultural and historical world, and assigned places in that world. This self beyond itself makes ethics possible, because the boundaries between self and others and between self and environment are loosened or permeable and even shared. The neuroscientist Donald Pfaff of Rockefeller University identifies the loosening of personal boundaries as the only or principal mechanism that makes ethics possible. This solves the problem of altruism, of why an individual might help somebody else, even a nonrelative, and even to his or her own detriment. A neurological mechanism of shared selves—or as a recent conference on Buddhism and science puts it, "the neurobiology of we"—really redefines the problem of altruism because it changes the unit of benefit from the individual to a blending of individuals in groups, small and large.[12] Scholars of East Asian cultures and religions have reminded me that the unit of analysis in Buddhist ethics is not the individual but interactions and interrelationships.[13]

Other recently discovered neural mechanisms also contribute to making extended and shared (or distributed) selves: mirror neurons are everywhere in the news these days—they make possible emotional contagion, other

forms of unconscious imitation and joint action, and a grasp of others' emotions. Another relevant neural mechanism is self-maps, which is the mapping into the bodily neural moment-by-moment representation of the self
of all kinds of biographical information and group identifications. In this
chapter and the following ones I discuss how these and similar mechanisms
operate and contribute to ethics—but I also show how they make moral life
precarious and difficult. The self beyond itself all too easily devolves into the
group self; the loosening of the boundaries of the self makes it all too easy to
lose the self in others. More often than not that absorption of self poses a
moral problem, as much as or more than does the individual self.

Our fundamental and overwhelming competitive and self-serving desires,
rather than expressing themselves primarily in a dog-eat-dog endless fray, as
we tend to presume, actually play out via what neuroscientists are now identifying as our *extended self* and our *shared* (or *distributed*) *self*. By the extended
self I mean all the tools and other things we use to do what we do, from
screwdrivers to cars. By the shared or distributed self I mean our social biographical, historical, and cultural selves; these are the social identifications
and situations that the self projects itself into and pursues as self. Yet they
can lead to the loss of self in and to the group. The extension or expansion of
the self to include its tools and the like and the distribution or extrusion
of the self into the groups and projects to which one is committed are intertwined, however, as they play out: introjection is followed by self-extrusion
and further introjection, and on and on. It is only in theory that we can
separate the strands and untangle them for clarity's sake.

When we put together the vastness of possible interpretations, the extraordinary degree of human cooperation and sociality, and the self beyond
itself, we find moral danger as well as extraordinary promise. It was the
danger of situation and social identity (the self beyond itself), rather than of
individual character, as we tend to assume, that was so devastating in Nazi
Germany and even in the Stanford Prison Experiment and more recently in
the Abu Ghraib torture cases. Neuroplasticity enables, in an extraordinarily
variable way, the social shaping of the individual, and the channeling of
individual self-servingness (what I like to call, tongue in cheek, *selfiness*) toward group purposes. What this means is that our sociality and empathy,
but also our basic sense of self, self-protection, self-furthering, and even self-
servingness, are all filtered through, shaped, and channeled by our complex
world of social relationships, institutions, power relations, and group meanings. So what we human beings bring to the table, I think, turns out to

hinder the smooth and direct functioning of empathy, reciprocity, and the like. The vast neuroplasticity of the human neocortex that gives us the capacity for imagining all kinds of behavior, and especially for believing almost anything about that behavior, can enable an almost infinite possible extension of the range and sensitivity of moral concern—but it can also lead us to skew and corrupt moral action toward selfish and group-selfish desires.

Openness: The Case for Neuroplasticity

The brain changes with anything you do including any thought you might have.

—ALVARO PASCUAL-LEONE

Psychiatrist and researcher Norman Doidge offers in his book *The Brain That Changes Itself* an overview and history to date of the emerging revolution in neuroplasticity—the discovery of the lifelong adaptability of the brain. Discoveries in the neurosciences have opened windows into how the brain can be trained to rewire and reorganize itself throughout life. One review calls Doidge's book "an owner's manual for the brain."[14] And *Publishers Weekly* remarks that Doidge "turns everything we thought we knew about the brain upside down."[15]

For the past four hundred years the brain has been regarded by scientists and neurologists as more or less fixed: it has been thought to be unchangeable after childhood. Brain cells were held to be nonreplaceable, nor was the structure of the brain thought to be capable of change. The common experience that patients with brain damage rarely recovered completely seemed to make the unchangeability of the brain obvious to physician and laypeople alike. The presumption was that the brain is basically a machine that is hardwired and has permanently connected circuits that either work or don't work, each circuit performing a specific function.[16] However, scientists are now able to conduct experiments in which they use neuroimaging technology to observe the brain as it functions, and this has led to rethinking all those long-standing assumptions. Doidge says that in the late 1960s and early 1970s brain scientists unexpectedly began coming up with evidence that "the brain changed its very structure with each different activity it performed, perfecting its circuits so it was better suited to the task at hand. If certain 'parts' failed, then other parts could sometimes take over." This was the origin of the discovery of neuroplasticity, so named because neurons (nerve cells) were, unexpectedly, exhibiting modifiability.[17] Scientists were showing that children could sometimes improve the mental abilities they were born with; that damaged brains could reorganize themselves to side-step areas that were impaired and recruit other areas to perform the function of the damaged area; and, perhaps most surprising of all, it was discovered that "thinking, learning, and acting can turn our genes on or off, thus shaping our brain anatomy and our behavior." Doidge calls this last "surely one of the most extraordinary discoveries of the twentieth century."

Sensory Substitution

The first presumption of the standard model of the brain that Doidge challenges is the fixed and specialized localization of functions. He calls this the challenge to "localizationism," a belief that he points out still has many adherents. "A brain that is hardwired, and in which each mental function has a strict location, leaves little room for plasticity," he remarks.[18] The scientist and rehabilitation physician Paul Bach-y-Rita discovered that the wiring of our senses, rather than being completely specific, can be recruited to fill in to some degree a missing or destroyed sense. Bach-y-Rita called this process "sensory substitution." Bach-y-Rita recounted the story of a patient, Cheryl Schiltz, whose vestibular system (which maintains bodily balance and gives us a sense of location in space) was seriously damaged as a side effect of an antibiotic. Schiltz had the continuous feeling that she was falling, and she could not focus on objects. Tests revealed that she retained only 2 percent of her original ability to maintain balance and spatial orientation. Bach-y-Rita and his lab created a device that translated location and movement of the body onto the tongue, which in turn sent location signals to the brain. When Schiltz put on the device she not only felt normal for the first time in five years, but the residual effect lasted for twenty minutes after she removed the device. Eventually, after about a year of sessions in which she used the device, Cheryl no longer needed it at all—her brain had rewired itself. Bach-y-Rita described Cheryl's rehabilitation process as the "unmasking" of "secondary" neural pathways. This is different from the rewiring of the brain and can happen much more quickly because these pathways are there but unused. He described it using the analogy of driving: if your standard way of getting from point A to point B is blocked by the collapse of a bridge, you might initially take secondary roads that are not as direct and make the trip longer. Over time, however, you find shorter pathways. This unmasking of secondary pathways, Doidge reports, is "one of the main ways the plastic brain reorganizes itself."[19]

Perhaps one of the factors that enabled Bach-y-Rita to think outside the box is that he had extraordinarily interdisciplinary interests and expertise: medicine, pharmacology, ocular neurophysiology (the study of the eye muscles), visual neurophysiology (the study of sight and the nervous system), and biomedical engineering. At the age of forty-four he returned to medicine after a research career to do a residency in rehabilitation medicine in order to apply more directly his discoveries in neuroplasticity.[20] His concern was drawn to both rehabilitation and the plasticity of the brain when his father had a disabling stroke in 1959, from which neurologists at the time said he would never recover. Bach-y-Rita's brother took over care of their father and began, in a seat-of-the-pants way, to develop a series of exercises to teach him to crawl and then to walk, to exercise his weaker side and hand, and to speak. Their father improved enough in the course of a year to resume work, and he eventually recovered all his functions, yet an autopsy performed on him after his death showed that the brain damage from the stroke was catastrophic. Clearly his brain had rewired itself as a result of the rehabilitative exercises.[21]

Bach-y-Rita developed a device that enables "sight" in people whose damaged retinas had made them blind from birth. That device was reported on in the journal *Nature* in 1969. It was (at the time) an enormous device that a blind person sat in and which sent electrical signals of the image of scenes or words to a computer that then conveyed vibrations to the skin. "The stimulators functioned like pixels vibrating for the dark part of a scene and holding still for the brighter shades." Bach-y-Rita conjectured that skin could substitute for retinas as receptors because both are two-dimensional sheets with sensory receptors upon which pictures form.[22] Six blind experimental subjects were able to make out objects, see faces, perceive perspective, and learn to recognize common objects. They went from having a tactile experience to "seeing" things, albeit with fairly poor resolution.[23] As Bach-y-Rita said, "We see with our brains, not our eyes"—while our eyes sense changes in light, it is our brains that perceive.[24] Bach-y-Rita was one of the first scientists to reject localization of the senses and demonstrate the possibility of "sensory substitution." He considered the brain as "polysensory": the nerves are capable of processing electrical signal patterns from multiple senses rather than operating as discrete sensory units. "All our sense receptors translate different kinds of energy from the external world . . . into electrical patterns. . . . These electrical patterns are the universal language 'spoken' inside the brain—there are no visual images, sounds, smells, or feelings moving inside our neurons." The areas of the brain that process patterns of electrical signals, which are sent down the nerves, are all far more similar than scientists have generally realized.[25] Recent experiments on animals have confirmed the plasticity of the brain's sensory processors. And the late Susan Hurley, one of the foremost philosophers to rethink the mind in response to the neurosciences, argued persuasively that it is the body's activities in a sensory mode (looking around for vision, touching things for touch, etc.) that trigger and engage the rewiring of neurons usually dedicated for one purpose to another.[26]

Retraining and Rewiring the Brain to Overcome Cognitive Deficits

Doidge's second case is that of a woman, Barbara Arrowsmith Young, who was labeled "retarded" as a child although she had some remarkable abilities as well as cognitive impairments. Some of her difficulties were with connecting behavior with its consequences, with understanding logic and grammar, and with comprehending cause and effect; she could not visualize space mentally; she was awkward; she could pronounce words only with difficulty; she could put only a few letters together in reading. She could not decode a TV program in real time. On the other hand, her memory was spectacular. Her observations of children were wonderful and insightful and led to a career in education. Her determination was extraordinary. She was able to overcome much through sheer doggedness and memory. (She would read a paper or a book twenty times over to be able to get it.) She completed college and went on to grad school in education where, in her twenties, she discovered the work of Aleksander Luria, the founder of neuropsychology, about the cognitive and other deficits of stroke and war victims. In Luria's work she found descriptions of the precise brain functions that might be missing or impaired in herself. At this point Young came upon the research of Mark Rosenzweig on neuroplasticity in rats. Rats that had had more stimulating environments—objects to explore, toys to play with, and climbing and other devices to exercise on—had more neurotransmitters, heavier brains, a greater blood supply in the brain, and more neural connections than genetically identical animals reared in impoverished environments, and they were more capable of learning. Doidge says that Rosenzweig's research demonstrated that "stimulating the brain makes it grow in every conceivable way." These kinds of effects on brain anatomy have been shown in all kinds of animals, and evidence from human autopsies has shown that the same holds true for people: for those with higher education, and proportionately as education increases, the number of branches among neurons increases, as do the volume and thickness of the brain.[27]

Rosenzweig's discoveries sparked Young to begin on a self-directed program of mental exercises to retrain her brain in just those areas of cognitive deficits that she could now identify. It worked—she overcame her mental deficits and no longer suffers from them. She and a fellow student named Joshua Cohen, who also had had cognitive deficits and who eventually became her husband, expanded their techniques of retraining the mind, and in 1980 they founded the Arrowsmith School in Toronto, where they develop methods to help rewire the brains of children with various learning and cognitive deficits.[28]

The Evidence Against Brain Modules and for Flexibility and Openness

The conventional view sees the brain as made up of a group of specialized processing modules, genetically hardwired to perform specific functions and those alone, each developed and refined over millions of years of evolution. Once one of them is damaged, it can't be replaced.

—NORMAN DOIDGE

The idea that the brain can change its own structure and function through thought and activity is, I believe, the most important alteration in our view of the brain since we first sketched out its basic anatomy and the workings of its basic component, the neuron.

—NORMAN DOIDGE

The evidence for the astounding neuroplasticity of the brain challenges long-standing assumptions about the localization of functions and mechanisms in the brain. Those standard assumptions envision the brain as consisting of "modules," enclosed capacities that are not open to modification from other modules and are unable to take over other modules' tasks. So an ethical module, if there were such a thing, would be a uniquely dedicated brain capacity not capable of being influenced by other capacities—such as emotions, for example. On this view, the discrete module would underlie all ethical behavior. It is this standard modular account of higher capacities that neuroscientist Jaak Panksepp gives a knockout blow to in an article that has attracted quite a lot of attention. In "The Seven Sins of Evolutionary Psychology," Jaak and Jules Panksepp charge many who advocate the standard modular view, and especially "evolutionary psychologists," with irresponsibly attributing all kinds of human behavior directly to hardwired genetic modular mechanisms.[29] To make the claim of modularity goes way ahead of the available evidence, the Panksepps say. They argue that those favoring a modular view of the brain largely ignore the available and growing evidence about the actual neural mechanisms operative in behavior. "Real neural functions across a variety of species should provide definitive constraints on speculation about what evolution did or did not create within human and animal brain/minds," they say.[30]

No modular mechanisms of behavior, the Panksepps point out, have ever been shown to be a product of evolutionary adaptation of the human neocortex when it underwent massive development in the Pleistocene. There are modular emotional and motivational capacities in subcortical (more primitive) brain regions that humans share with other mammals, but there is no current evidence for discrete special-purpose mechanisms and functions in the human neocortex. Instead, the higher regions of the human brain appear at birth to consist "largely [in] general-purpose computational devices." This does not mean, however, that the neocortex of the human brain is like a digital computer. Instead, its organic brain processes are "spontaneous, self-organizing, and non-linear."[31] Regions of the neocortex, nevertheless, may acquire "special-purpose functions" as a result of life experiences.[32] So any special modularized functions in the human neocortical regions of the brain would be softwired, a result of the interactions of the subcortical regions with experience. Enter neuroplasticity. The Panksepps point out that it is not clear yet whether even language emerged from specifically human evolutionary adaptations or instead from the great growth in general-purpose symbolic processing that enabled more of the brain's capacity to be used for language. The long-standing view that language is a discrete and localized capacity is no longer warranted and has been crumbling.

Psychologists, and especially evolutionary psychologists, some engaged in trying to discover the source of the human ethical capacity, have been all too wedded to a model of the mind as a computer and are loath to give it up despite mounting evidence to the contrary. The commitment to a vision of the brain as a computer has made the discoveries in neuroscience of the actual embodied mechanisms seem all but irrelevant. Modularity is exactly the way a computer works, but, it turns out, it is not the way the neocortex of the human brain works. Modularity both assumes and mandates a much less flexible, less dynamic kind of system. It's true that both computers and the human brain utilize "action potentials," but that, the Panksepps point out, is a "surface similarity."[33] Much of what has been thought to be modular "may simply reflect our multi-modal capacity to conceptualize world events symbolically and to relate them to primitive affective feelings that reflect specific fitness concerns."[34] So it is due to the absence of modularity and the presence of neuroplasticity that "ancient emotional systems are able to imbue 'cold' perceptions with hot affective charge."[35] Thinking and beliefs are filled with emotion; thought and affect interpenetrate each other. "Creatures like ourselves have been endowed with a massive random-access type of general-purpose intelligence. To some yet unfathomed extent, we have been liberated from the crucible of a mindless biological emergence."[36] So "everything that mature humans do is filtered through their higher neural capacity for flexible 'intelligent' action. . . . The nature of the higher regulatory system in humans does permit many alternative courses of action."[37] This is the source of the real and vast openness in human behavior. We now turn to how plastically and changeably and ever responsively the brain maps the body.

The Plasticity of Brain Maps

The neuroscientist Michael Merzenich has been considered the world's leading researcher on brain plasticity, having worked extensively on the problem of "brain maps."[38] Before Merzenich, research on brain maps began with the idea that areas in the brain could be found that represent parts of the body and its specific actions. Early experiments with electrodes showed that electrically stimulating a specific area of the brain during brain surgery led to the same feeling in the patient's hand as touching the hand. That was true of other correlated brain-body areas as well. The brain had a sensory map of the parts of the body's surface. So too with movement—stimulating specific brain areas that control movement could trigger specific movements such as those we initiate when we move our hands or feet or other body parts. And adjacent areas of the body were discovered to be adjacent on the brain maps.

These brain maps were thought to be universal in a species and immutable. They seemed to lock in localization (and perhaps even point to hard-wiring) of brain function—but Merzenich discovered otherwise.[39] For he came to brain mapping research just at the moment of the invention of a new technique, micromapping, that made possible the investigation of the firing of individual neurons as communication to other neurons. It was at this level of specificity that neuroscientists were now capable of peering into the brain's operation and mapping. There are 100 billion neurons in the human adult brain, and a neuron firing lasts about one-thousandth of a second. Many fire at the same time. Just think of how many possible combinations of neurons firing together that makes possible. That's the source of the vast flexibility, the openness, of human learning, thinking, and behavior.

A basic description of the nervous system will help us understand what Merzenich's discoveries are really about. The nervous system is divided into the central nervous system, which consists of the brain and the spinal cord, and the peripheral nervous system. The central nervous system commands and controls, whereas the peripheral nervous system relays messages between the central nervous system and the rest of the body. The nervous system as a whole monitors both internal and external environments and responds to them. The peripheral nervous system had been known for a long time to be plastic—a cut nerve in a person's hand, for example, can regenerate itself. But the central nervous system had long been thought to lack that capacity and to be set.[40] Merzenich and colleagues did experiments with cutting a peripheral nerve to the hands of monkeys and sewing the two severed ends near but not touching each other. They anticipated that this would cause havoc in the brain maps of these monkeys since the signals back to the brain would be confused. What they discovered, however, was that the monkeys' brain maps had accommodated themselves to the changes and were roughly normal. They had discovered neuroplasticity in the central nervous system.

In a later experiment on monkeys in which he cut the median nerve that conveys sensation from the middle of the hand, he found that the nerves that convey the sensations of the outsides of the hand had taken over conveying the sensations of the middle. Looking at the brain maps of these monkeys, he discovered that the brain maps for these two outside sections had nearly doubled in size and taken over the function of the disabled middle nerve. The unused median nerve space was taken over by the two other nerves, which had invaded its map space and was using its capacity to process their inputs. Map space was apparently valuable and functions competed to get more processing capacity. "Use it or lose it" seemed to be the rule—competitive plasticity explains a great deal of why we come to excel at some things but are weak in little-used areas and capacities. It also explains how bad habits can take over lots of map space and be hard to dislodge.[41]

The maps, Merzenich now surmised, were arranged topographically, which is to say, the map had its own internal organization that is like that of the body, but there is not a point-to-point correspondence between the body and the brain map. That is, brain maps were not like the old telephone wiring systems, in which operators plugged in wires that connected one person directly to the person he or she was calling. Maps were also constantly changing to reflect changes in the body and activity. Merzenich was showing that "maps could alter their borders and location and change their functions well into adulthood."[42] Maps that could reorganize themselves in response to changes in the body, Merzenich conjectured, would confer an incredible evolutionary advantage.[43]

Neural maps also change through experience. In 1949 the Canadian behavioral psychologist Donald O. Hebb proposed that when neurons fire at the same time repeatedly, the two develop a strong connection; the more they fire together, the stronger the connection. Here is a principle of neuroplasticity that wasn't just about sensory capacities but broadly applicable to all kinds of learning and experience. The opposite principle was also demonstrated to hold. While Merzenich was doing his doctorate at Johns Hopkins in the early 1960s, some other scientists in the same group discovered that the brains of animals had a "critical period" of extraordinary plasticity in which they were shaped by early experience. Newborn kittens, for example, had to have visual experience in the first few weeks of life or they would be blind. If a kitten had one eye closed by the researcher and one eye open, the visual map of the closed eye would not develop and the cat would be blind in that eye for life. We recall the "imprinting" experiments of Konrad Lorenz in which ducks brought up without a mother in a critical early period bonded with a person as a mother substitute. And Freudian theory is all about critical periods of openness in which a certain development must take place for psychological normalcy. The kitten experiments revealed something else: the parts of the brain map that would have been operative in relation to the closed eye became connected and functionally contributory to the good eye.[44] That there is a critical period of plasticity in infancy when the brain is in a state of rapid development began to be accepted. Merzenich, however, began to discover adult plasticity as well.

Merzenich wondered how the topographic order of the maps emerged. He found that repeating sequences of movements in a fixed order link certain functions that are then mapped together. Often these movements are of contiguous areas of the body, such as the fingers of a hand. So the thumb and the index finger are closely mapped because they are repeatedly used together as we grasp things, for example. Another example: low notes tend to occur together, so low sounds are mapped together, as are high sounds. Merzenich and his colleagues developed experiments in which they taught monkeys new skills in order to see what would happen to their neural maps as a result. One experiment had a monkey learn how to spin a disk in a very nuanced way that involved using its fingertips with great sensitivity in order to get a reward. The monkey had to pay very close attention to reproduce the exact spin the researchers demanded. The result was that the area of the monkey's brain map devoted to fingertips expanded as it learned to exert just the right amount of pressure. A similar effect was also demonstrated for various cognitive and sensory skills such as the recognition of sounds. It turned out that both paying close attention and repetition were crucial factors in initiating and maintaining plastic change. Initial close attention predicted long-term change, and without that initial concentration there was a failure to retain much change in the brain maps.

An even more interesting result was that as the neurons became more efficient from practicing the task, the expansion stopped and eventually began to recede, for at that point fewer neurons were needed to perform the task. The researchers discovered from this and other experiments like it that when neurons are trained through practice and become more efficient, they also process faster and use less brain space. Hence there is plasticity in the very speed in which actions are performed.

Merzenich regards the brain as "structured by its constant collaboration with the world." All kinds of experiences—sensory, motor, cognitive, emotional—shape and reshape the different functions of the brain. While the cortex has more neurons than other parts of the brain and may be more plastic, plasticity is nevertheless a capacity of all brain tissue, Doidge holds—even the amygdala (a seat of emotion) and the hippocampus (which controls the conversion of short-term memories to long-term ones). "Research has shown that neuroplasticity is neither ghettoized within certain departments in the brain nor confined to the sensory, motor, and cognitive processing areas," Doidge writes. And Merzenich argues, "You cannot have plasticity in isolation. . . . It's an impossibility." His research has demonstrated that if one brain system changes, the systems connected to it also change. This is because brain systems function together.[45] "Thinking, learning, and acting actually change both the brain's physical structure (anatomy) and functional organization (physiology) from top to bottom."[46]

But all this speaks to the dark side of neuroplasticity as well as its tremendous potential to be used for human benefit. The dark side includes addictions, obsessive-compulsive behavior, and various other kinds of resistance to change, both cognitive and behavioral. We train ourselves in bad habits through repetition and intensity of focus, and the environment can lock us into patterns of behavior and meaning. Now we can better understand groupthink and the tyranny and danger of habitual interpretations of the world. This is why free will is so deceptive and otherworldly an ideal. It's not possible for creatures like us. We can't stand above our brains and rewire them by hand or by will. We don't choose our beliefs and actions; instead, we *are* those actions and beliefs because they are written into our brains. We have written them into our brains, not by will but by experience and unconscious osmosis, so to speak. And the neural maps are also one of the sources of the self and also of a feeling of self, as I will discuss below.

Retraining and intense rethinking, however, can give us some freedom from the tyranny of our deeply ingrained patterns of engagement with the environment, and even help us gain new perspectives on ourselves. Now we may begin to have a sense of where freedom lies. Individually, it is in coming to understand how our brains work and seeking interventions that reshape our habits and broaden our thinking. On the societal scale, our freedom lies in developing institutions and cultural beliefs and practices and families that shape our brains toward the broadest good rather than toward narrow interests, and toward health rather than addictive habits and other limitations, starting early in life. We can use the growing understanding of how the brain works to reshape our environments—and thus develop contexts and institutions that foster habits of thought and behavior that further both the personal and the general benefit—and we as individuals can also use that knowledge to broaden our thinking about the world and ourselves within it by intensive and enhanced learning.

The Neuroplasticity of Thinking

Alvaro Pascual-Leone, chief of Harvard Medical School's Beth Israel Deaconess Medical Center, was the first to use transcranial magnetic stimulation

(TMS) to investigate brain maps.[47] The TMS emitting device can directly make neurons fire in the brain and hence directly influence behavior; TMS can also be used to block temporarily an area of the brain from functioning so that the precise function of a particular brain area can be identified. TMS can also be used therapeutically, for example, to excite areas of the brain that are inhibited in depression. Pascual-Leone has identified what he believes are two neuroplastic pathways, one that brings about a short-term change and another, long-term one that consolidates change. Early in his career he studied the brain maps of blind subjects learning to read Braille and found that the neural motor maps for the Braille reading fingers became larger as the subjects learned to read. The learning occurred during the week from Monday to Friday and then subjects had weekends off. By Friday the subjects' maps were much larger than on Monday, but by the end of the weekend, the maps had returned to normal size. What was happening? The size of the Monday maps started to change only after six months had gone by. The Monday-to-Friday map changes, Pascual-Leone surmised, were dramatic short-term changes that strengthened existing neural networks, whereas the long-term changes evident at six months were evidence of "brand-new structures, probably the sprouting of new neuronal connections and synapses."[48] Doidge suggests that this two-phase process is probably why we can quickly learn something by cramming for an exam but then forget it just as quickly. Enduring change, truly mastering new skills, takes longer. (In his book *Outliers*, Malcolm Gladwell argues that becoming an expert at anything whatsoever, from piano playing to sports to computer programming to being a rock star, takes almost invariably ten thousand hours or about ten years of intensively focused work. Not that putting in ten thousand hours necessarily makes you a star; it's a necessary but not sufficient condition for expertise. There's also a lot of luck, privilege, and social and cultural context involved, according to Gladwell. It's a matter not of will but of opportunities taken.)[49]

After his work on the motor and sensory cortex, Pascual-Leone turned to investigate the way thoughts change the brain. His research involved two groups who learned a piano piece: one group practiced the piece on the piano for two hours per day for five days, and the other group merely sat in front of a piano and imagined playing the piece in their minds for the same length of time. Mental practice, it turned out, produced the same physical changes in the motor system of the brain maps as the actual practice. Both groups played the music equally well at the end. This finding and those like it are enabling the development of devices to help the paralyzed regain function by connecting their thoughts to technological products that can replace their paralyzed limbs, for example. They can learn to move devices with their thoughts, it turns out, by going through the exact thinking sequence that moving the limb would entail and then linking that sequence to a computer that recognizes it and is attached to a device that can produce the motion. Experiments with rats have borne out the feasibility of such an invention. It worked with rats: through a series of steps rats learned to merely imagine their paws pressing a bar for water, and a computer that recognized the neural pattern that precedes the action delivered them water.[50] Thinking is not a passive process, as we tend to assume, nor is it emotionally detached. It would appear to be as engaged and active in terms of our brains as learning a piano piece through mental practice.

Thinking is a form of action, it turns out. Pascual-Leone likens our brain to Play-Doh. "I imagine," he says, "that the brain activity is like Play-Doh one is playing with all the time." Everything we do changes the shape of the Play-Doh.[51] Neural pathways are always being laid down or strengthened. It is these stable patterns that give us a sense of identity as the same person over time. We are our brain maps and our brain mapping.[52]

The Learning Brain

When we learn, the structure of neurons is changed, and synaptic connections are formed and re-formed. These findings won Eric Kandel, Norman Doidge's mentor at Columbia, the Nobel Prize in 2000. Subsequent research by Kandel and others showed that learning alters which genes in our neurons are expressed or turned on and which are not.[53] This is a process that is implicated in psychotherapy as well as other kinds of learning, according to Kandel. In a way this turns our standard assumptions about the direction of genetic causality on its head: not only do genes cause behavior, but behavior influences how our genes operate. Nature and nurture are far more deeply interwoven and mutually implicated than we had ever imagined. There are thirty billion neurons in the human cortex, which means that our brains have the capacity to make one million billion synaptic connections; the possible neural circuits made up of these connections reaches a number impossible to even grasp: 10 followed by a million zeroes![54] We need to begin to come to grips with the extraordinary vastness of our human neuroplasticity, of the extraordinary openness of the range of our possible thinking and behavior. That ungraspably enormous range is to think bio-logically about ourselves as a species. To think biologically about our hu-manness is not to invent narrow, evolutionary psychological tales about who we are as a species and how we must therefore act—for example, that we are inevitably promiscuous or violent or monogamous or hierarchical. Our neu-roplasticity blasts open our possible ways of being in the world to propor-tions unfathomable and, for all practical purposes, infinite.

Neuroplasticity and Cognitive Framing: The Promise and the Danger

> We are neural beings. . . . Our brains take their inputs from the rest of our bodies. What our bodies are like and how they function in the world thus structures the very concepts we can use to think.
>
> —GEORGE LAKOFF

> What the world is like can be part of what enables us to experience what it is like. . . . Evolution has no reason of principle to respect the [boundaries of the] skin in enabling experience.
>
> —SUSAN HURLEY

Given the unimaginable vastness of neuroplasticity, how do we become distinctly who we are rather than mere harborers of infinite possible actions, beliefs, and selves? With the kind of openness that constitutes us, how can we ever decide on anything? Become a stable and recognizable personality? Have any continuity across time? If we, our "selves," are our brain maps, how do these maps get drawn up and pinned down? In what ways do they remain open as well as set? How do our experiences and interactions and engage-ments in the world—our family, where we were born and grew up, our auto-biography, our place in the biographies of other people who were close to us and in the larger immediate and historical world, the great historical events of our lifetime, our roles in the family and at school and in all kinds of insti-tutions and groups, and on and on—come to be pathways in the neuroplas-ticity of our brain maps?

A significant part of the answer to this question is what cognitive scien-tists call *cognitive framing*.[55] Our minds are not receptacles of isolated pieces of information but rather built of cognitive frames through which we organize and interpret our experience and the world around. These frames are largely unconscious—that is, we are unaware of how they arose, we are aware (or conscious) of them in the use of them, and we can become self-consciously aware of them as well. They are also emotionally laden; we absorb them from the cultural, familial, institutional, and historical environments we grow up in. George Lakoff, a linguist and cognitive scientist at the University of California at Berkeley, has been one of the foremost contributors to rethinking language and philosophy in the light of findings in the cognitive and neurosciences; he has also applied these new fields to rethinking politics and political language. Our lives, and the stories we tell about them, shape the basic beliefs through which we perceive our environments. "Our brains and minds work to impose a specific understanding on reality," Lakoff writes, so "not everyone understands reality in the same way."[56]

Lakoff points out that language uses "frames, prototypes, metaphors, narratives, images, and emotions" in shaping our understandings or interpretations of ourselves and our contexts. But language, he warns, "is the surface, not the soul, of the brain."[57] Deeper things are going on that get expressed and modified in and by language. For example, the unconscious character of 98 percent of our thinking (as the cognitive psychologist Michael Gazzaniga and others estimate) gets expressed in covert ways via language. The emotional ladenness of our thinking gets expressed in and through language. The narrative character of life or of experience generally—that our experiences have a pattern of beginning, middle, and end that we impose upon them—gets expressed in language. We use metaphors based in bodily experience to conceive ourselves and our lives, and this shapes language.

Lakoff tells us that cognitive frames are "scripts" that "are among the cognitive structures we think with" and that "frames tend to structure a huge amount of our thought."[58] An illustration he offers is of a murder mystery, a frame we all know. There is a murder, a detective, a process of investigation and discovery, the catching of the murderer, and so on. So "each frame has roles (like a cast of characters), relations between the roles, and scenarios carried out by those playing the roles."[59] And such frames even provide the underlying structure of key institutions. We all know, for example, that in the United States, public schools are regionally distributed, have teachers, a principal, janitors, classes generally divided according to age, sports activities, school colors, and summer vacations. The school frame enables us to grasp any given school quickly and know generally what it's about, how it operates, and what the basic roles within its structure are. Events also have cognitive frames.[60] Take a baseball game, for example: we know it has innings and teams, we know the arc of how the game progresses, we know how a game is won, and we know the rules that govern the process to completion. We can tell the story of the Red Sox winning the World Series for the first time in eighty-six years in 2004 because we have in our minds, especially as Americans, the general cognitive frame of how baseball games work.

Even simple actions that we perform, such as taking a drink of water or tying our shoes, are dependent upon event structures that transpire over time. These are simple narratives or stories about first lifting your arm to the sink, turning on the tap, and so forth.[61] It turns out that the same narrative structures are engaged both when we perform a particular action or live out a narrative and also when we understand someone else's action or series of actions or hear their story.[62] So our mind, through its cognitive frames, is a social mind both in its comprehension of its environment and in its active engagement with that environment. It is also broadly cultural and historical as well as autobiographically nuanced, which involves a smaller social arena of history, culture, and location. We *are* our cognitive frames, so we have selfiness (feelings of ownership and desire for their furtherance) toward them. Those cognitive frames are also "groupy"; that is, they arise in all kinds of socially and historically shaped contexts and express shared under-standings that we grasp and enact. It is via cognitive frames that basic social mammalian moral feelings and tendencies are embedded and played out—and it is within these frames that they arise, for that matter.

Cognitive framing exposes how neuroplasticity, at the very least, links meanings to each other and to context and to our individual biographical selves. We don't have words all jumbled in our minds or objects in little cubbyholes. We don't pick out isolated words or isolated objects to create scenes and meaning. Instead, we draw on highly contextualized, overlapping meanings, language, and experience frames when we think, talk, learn, and encounter new experiences. We constantly enrich our neural maps by filter-ing our experience through interpretive frames, which in turn are modified in the process. Lakoff points out that the evolutionarily oldest part of the brain, the limbic system, one of the brain's basic emotional regions, has two emotional circuits, one producing positive emotions and the other negative emotions, and there are pathways in the brain linking these circuits to the forebrain, where it is thought that cognitive framing and interpretation take place, binding rapid and unconscious emotional reactions to narrative sequences. This "allow[s] the right emotions to go where they should in a story. They are the binding circuits responsible for the emotional content of everyday experiences," Lakoff says. "Narratives and frames are not just brain structures with intellectual content," he continues, "but rather with integrated intellectual-emotional content."[63] The highly emotional quality of cognitive frames makes them feel very personally owned. They are *selfy*. They offer and anchor a personal point of view.

We all know the rags-to-riches narrative frame. That frame has special cultural resonance in the United States, as does the political narrative frame of being born in a log cabin and rising to become president. These are two special narrative frames that evoke "cultural prototypes, themes, images, and icons" and, as a result, powerful emotions and national group ideals and attachments.[64] Other examples of narrative frames that have special American cultural resonance and attachment involve the reinvention of the self and the redemption of the self. Bill Clinton's "Comeback Kid" nickname exemplified the redemption story as well as the born-in-a-log-cabin and rags-to-riches stories. George W. Bush's dramatic turning away from alcoholism in his forties evokes the redemption and reinvention stories; he had to play down the born-with-a-silver-spoon story and invoke the just-a-regular-guy-to-go-have-a-beer-with story to get the American people to feel he was presidential material; by contrast, John Kerry didn't evoke that story and he didn't play down the silver-spoon story enough to appeal to Americans' sense of collective ideals and identity. Who we are, or who we think we are, and the narratives and images that evoke those senses of self have enormous emotional power to shape our choices and decisions—and they do so beneath the level of conscious and rational thought.

The stories of people's lives can be told in a vast number of ways. In *Outliers*, Malcolm Gladwell chronicles the rise of the very famous and highly successful, beginning with the standard cultural narrative of great success as the result of special talent and even genius. But Gladwell seeks to undermine this conventional narrative, retelling the stories of software billionaires like Bill Gates and Steve Jobs, renowned Canadian hockey players, and even the Beatles, as stories that reveal success as the product of extraordinary opportunities encountering cultural and familial legacies. Those who were lucky enough to have both and who could and did take advantage of them through obsessive focus and work succeeded in a big way. Gladwell's purpose in reframing how we think about success is to induce in us changes in our cognitive framing of success—to change the stories we tell ourselves and give as explanations of success, decisions, and social and cultural trends.

This is possible because of neuroplasticity. For example, he wants us to think that extraordinary focus and hard work are more important and effective in success than any natural talent or even genius. The neuroplasticity of cognitive frames suggests that the best intervention to get people to change their behavior may be to change the way they see the problem. It's surely a necessary first step. Now we can understand why Republicans and Democrats spar endlessly over the way problems are defined. Are welfare recipients cheats, or are they deserving poor humbled by circumstances we could all fall into, such as losing a job or becoming ill or losing health insurance? Is buying one's own health insurance a way of standing tall, or is it the insurance industry and the pharmaceutical companies exploiting the vulnerable and sick for profit? Which story wins out over the other in these cases determines a lot of our future. Narrative frames are in competition, and embedded within them are personal and group interests. They are neither neutral nor innocent.

Lakoff holds that there are three types of cognitive frames: universal ones, group and cultural cognitive ones, and individual and highly personal ones. These, of course, all intersect and interact in complex ways within us. But the personal one is perhaps the most emotionally charged and motivational for us. It is, Lakoff says, a "personal narrative," a life story that is at the center of who we are. Each of us has an overall cognitive frame that is our life story; that phenomenon is the most up-to-date way that psychologists define personality, according to Lakoff. We live out and from our personal stories. But these stories are largely implicit, an unconscious background that structures who we are and where we are going. "The roles in narratives that you understand yourself as fitting give meaning to your life, including the emotional color that is inherent in narrative structures," Lakoff writes.[65] The roles we implicitly assign ourselves (or have assigned to us by family and other contexts) include hero, victim, and helper, he says. So Lakoff proposes that a nurse may see himself as the helper to the hero-doctor. A president may see herself as hero-rescuer of a victim-nation. These are stories that we are not born with but born into. Lakoff points out that many deep narratives are activated together—personal, cultural, situational, familial—and that these complexes affect how we fit ourselves into social structures, how we understand our world and place ourselves within its patterns and meanings. These patterns of simultaneous activation of cognitive frames become fixed in our neural pathways. They unconsciously shape our brains, selves, and worlds. They enable us to understand our environments. They determine our choices.

Conclusion

The Greeks understood that moral character was not just about imbibing moral rules or even virtues but about understanding, rethinking, and re-imagining the trajectory of one's life. That the unexamined life is not worth living was Socrates's moral insight, the philosophical elaboration of the Delphic maxim to know thyself. For Aristotle, ethics was not just about acting according to the golden mean or the acquisition of discrete virtuous habits (as his theory is generally presented in America today) but also about developing the self-understanding and understanding of the human world—and even of the cosmos and of our place within it—needed to discover the ideal life and the best trajectory of a life for a human being. That best human life, in Aristotle's estimation, was the life of the mind, engaging in a lifetime of wide-ranging investigation and learning.[66] What could be more central to rethinking ethics than becoming aware of how our lives follow available narrative patterns gained from our biographies and social contexts and cultural meanings, which we also project into our futures and invoke to explain our pasts? Yet if these frames are largely unconscious (by which I mean implicit) social, cultural, and biographical interpretations that we enact rather than self-consciously construct, then ethical engagement has to take place at a different level from self-conscious deliberation about discrete moral decisions or principles. It is really something else entirely.

Moreover, cognitive frames that pin down interpretations personally and culturally must be understood against the background of our human enormously vast neuroplasticity, with its infinite possibilities for meaning and action. When we recognize that interpretation, belief, and understanding are so open-ended that they can take almost any form, from the most anchored and literal to almost infinite flights of complex metaphorical, symbolic, highly culturally specific, and technically specialized meanings, and that they can also be easily hijacked by self-serving cognitive framing, we see both promise and danger. Action, including ethical action, can be managed and directed via interpretation, which is to say at the cognitive level of belief and understanding and their accompanying emotions and motivations. How do we ensure that we embrace the cognitive frames of the saintly rescuers and not those of the Nazi perpetrators? If our cognitive frames, our well-worn paths of understanding, associated emotion, and action, are mostly implicit rather than available to self-conscious awareness, inherited rather than chosen, enacted rather than explicit, largely social and cultural and institutional rather than individual and self-made, how can we develop and ensure a reliable ethical life for ourselves, for our children, and for the society and world in which we live? The flexibility provided by neuroplasticity gives us hope. Intervention and change would seem possible, in fact, vastly possible. But how? It seems that ethical changes in our neuroplastic pathways must come principally from changes in social and cultural beliefs and practices. Yet is there also the possibility of developing a more individuated self, a self whose pathways of understanding, belief, motivation, and action differ from those of the crowd and are oriented toward a more independent moral sense and sensitivity? It is to these questions we now turn.

8

The Self in Itself: What We Can Learn from the New Brain Sciences About Our Sense of Self, Self-Protection, and Self-Furthering

Selfiness: The Urge

> The self is a unit, in the sense that organisms go to great pains to keep themselves alive and well. . . . But self-preservation is a universal motive, independent of whether an organism is aware that it is working toward this goal.
>
> —JOSEPH LEDOUX

> The ego is first and foremost a body ego.
>
> —SIGMUND FREUD

This chapter has as its theme the self. I recount here some of the salient and growing evidence of how self-promoting and self-serving we humans are as a species. I identify and provide lots of examples of the ways in which we exhibit the pervasive desire for self, so obvious and ubiquitous that we tend to take it for granted and thereby (ironically) render it almost invisible. Our selfiness, as I call self-interestedness, is better thought of as an urge, a process, a system, and a point of view, rather than as a thing. It is the urge to maintain, promote, protect, and further our embodied organic selves, body and mind. After flooding us with much bad news for ethics of our selfishness, manipulativeness, self-deception, hypocrisy, and the rest, I end the chapter with a light at the end of the tunnel, proposing a way that from all this self-absorption and self-focus we can nevertheless begin to discern the possibility of where a sense of self-awareness and responsibility might come from. But first the bad news: our selfiness.

In *A Mind of Its Own: How Your Brain Distorts and Deceives*, Cordelia
Fine presents with great humor a compendium of recent research that shows
the overall urge of each of us to protect ourselves, body but also mind. There
are multiple "protective layers encasing our self-esteem" as well as "a multi-
tude of strategies our brains use to keep our egos plump and self-satisfied,"
she writes.[1] And there is a vast and growing compendium of evidence that
the emotions, which are self-promoting and self-protective but hardly rationally so, drive and inform our thinking and beliefs. Fine puts her summary
assessment quite humorously, but the news is still devastating: "The truth of
the matter . . . is that your unscrupulous brain is entirely undeserving of
your confidence. It has some shifty habits that leave the truth distorted and
disguised. Your brain is vainglorious. It's emotional and immoral. It deludes
you. It is pigheaded, secretive, and weak-willed. Oh, and it's also bigoted."[2]
The evidence Fine has amassed and presents so entertainingly should make
not only philosophers but all the rest of us stop in our tracks. She does not
offer a solution for our quandary but lets us stew in the problem of where we
should turn for a reliable moral capacity (what the philosophers call "moral
agency") and what direction ethics ought to take to neutralize or work
around human self-deception, self-servingness, and corruption.[3]

Fine starts out by discussing what psychologists call the self-serving bias,
which leads us to cast ourselves in the most favorable light. She presents evidence for how the self-serving bias even enlists the two great capacities we
think will set us straight and give us the bare facts: memory and reason. For
example, we remember only good things about ourselves, whereas the negative feedback slips away. She remarks that "it is easier for a camel to pass
through the eye of a needle than for negative feedback to enter the kingdom
of memory." To illustrate the point Fine cites a study by two Princeton researchers who gave student subjects two fake scientific articles. One group
read an article that claimed research showed extroverts do better academi-
cally, while the other group read an article claiming that introverts do better. Sure enough, on subsequent self-assessments the first group brought to
mind memories that indicated to them that they were indeed extroverts, and
the second group brought up memories that indicated the opposite.

Reasoning works in a similar way, Fine says. Evidence favorable to oneself
and to one's opinions, habits, or tendencies is quickly accepted, while coun-
terevidence is brought under harsh and skeptical scrutiny, and most often
dismissed. This phenomenon has earned a title: it's called motivated skepticism. Fine cites another study in which experimental subjects were given

an article to read about the danger to health of coffee drinking to women but not to men. Sure enough, men and non-coffee-drinking women found the article extremely convincing, while women coffee drinkers were highly skeptical, finding the link between coffee drinking and health risks to women unconvincing. Fine succinctly puts it this way: "Judgment is swayed by desire." It's not just a nuisance at the margins but central to many aspects of our everyday lives.

In the chapter she calls "The Immoral Brain: The Terrible Toddler Within," Fine provides a great deal of evidence that we are all much like her two-year-old son: "The path of righteousness is as plain as day—it corresponds exactly to what my son wants." The evidence she provides suggests that young children and some higher mammals may even be at a moral advantage in some respects compared to most normal adult human beings, for it turns out that our big brains generally are brought in not to make an informed, judicious moral assessment but to rationalize and justify our self-serving gut responses we have already put into effect. "Thanks to the emotional brain's clever deception," Fine writes, "it normally seems—both to ourselves and others—that we engaged in our skillful cogitations before, rather than after, forming our moral verdict." Fine argues that two-year-olds turn out to be generally more honest in their self-valuations than adults:

> Currently, one of my son's favorite misdeeds is to roll his baby brother from his tummy onto his back. . . . What he does next starkly exposes his childish lack of understanding of grownup notions of right and wrong. He does not pretend that the baby deserved it, nor blame the baby for being so seductively rotund. He does not excuse himself by calling attention to his tender age. He makes no claim that other, less pliant toddlers would flip the baby over much more frequently. Nor does he even appear to consider suggesting that the baby's tears stem from joy, rather than shocked bewilderment at finding himself unexpectedly staring at the ceiling. Instead, he does something that no self-respecting adult brain would ever permit. He chastens himself with a cry of "Naughty Isaac!" and, with genuine humility, places himself in the naughty corner.[4]

What her two-year-old hasn't yet learned is how to excuse and justify bad behavior—a standard adult capacity and strategy, according to Fine, and she presents research on a number of cognitive and emotional mechanisms that drive pronounced tendencies in adults to engage in self-serving behavior and subsequent moral rationalization, rather than moral honesty, rational and conscious choice, and upright behavior.

The first psychological tendency she points to that influences, corrupts, and distorts moral judgment has to do with the unconscious influence of a person's overall mood or of a seemingly extraneous emotional state at the time of making an unrelated moral assessment. Moral judgment is not locked in a safe compartment but instead is porous to the overall state of the self and also to discrete events and circumstances. For example, experiments have shown that if a person is angry about something entirely unrelated (in this case, the subjects were exposed to a video displaying a brutal act of violence against a teen), his or her moral judgment in a hypothetical moral situation test administered immediately afterward becomes far more severe and punitive than that of control subjects who did not see the video.[5] Another source of moral misjudgment is a result of the need to believe we live in a just world and that our own good behavior will be rewarded, both socially and cosmically. It also serves us as a way to reassure ourselves that since we are good, no such bad thing could befall us. So we tend to blame the victim: he must have done something to deserve his sorry fate. We persuade ourselves, as Fine puts it, that bad things happen only to bad people. Or otherwise we, too, "through no fault of our own . . . might lose our job, our home, our health, our sanity, our child."[6] Fine goes on to cite an experiment in which volunteers, individually and consecutively, watched as a purported volunteer student, but actually a confederate of the psychological experimenter, was shown on a TV screen hooked up to electrodes that purported to deliver a painful shock whenever the student failed at a learning task. The confederate expresses her terror in anticipation of being shocked and when she is seemingly shocked she writhes in pain.

The experiment was conducted in several versions. In the first, the volunteers are told that the shocked student will receive no payment for her participation; in the second, she is to receive modest payment; and in the third, substantial payment. And in a final version, the researcher tells the confederate, while the volunteers look on via the TV screen, that the volunteers will not receive their monetary compensation if she does not complete the experiment. So in that version the confederate is seemingly martyring herself for the benefit of the volunteers. At the completion of all versions of the experiment, the volunteers were asked to rate the personality of the student who was being shocked. In all cases the volunteers saw exactly the same tape of the purported shocks administered to the confederate as punishment for learning mistakes. Only the background stories differed from one experimental version to the next. The results from the experiment are pretty depressing. It turned out that the volunteers' dislike for the student being shocked *increased* the less they thought she was being compensated. So the less just the experiment was, the more they disliked the victim. And the most disliked of all was the student whom the volunteers thought was martyring herself for their own financial benefit. *The less a person was thought to deserve the suffering, the more despised she was.* Rather than sympathizing and identifying with victims of undeserved suffering, there seems to be a tendency to despise them. If they are unworthy, then we don't have to help them, and anyway, it won't happen to us since we, unlike them, are the good people.

Another marked human tendency that Fine points to is a divide between our moral assessments of ourselves and our assessments of others. Experiments show that when we morally evaluate ourselves we give ourselves all kinds of excuses and refer to extenuating circumstances. But when we evaluate others, we are hard-nosed, not giving them any of the wiggle room we so gladly provide for ourselves. Our own failures are due to extenuating circumstances, whereas other people's are squarely their fault. We extend to those closest to us a similar benefit of the doubt, this "benevolent fog." Similarly, we exaggerate our own good intentions and ignore those of others. We judge ourselves by our good intentions but others by their actual actions. Perhaps most significant is the effect of hierarchy and social position upon moral judgment. "So powerful are the psychological pressures of a hierarchal environment," Fine writes, "that a co-pilot may sacrifice himself—as well as passengers and crew—rather than question his superior's authority."[7] The deference owed a superior can trump the realistic and truthful grasp of situations even when survival is at stake. Saving face trumps saving lives. (I will return to the effects of hierarchical social position on the objectivity of moral judgment in the next chapter.)

Fine calls the overall phenomenon she is pinpointing "the brain's narcissism." When it comes to ourselves we can't really believe that we are this morally frail—although we are perfectly willing to think other people are. This point was poignantly illustrated by an experiment with psychology students who had just been taught about the Milgram experiments and their extensions over the years. After hearing about those famous experiments, they were asked to predict how they would act in a Milgram situation. Even with all they knew about the pressures to comply, they overwhelmingly predicted of themselves that they, unlike the overwhelming majority, would

Mil for law

not deliver painful shocks. Fine concludes: "The problem is that you may know, intellectually, that people's moral stamina is but a leaf blown hither and thither by the winds of circumstance. You may be . . . comprehensively informed about the self-enhancing distortions of the human brain. Yet this knowledge is almost impossible to apply to oneself. Somehow, it fails dismally to penetrate the self-image."[8]

Another narcissistic brain tendency is that we have an illusion of control in situations where we in fact have little or none. Studies show that we imagine we have control when things turn out the way we wish they would. A cleverly designed experiment had volunteers push a button that they were told might turn on a light, but the light actually was set to go on and off randomly. The experimental subjects, while having no influence over what the light did, nevertheless experienced an illusion of control, and the more the light was scheduled to be on, the more control the volunteers thought they had over it. This tendency contributes to our illusion of free will and helps mask the absence of it. It also highlights a point I will bring up later about the predominantly unconscious and even nonconscious character of many mental operations, especially our procedural learning and knowledge: we do not in fact have access to our own choice or decision-making processes, so we make guesses about our motives based on externally visible associations. We watch ourselves largely as outside observers (rather than being introspectively transparent to ourselves) and draw inferences about ourselves therefrom. Here those externally observable associations were the light going on and the finger pushing the button. Any observer could have made just as valid—or invalid—a guess. Perhaps a neutral outside observer in the lightbulb experiment would have been less likely than the experimental subject to jump to the conclusion of actor-controlled agency rather than the true externally controlled random agency.

Fine points out that all this makes the few who defy the self-serving self in the interest of moral integrity all the more admirable for it.[9] For we are greatly prone to hypocrisy as well. We all too easily change our beliefs ex post facto to match our actions. "And really the situation had no moral content to it at all anyway," we tell ourselves. Given all these selfy tendencies, it's remarkable that anyone engages in independent moral thinking, judgment, and behavior at all.

Despite all this bad news, Fine points out that human beings are not completely delusional but rather read the evidence in self-serving ways. If we were completely impervious to reality, as she puts it, there would go our san-

egoism arrogance

ity. Most of us don't go that far, thankfully. She says of herself that even though she might not generally acknowledge to herself how mathematically challenged she is, she's not about to apply for a professorship in mathematics, either. We generally censor and distort the truth about ourselves but don't give up on it entirely. The few of us who think we're Napoleon or Jesus are in trouble. So this is a matter of degree. It turns out those who tend toward arrogance are less connected to reality than those who are humbler. Fine reports on a psychological experiment conducted on stock traders that revealed that the more arrogant a trader was, the more he thought he could control changes in the value of a stock index, when in fact the changes were controlled randomly by a computer program. The more arrogant the trader, the higher was his illusion of control on the psychological scale and, as a result, the less he earned. The most arrogant, the experimental psychologists determined, earned about $100,000 less than the averagely arrogant.[10] On the other hand, there are folks who see themselves more realistically, perceiving the warts and not just the dimples. Who are these lucky, clear-sighted philosophers who observe themselves and the world realistically and fairly, assigning responsibility largely with impartiality to themselves and others? They are people who suffer from depression. Fine comments that "they are living testimony to the dangers of self-knowledge." For normal people there is what psychologists call a "window of realism": when a decision is to be made, a short period of more realistic self-assessment may set in, but it quickly shuts down once the decision has been made. That's good news for most of us. Psychologists have come up with a name for analyzing the standard state we're in and our self-serving biases: "terror management theory." Life without self-deceiving and self-serving illusions would be unbearable; no wonder the psychologically depressed don't want to get up in the morning. But that is not good news for our reliance on individual moral judgment and a presumption of rational free will agency.

The Evidence from Cognitive Dissonance

Most people, when directly confronted with evidence that they are wrong, do not change their point of view or course of action but justify it even more tenaciously. Even irrefutable evidence is rarely enough to pierce the mental armor of self-justification.

—CAROL TAVRIS AND ELLIOT ARONSON

The various cognitive biases that Fine identifies can be chalked up to the overall tendency of our minds to avoid what psychologists call *cognitive dissonance*. That term was invented by Leon Festinger, a Stanford psychology professor, in 1956 to describe the psychological discomfort of having a belief brought into question by mounting empirical evidence that challenges its validity. Psychologists Carol Tavris and Elliot Aronson make the case that our response to such dissonance is to banish from view the conflicting evidence rather than revise our beliefs; this is "the hard-wired psychological mechanism that creates self-justification and protects our certainties, self-esteem, and tribal affiliations."[11] More than three thousand experiments have been conducted exploring the particulars and bounds of cognitive dissonance, and these, "taken together, have transformed psychologists' understanding of how the human mind works."[12]

Overwhelming evidence challenges the notion that human beings act rationally for their own practical (economic or other) benefit; it challenges the behaviorist account of motivation as a response to rewards; it even challenges that seeing is believing. What is more true, they say, is that "believing is seeing."[13] That phrase captures well one aspect of cognitive dissonance called *confirmation bias*: we consider salient any evidence that confirms our presuppositions and belief commitments, whereas disconfirming evidence pales. Perceptions follow beliefs, rather than vice versa, more often than we care to admit. Politics is a great example: it should hardly be a surprise to any of us that experiments testing people's perceptions of how their own political party's candidate performs in a debate against the other party's candidate confirm that perceptions follow party lines. The neural basis of the confirmation bias has recently been exposed in an experiment on political perceptions. Drew Westen, an Emory University psychologist, has demonstrated that the brain's loci of cognitive reasoning ceased to function when subjects were confronted with negative, cognitively dissonant information about their preferred candidate. And, conversely, when the dissonant information was withdrawn, emotional areas of the brain signaling happiness lit up.[14] Other experiments show that disconfirming evidence may even serve to strengthen rather than weaken one's beliefs, as we would expect. This is particularly true of decisions that we have already made, especially true if those decisions were irrevocable, and even more so if the decisions were costly. Self-justification is our standard operating procedure.

Experiments testing how cognitive dissonance affects the treatment of victims of violence by aggressors are particularly chilling. For it turns out that "aggression begets self-justification, which begets more aggression."[15] Once aggression against victims is set in motion, justifications that blame the victim emerge, and the more those beliefs are strengthened, the more aggression ensues. This is not good news about the strength of our moral capacity. But on the other hand, good actions induce positive feelings and hence positive judgments about the goodness of the character of the object of such kindness; these snowball as well. So kindness, too, has a reinforcing effect upon beliefs about the objects of one's charity. Both violence and kindness spiral. The spiraling is particularly significant because there is a slippery-slope dimension to our responses to cognitive dissonance. The initial action taken in a morally ambiguous context may be in a gray area, the consequences not at all clear, for the process often plays out gradually. After the initial decision is made, however, the consequences begin to hook one in, and self-justification gradually sets in. This is what happened in the Milgram experiments, according to Tavris and Aronson: "The Milgram experiment shows us how ordinary people can end up doing immoral and harmful things through a chain reaction of behavior and subsequent self-justification. When we, as observers look at them in puzzlement and dismay, we fail to realize that we are often looking at the end of a long, slow process down that pyramid."[16] A seemingly inconsequential initial moral decision to give an apparently small electric shock to the subject sets in motion a cascade of actions and then ever greater self-justifications as the stakes get higher. This, they say, is a form of "entrapment" that "increases our intensity and commitment, and may end up taking us far from our original intentions or principles." We get "hooked."[17] The slippery-slope phenomenon is widely applicable, they say, and Tavris and Aronson's book ranges over a broad array of situations in which it has been documented, from the personal to the political to the professional.

We believe that we have independence of judgment, but it turns out, for example, that our actions toward others more often than not define them for us, rather than vice versa, as we generally assume. Experiment after experiment reveals that the denigration of victims is a *consequence*, rather than a *cause*, of their victimization.[18] It is a self-justification effect. That pain felt directly is more intense than the pain we experience in inflicting it upon others contributes to the ease of victimizing others. In investigations of prisoner abuse and in studies of the use of torture as a technique embraced at the highest political level, studies and interviews show that the perpetrators, to a person, claim that their use of torture was far less than the use of it by their enemies.[19] The authors conclude that "even the most awful guys think they are good guys." Again and again studies have shown that despots see themselves "as self-sacrificing patriots." The authors cite a particularly ironic example, a poster that the tyrannical dictator and human rights violator Jean-Claude "Baby Doc" Duvalier put up all over Haiti. It read, "I should like to stand before the tribunal of history as the person who irreversibly founded democracy in Haiti," and it was signed, "Jean-Claude Duvalier, president-for-life."[20]

We might apply the term *corruption of judgment* to the human tendency to reduce cognitive dissonance by self-serving self-deceptions. Tavris and Aronson pinpoint that what drives the phenomenon they document so ubiquitously is our highly emotionally charged self-relation, what I have called selfiness. It stems from the emotional attachment we have to ourselves and our self-concepts or sense of ourselves—which includes beliefs about ourselves and about the world and especially how we fit into the world. "Dissonance," they write, "is bothersome under any circumstance, but it is most painful to people when an important element of their self-concept is threatened—typically when they do something that is inconsistent with their view of themselves."

The Totalitarian Dictator in All of Us

In an influential essay in the *American Psychologist* in 1980, psychologist Anthony Greenwald proposed that recent discoveries in the cognitive sciences, when taken together, exposed that there is what he called a "totalitarian dictator" within each and every one of us. The discoveries about normal cognition, he said, offer a portrait of a human nature that looks a lot like a totalitarian dictator's control and suppression of information. He titled his article "The Totalitarian Ego." Greenwald argued that the mind's cognitive orientation falls into three categories: (1) egocentricity—the focusing of knowledge and memory upon the self and the overestimation of one's own role in what happens; (2) "beneffectance"—a term that Greenwald coined to refer to the tendency in people to take credit and responsibility for their successes but disavow responsibility for their failures; and (3) cognitive conservatism—the persistence of belief and the reluctance to modify or relinquish beliefs in the light of countervailing evidence. This last category is what Tavris and Aronson document as the ubiquitous response to cognitive dissonance: people generally read the data and remember the data selectively to confirm the opinions they already hold. Greenwald points out that this is true not only for controversial opinions and attitudes but also for factual knowledge. Memory is fabricated, and the data are constructed and reconstructed to confirm standing beliefs.

These three cognitive biases, Greenwald concludes, are "pervasive and characteristic of normal personalities." The three biases capture the self-servingness I have dubbed *selfiness* (rather than "the self" or "a self" or even "the ego," as Greenwald calls it; I prefer to avoid the whole host of psychoanalytic assumptions that go hand in hand with the idea of "ego"). These tendencies of thought give us a window into a self behind or beneath them. Yet that self is not or need not be a thing. Rather, it is an urge we have to make a self-serving picture of ourselves and the world that is coherent rather than piecemeal and contradictory. The self-organizing urge toward systematic coherence in beliefs produces a coherent self-process rather than leaving us with a collection of disconnected and momentary selfy episodes. Moreover, it is the beliefs and thoughts that give us a jolt of pleasurable emotion that we allow into our picture of ourselves.

Selfiness: Thinking and Believing Are Always Emotional

> Sometimes you have to work with the brain you have, not the brain you might have wished you had.
>
> —DREW WESTEN

The cognitive biases that I have been presenting have a common selfiness, or self-serving motivation. *Motivation* is the key word here, for what is being exposed to us is how emotional, and emotionally driven, our thinking and beliefs really are. What I have been unmasking, in sum, is the *desiring* nature of our mental functioning. The human brain is the brain of an organism keeping itself alive, pursuing its survival, competitively and cooperatively. While we may think that having a special ability to separate our thinking from our imperative to survive, our reason from our emotions, is what makes us human, in fact it turns out that one dimension of what makes us human seems to be our extraordinary *integration* of cognition and affect. Science writers Sandra and Matthew Blakeslee detail how deeply integrated our ideas, beliefs, and thoughts are with our emotions, feelings, and motivations.

They report that neuroanatomist Arthur Craig has found that "in every brain imaging study ever done of every human emotion, the right frontal insula and anterior cingulate cortex light up together. . . . He takes this to mean that in humans, emotions, feelings, motivations, ideas, and intentions are combined to a unique degree, and that this is a key element of our humanity."[21] In primates in general and in human beings supremely, the internal perception of the state of the body, called *interoception*, is extraordinarily rich, the Blakeslees point out. The right frontal insula of the brain is active both when we feel strong physical sensations and also when we feel emotions. The Blakeslees say that it would be difficult to overestimate the significance of the extension of physical feeling to encompass the emotional as well:

> This dual physical-emotional sensitivity is not just a coincidence. The right frontal insula is where conscious physical sensation and conscious emotional awareness coemerge. Consider this amazing fact: The right frontal insula is active both when you experience literal physical pain and when you experience the psychic "pain" of rejection or the social exclusion of being shunned. It lights up when you feel someone is treating you unfairly. Scanning experiments have proven all this, and the results are profound.[22]

Robert Solomon, the late philosopher of the emotions, connects the dots among emotion, cognition, and self in this way: he says that emotion "is a reflection of a *self*," "show[ing] . . . who one is." Emotions are "our engagements with the world" and give us a "basic orientation to the world and to one another."[23] Human thought and belief are profoundly emotionally influenced and driven, and conversely, our emotions are infused with our ideas and presuppositions to a very high degree. Cognition modifies and influences emotion as well as vice versa.

It is the right frontal insula that makes the connection between our physical states and our mental states, both cognitive and emotional-physical. Internal information or readouts of our body and of our brain are integrated there. It is there that "you detect the state of your body and the state of your mind together," the Blakeslees say. "It is here that body and mind unite."[24] The right frontal insula is capable of integrating mind and body in this way because it also forms strong connections with three other relevant areas of the brain: the amygdala, the orbitofrontal cortex, and the anterior cingulate cortex. The amygdala links emotions and experiences, the orbitofrontal cortex is involved in planning and responding to rewards and punishments, and the anterior cingulate cortex is crucial in self-monitoring and agency.[25] *Our emotions are the site where mind and body connect.* So just as we have had little reason to doubt that our bodily sensations are our own and that we respond to them in self-serving ways, we can now begin to see that the same selfiness ought to be attributed to our mental life—to our emotions, of course, but also to our beliefs, ideas, and thinking generally. For they are built upon the bodily sensations that we do not doubt are our own, nor do we question why they are motivating of self-protective responses. (How could Descartes have thought it intuitive that our minds are our own but that we could doubt that about our bodies? Nevertheless, Descartes's attribution of a sense of self to our thoughts and other mental phenomena, in retrospect, has a kind of prescience—although he was wrong that the mind (thoughts and emotional feelings) are amenable to direct voluntary control, to free will.) We are beginning to discover why our cognition works in the selfy ways that it does.[26]

Further evidence of the emotional drivenness of perceptions and beliefs comes from psychologist Drew Westen's research into political attitudes and judgments.[27] While his focus and agenda are political, his argument and data have broader significance and application. He exposes pervasive selfiness in our thinking, a selfiness that is not confined to the political arena. Westen states his basic point simply and clearly: "The central thesis of the book is that the vision of the mind that has captured the imagination of philosophers, cognitive scientists, economists, and political scientists since the eighteenth century—*a dispassionate mind*—that makes decisions by weighing the evidence and reasoning to the most valid conclusions—bears no relation to how the mind and brain actually work."[28] Recent research, including Westen's own, demonstrates unequivocally that our minds are inherently what he calls "partisan." It is clear that partisanship, as he describes and explains it, applies not just politically but in all kinds of ways, from the individually idiosyncratic and biographical (what I call *selfiness proper*) to various forms of social identity and identifications (what I call *group selfiness*). Westen and his colleagues used functional magnetic resonance imaging (fMRI) to study the brains of political partisans during the 2004 election. They presented strong Democrats and strong Republicans with statements about their party or candidates that pointedly contradicted their partisan inclinations. Their aim was to test if the clear information they presented would lead the volunteer partisans to reassess their positions as a consequence of assimilating the new data fairly and dispassionately and come to the obvious conclusions, albeit against their inclinations, or not. Would desire and emotion influence the partisans' reading of the data enough to corrupt their judgment? Or would they stand firmly by an honest reading despite inclinations to the contrary? Would the volunteers be rational in their decision making? What Westen and his colleagues suspected turned out to be the case: desire and emotion triumphed over reason and truth.

During the election season of 2004 Westen and his colleagues gave their fifteen committed Democratic and fifteen committed Republican volunteer partisans six sets of quotations (some real and some invented) in which George W. Bush and John Kerry were presented as contradicting themselves. As a control, the research team also presented famous but politically neutral figures (such as Tom Hanks and William Styron) contradicting themselves. The quotations from the politicians and the politically neutral figures were partly fictionalized but embedded in real events to make them believable. The researchers had the experimental partisan subjects lie down in brain scanners while they were shown a series of these quotations. Then the volunteers were asked to what extent (on a scale from 1 to 4) they agreed that each candidate had made contradictory statements. As anticipated, the partisans were easily able to identify the contradictions of the *opposing* candidate and rated his contradictoriness at 4, the highest level. But their own candidate's contradictions were rated on average as 2, which is low. The contradictory statements attributed to each of the neutral figures inspired no such partisan discrepancies. At the neural level, Westen and his colleagues found, dissonant information activated neural circuits associated with negative emotions. Areas of the brain that are known to be highly involved in the regulation of the emotions were activated. An area of the brain involved in conflict resolution was also activated. And finally the outcome was a decision primarily motivated by the desire to reduce the cognitive dissonance and emotional pain of one's own candidate's seeming contradictory statements.

Hence the experimental subjects' analyses and moral judgments about the merits of the cases turned on gut emotional reactions. Areas of the brain that were activated in prior studies of reasoning were not strongly activated in the partisan political judgment process even though the task, as Drew Westen puts it, was "a reasoning task."[29] The reasoning engaged in consisted largely of subsequent rationalizations of the gut feelings that drove the analysis, rather than rational judgments and decision making. In addition, the emotionally driven judgments that resolved the conflicts between desire and data took place very rapidly—and perhaps beneath explicit self-awareness, Westen and colleagues concluded, because the "partisans mostly denied that they had perceived any conflict between their candidate's words and deeds."[30] Something even happened that Westen had not predicted: not only was distress avoided in this way but also the brains of the partisans were concurrently flooded with pleasure, because the reward circuits had been activated. Westen comments, "The results showed that when partisans face threatening information, not only are they likely to 'reason' to emotionally biased conclusions, but we can trace their neural footprints as they do it."[31] The basic motivation of the volunteer partisans turned out to be how to get rid of the painful emotional dissonance between knowledge of the evidence, on one hand, and the desire to have one's candidate look good, on the other. And this is what happened in the brain:

> The neural circuits charged with regulation of emotional states seemed to recruit beliefs that eliminated the distress and conflict partisans had experienced when they confronted unpleasant realities. . . . Not only did neural circuits involved in negative emotions turn off, but circuits involved in positive emotions turned *on* . . . activating reward circuits that give partisans a jolt of positive reinforcement for their biased reasoning. These reward circuits overlap substantially with those activated when drug addicts get their "fix."[32]

A significant implication of these findings, Westen concludes, is that "the political brain is an emotional brain."[33]

Thinking and feeling, Westen points out, evolved together and work together. Emotions both fuel action and also put brakes on it. Emotions are part of our social equipment and regulate social interactions. Emotions are communicative. Emotional intelligence includes the ability to perceive, respond to, and influence others' emotional signals and states. They are also a source of motivation and hence of behavior. Emotions lead toward things and away from them, and cognitively meaningful associations built up from experience inform our patterns of approach and avoidance. In both capacities, the social and the motivational, emotions have survival value.[34] As I sit on my terrace under the grape arbor with my computer, my cat Moxie is meowing vociferously and plaintively at the screen door and looking at me insistently. I feel the emotional insistence and distress in her voice and pick her up to comfort and play with her a moment. She doesn't manage to get free this time to run outside to chase butterflies or scamper down the garden path, but she is mollified. Her cry has communicated her distress to me without any need for language and has led to an attachment response on my part, a social interaction, which we both capture and understand.

Studies of people who have been neurologically injured and have their emotional capacity impaired show that they cannot assess the emotional meaning of actions and people's expressions; they approach danger, for example, even when they know by rational assessment that it's a bad thing to do. Moreover, they cannot come to decisions, and they fail to make moral judgments. Antonio Damasio, in his popular book on affective neuroscience, the study of the neurobiology of the emotions, has coined the term *somatic marker hypothesis*. The term is meant to capture how beliefs and thoughts have to have emotions attached to them in order for them to motivate action. A somatic marker is a remembered link between an experience and an "associated physiological affective state," which is to say, it is the remembered emotion you got in the past from making a certain decision or performing a certain action. The inital linkage is largely beneath conscious self-awareness.[35] If, for example, you got stung by a bee when you went apple picking when you were three years old, apple picking will have an associated negative emotional valence that weights the decision about whether to go apple picking again. As Cordelia Fine puts it, "Without these emotional tags even the most encyclopedic knowledge or powerful intellect cannot help us to pluck a bottle of shampoo off the supermarket shelf."[36] A number of years ago I wrote about Damasio's research on the affectivity of thought.[37] In his first book for the general public, *Descartes' Error: Emotion, Reason, and the Human Brain*, Damasio reports that he had come to the conclusion that all thinking is affectively laden. He had initially accepted the standard notion of the separation of reason from the emotions, but as a result of his experience with patients whose neurological defect kept their reasoning power intact whereas their emotions were severely truncated, he changed his mind.[38] He discovered that, oddly enough, those who had sustained damage to their ability to experience emotions but not to their cognitive capacities were incapable of the kind of rational decision making (including ethical decisions) and the carrying out of those decisions that philosophers standardly (and wrongly) ascribe to reason and thinking alone. Rational decision and action, he concluded, require emotion and cannot take place in its absence. "Feeling," he realized, "was an integral component of the machinery of reason."[39] To say that feeling has a strong role to play in reason is to claim not only the affectivity of thought but also the incorporation into thought of the regulatory mechanisms that drive emotions, Damasio argues. For emotions are not merely feelings but "expressions" of the "mechanisms of biological regulation."[40] It is thus the feedback loops of biological regulation that are operative in reason through the infusion of thought with emotion. Damasio proposes that

> human reason depends on several brain systems, working in concert across many levels of neuronal organization, rather than on a single brain center. Both "high-level" and "low-level" brain centers, from the prefrontal cortices to the hypothalamus and brain stem, cooperate in the making of reason. The lower levels in the neural edifice of reason are the same ones that regulate the processing of emotions and feelings, along with the body functions necessary for an organism's survival. In turn, these lower levels maintain direct and mutual relationships with virtually every bodily organ, thus placing the body directly within the chain of operations that generate the highest reaches of reasoning, decision making, and, by extension, social behavior and creativity. Emotion, feeling, and biological regulation all play a role in human reason. The lowly orders of our organism are in the loop of high reason.[41]

The most primitive sensations (those in the brain stem, a structure common to us and to the most primitive mammals), such as taste or smell, arouse emotional feelings that motivate approach or avoidance. Associations between experience and such sensations are etched into neural networks, so the repetition of the associated stimulus initiates the feeling and its concomitant reaction—for example, in an experiment rats avoided a taste they came to associate with a concurrent painful dose of radiation, even though the taste was not bad in itself. Other sensations also work on emotional associations, such as sight and sound, and especially music. The evolutionarily primitive amygdala (yet not as primitive as the brain stem) scans the environment for emotional stimuli beneath conscious awareness. Feelings and their associated memories motivate and guide responsive actions more rapidly than they can be consciously brought into awareness. The amygdala has a vital role to play in emotional processes and emotional memories, as it forms and stores memories associated with emotionally charged events.[42] When it is damaged, classical conditioning is impaired. The amygdala's functions include "identifying and responding to emotional expressions" in other people, "attaching emotional significance to events," and "creating the intensity of emotional experiences."[43]

The cerebrum's neural pathways broaden the capacity to associate pain and pleasure with thoughts and beliefs. The cerebral cortex is the surface of the cerebrum and amounts to 80 percent of our human brain mass. Although brain function is widely distributed, certain areas nevertheless contribute certain capacities. The dorsolateral prefrontal cortex, the part of the cortex at the top and sides of the frontal lobes of the brain, enables us to have information self-consciously in mind, and it is active when we deliberate and make conscious choices. When partisan thinking is at stake—and I'm arguing that this includes all selfy thinking, since partisan behavior is one form of selfiness—this part of the brain, according to Westen, "isn't typically open for business when partisans are thinking about things that matter to them."[44] The ventromedial prefrontal cortex, which is located behind the eyes and forehead, has an important function in linking emotion and cognition. It is "involved in emotional experience, social and emotional intelligence, and moral functioning" and is crucial in "using emotional reactions to guide decision making."[45] It has many neural linkages to the amygdala. Westen describes how basic emotional systems are linked with neural networks that together capture our personal desires and biographies, our patterns of mental associations and their accompanying feelings, and connect to larger patterns of cultural meaning and interpretation as well.

> The brain gravitates toward solutions designed to match not only data but desire, by spreading activation to networks that lead to conclusions associated with positive emotions and inhibiting networks that would lead to negative emotions. Positive and negative feelings influence which arguments reach consciousness, the amount of time we spend thinking about different arguments, the extent to which we either accept or search for "holes" in arguments or evidence that is emotionally threatening, the news outlets we follow and the company we keep. In short, as suggested by the neuroimaging study with which this book began, our brains have a remarkable capacity to find their way toward convenient truths—even if they're not all that true.[46]

"What tends to 'drive' people . . . are their *wishes, fears,* and *values,*" which Westen defines as "emotion-laden beliefs about how things *should* or *should not* be." All of these are emotionally charged.[47] We have "emotional agendas," agendas that are not consistent with direct rational self-interest but instead are inextricably linked with our identities. Studies over decades have consistently shown that "people's [material] self-interests often show surprisingly little connection to their voting patterns."[48] The linguist and philosopher George Lakoff has commented (in his book *Moral Politics*) that people vote their *identities*, not their personal interests. Our identities are biographical and personal and also influenced by our identifications. The important point is that in large part emotions influence beliefs and those emotions emerge from and express *who we are* and *where we are*. They are built out of the connections made between our affective responses to our salient experiences—all of which occur within contexts of cultural meaning, history, present situation, and relationships.

There is a covert association between my invented category of selfness and satirist Stephen Colbert's notion of "truthiness." It's that the feeling of validity or truth derives from emotions that promote and protect who we are at a deep gut level rather than from disinterested knowledge and open and fair assessment of the facts. Selfness, the emotional ownership we feel toward our body-mind, is what's behind our pervasive truthiness. And truthiness is exactly the focus of Westen's examination and explanation, as Westen himself recognizes when he includes a wonderful quotation from Colbert that gets to the heart of what needs explanation in terms of the new brain sciences:

> I'm no fan of dictionaries or reference books. They're elitist, constantly telling us what is or isn't true. . . . Doesn't taking Saddam out *feel* like the right thing? Right here, in the gut? Because that's where truth comes from, ladies and gentlemen: the gut. . . . Next time, try looking [something] up in your gut [rather than a book]. I did, and my gut tells me that's how our nervous system works.
>
> The truthiness is, anyone can *read* the news to you. I promise to *feel* the news *at* you.[49]

Westen qualifies the pervasiveness of truthiness over truth with the following proviso: truthiness prevails over truth particularly when "the stakes are high."[50] I take that to mean when a person's emotional involvement in the matter at hand is high, and I would say that's the case when a high degree of selfness, of identification, of what is included within one's sense of self or identity, is at stake. Westen and his team of researchers concluded that "the dispassionate mind of the eighteenth-century philosophers allows us to predict somewhere between 0.5 and 3 percent of the most important political decisions people will make over the course of their lives"—though for the politically more engaged, these figures are probably an overestimate.[51] So "the results are unequivocal that when the outcomes of a political decision have strong emotional implications and the data leave even the slightest room for artistic license, reason plays virtually no role in the decision making of the average citizen."[52]

We Cannot Act from Reason Alone

Kant . . . wrote against the scholars in support of popular prejudice, but for scholars and not for the people.

—FRIEDRICH NIETZSCHE

In a wonderfully ironic essay titled "The Secret Joke of Kant's Soul," Joshua Greene, a self-described "neurophilosopher" at Harvard, argues against Kant's moral psychology.[53] Kant is famous for advocating what philosophers call a deontological theory of ethics, that is, a theory with a focus on rules, which themselves are defined in terms of rights and duties. This is in contrast with consequentialism, an account of ethics that defines moral actions as good or bad by virtue of their (moral) consequences. Deontology precludes an appeal to (moral) sentiments, emotions, as morally legitimate reasons for actions. Legitimate deontological judgments are those based strictly on respect for duty and respect for law. They are those that rationally recognize and respond to obligation as such. So on this account what makes moral action moral, according to Kant, is its appropriate rational motivation, namely, to act from duty alone. In this essay, Greene argues not only that such a moral demand of human nature is impossible for creatures like us, but, ironically and paradoxically, it is precisely when we think we are acting from duty alone that we act most completely from emotional motivations that we consciously feel but do not generally allow to reach self-conscious awareness, rather than from rational considerations. The one way that deontological moral judgment cannot function, according to Greene, is exactly the way it's supposed to function, namely, as actions taken primarily from rational motives— let alone from reason alone (as Kant would have us do). Instead, reason comes in, Greene says, mostly ex post facto, as rationalizations for the moral judgments and actions already taken, which were, in fact, emotionally motivated and chosen.[54] The judgment of an action as right or wrong in itself is necessarily both unconsciously decided upon (and because it is procedural, and that is beneath not only self-conscious awareness but even beneath consciousness, according to the neuroscientist Valeria Pulcini) and also emotionally driven. The reasoning enters in, in almost all cases and circumstances, as a subsequent rationalization of an already made unconscious decision that was necessarily emotionally motivated. (The motivating emotion is present in consciousness but may be unacknowledged or even disavowed at a self-conscious level, according to Jaak Panksepp.)

Research on how the brain works supports the notion that cognition in and of itself cannot result in action. Action is possible only to the extent that thinking is affectively charged and experienced.[55] I have remarked earlier that the evidence reveals that those who have lost the affectivity, the emotional charge, of their moral cognition through neurological injury or disease also lose the ability to come to moral decisions and act upon those decisions. Some can know in theory the moral path but cannot in fact act upon it. They have lost the capacity for moral concern and with it any motivation to act morally. In normal people the degree or intensity of emotional motivation in moral decision making and actions differs according to what is at stake in the decision. The research reveals that judgments of right and wrong about actions (murder, for example, or rape) are highly emotionally charged, whereas the mental acts of considering the consequences of actions are emotionally less intense (but not emotionally neutral) and hence more detached. "When harmful acts are sufficiently impersonal," Greene comments, "they fail to push our emotional buttons, despite their seriousness, and as a result we think about them in a more detached, actuarial fashion." When harm is personal and up close, however, it triggers "an alarmlike emotional response" of great neural activity "in brain regions associated with emotional response and social cognition." Researchers have called this "the identifiable victim effect." The mental weighing of overall consequences, on the other hand, produced greater activity in classically cognitive brain regions.[56] However, the inclination to retribution (a deontological immediate intuition) is highly emotionally driven. Moral violations make people angry and outraged, and the degree of outrage determines the nature of the punishment they wish to inflict. Research indicates that people tend to inflict punishment for its own sake. They rarely choose to punish based on assessing the consequences.[57] We see this phenomenon play out, for example, with morally harmless acts that nevertheless strike people as disgusting, violating their moral sensibility, such as eating the family dog or using the national flag to clean the car.[58]

Further research has exposed the post hoc character of the reasoning entailed in people's descriptions of why they make the decisions they do. Neuroscientist Jaak Panksepp suggests that all thinking is emotional and value-laden, but often covertly so. He says the neural evidence suggests that "cognitive states embed affective value(s) and that all sustained cognition is affectively directed and motivated, often invisibly in a fashion that promotes the illusion of cognitive autonomy from emotion."[59] So we human beings "confabulate," which means that we make up plausible (and self-serving, self-protective) stories after the fact about why we did what we did. Besides the selfiness involved, we confabulate because we do not have introspective mental access to much of our motivation. A lot of our motivation, it turns out, is unconscious —which is to say that we consciously feel our emotional motivations but often do not have *self*-conscious awareness of them. Moreover, we are largely unaware that we *are* unaware. We are not even self-consciously aware that the stories we make up to explain our own actions are plausible reconstructions rather than actual memories. They are largely invented narratives that make sense (after the fact) of our actions. An experiment conducted by Donald G. Duton and Arthur P. Aron captures well the phenomenon of confabulation. Duton and Aron had male experimental subjects cross a footbridge over a deep gorge. At the other end of the bridge was an attractive woman who was one of the psychologists conducting the experiment. Control subjects did not cross the bridge but sat on a bench. It turned out that twice as many men who crossed the bridge than who sat on the bench later called the attractive experimenter for a date. They had misread their own internal state of arousal (of sweaty palms and hearts pounding) as sexual attraction rather than as fear.[60] The problem here was the inaccessibility to them of their actual motives as a result of the general, nonspecific nature of arousal: "Because the arousal is the same whatever the emotion (it only varies in intensity), your brain has the job of matching the arousal with the right thoughts."[61] The phenomenon of confabulation has been shown by researchers to be obviously and tragically true in people whose left and right brain hemispheres have been severed (in some cases surgically to relieve the symptoms of epilepsy and in others by neurological injury or illness), but it is also standard in all of us to a much greater extent than we are aware or generally admit.[62]

Given both the tendency to confabulate and also the emotionality of strong moral judgment, Greene asks rhetorically: "What should we expect from creatures who exhibit social and moral behavior that is driven largely by intuitive emotional responses and who are prone to rationalization of their behaviors?" He concludes that "the answer is . . . deontological moral philosophy," that is, ethics that relies only on rational rules for guidance. For "deontology . . . is a kind of moral confabulation," which is to say that deontology is itself a theory that rationalizes our urgent and immediate emotional moral reactions.[63] We act from emotion, but we fool ourselves that we are acting entirely or principally from rational considerations alone. We falsely believe that we act upon the reasons we come up with ex post facto to explain and justify our actions. Brain science, however, exposes that those reasons are not why we perform a given action but instead turn out to be how we prefer to see ourselves as having acted. We give ourselves a pat on the back and tell ourselves we acted from the best of motives, moral motives—duty, responsibility, universal moral rules, principles, and truths. But we were in fact driven almost without exception by emotion, and emotion is at bottom selfy. Which brings us to a deeper question: what are emotions anyway, and how do they operate?

Emotions: The Source of Agency in the Brain, a Source of a Deep Self

Jaak Panksepp, one of the foremost neuroscientists in the world and coiner of the term *affective neuroscience* to designate the neuroscientific study of the emotions, defines emotions as the "action systems" of the brain. So action and emotion go hand in hand. He calls the complex the "emotion action systems" and further remarks that neurobiological evidence reveals that "emotional feelings . . . are fundamentally experienced action systems."[64] There is little wonder, then, that thinking in and of itself cannot result in action. It is precisely because "all thinking is emotional and value-laden" that it can result in action. Jorge Moll, a researcher at the National Institutes of Health, explains the emotionality of all thinking this way:

> According to our view, emotions are not dissociable from cognition, and neither compete with nor are controlled by rational cognitive processes. Instead, they are integrated into cognitive-affective neural assemblies, or representations. A process can be defined as the dynamic engagement of neural representations in a given behavioral context. . . . Behind the engagement of any process there lies a representation, encoded in neural assemblies. . . . It is reasonable to assume that interpreting highly complex social situations as morally relevant depends upon the existence of complex cognitive-affective associations stored in distributed, large-scale neuronal assemblies. . . . [Social] attachment is a component required for certain moral emotions.[65]

Panksepp elaborates that all thinking and acting are emotional in that they are informed by a number of basic emotions ("primal affects," also referred to as "primary-process affects") that all mammalian brains have, and even all vertebrates have some of them. These basic emotions are "intrinsic brain value systems that unconditionally and automatically inform animals how they are faring in survival." They provide consciousness, and both humans and animals that sustain damage to these primitive brain regions lose the capacity for consciousness and feeling, according to Jaak Panksepp.

> Affective feelings come in several varieties, including sensory, homeostatic, and emotional. . . . Primary-process [i.e., basic] emotional feelings were among the first subjective experiences to exist on the earth. Without them, higher forms of conscious "awareness" may not have emerged in primate brain evolution. . . . All vertebrates appear to have some capacity for primal affective feelings.

The ancient brain mechanisms present in mammals still play crucial roles in human beings, in whom they are nuanced by subsequent evolutionary systems (as they are in other primates and to varying extents in other mammals), especially by neocortical cognitive ones. The causal direction of influence goes both ways, Panksepp says, for "the brain seems to be the only organ of our bodies that is clearly evolutionarily layered, albeit all levels functionally interdigitate." Panksepp identifies three layers of emotions. "It is useful," he says, "to divide evolved brain functions in terms of primary processes (tools for living provided by evolution), secondary processes (the vast unconscious learning and memory mechanisms of the brain), and tertiary processes (the higher-order functions of mind permitted largely by the cortical expansions that allow many thought-related symbolic functions)." The most basic layer, the primary-process emotions, he continues, "are among the most important aspect of our mental lives"; they "bring us great joys and sorrow, and intrinsically help anticipate the future."

Panksepp has identified a number of evolved emotional circuits or mechanisms that he designates as basic in that they are both built upon by experience (top down) and inform experience (bottom up). These systems underlie all emotions and hence also underlie action. If we are to understand agency—how and why and when we come to act—and moral agency in particular, this is the place to start. This is rock bottom. It is not the whole story, of course, but it is the crucial beginning of the story, the scaffold upon which all kinds of implicit ways of acting in the world—when to get up in the morning, how to brush our teeth, and an infinite number of other skills, for example—as well as fully cognized, culturally nuanced, socially constructed motivations, are grafted. What Panksepp refers to as the secondary and tertiary layers of emotion provide elaborations of the emotions, modifying agency. The secondary process consists of "mechanisms of learning and memory, deeply unconscious [or really nonconscious] brain processes, [which] are regulated by more primal emotional systems." This secondary layer includes all the nonconscious, embodied, and implicit procedural kinds of knowledge that inform how we get around in the world, for example, while the tertiary consists more in all the kinds of linguistic and cultural knowledge that shape our lives. This level provides self-conscious awareness. So it now seems that the primary level provides conscious emotional and affective feeling, the secondary level of learning and memory is largely procedural and its mechanisms transparent and hence unconscious, while the tertiary neocortical level provides self-conscious awareness. Tertiary-process mechanisms "remain tethered to what came before"; like the secondary-process mechanisms, they "rely on unconditional [basic] networks that evolved earlier." Furthermore, "it is only the tertiary-process level that cannot be well fathomed scientifically through current animal brain research—[these are] our cognitive thoughts and emotional ruminations, the sources of our art, beliefs, creativity, dance, fantasies, literature, music, theater . . . the cognitive aspects of schizophrenias, obsessions, manias, and depressions." All these "are constructed from our vast capacity for learning languages. . . . In contrast, our primal emotional urges can best be understood through animal brain research." Moreover, "practically everything that emerges in our higher neocortical apparatus arises from life experiences rather than genetic specializations." The cortex, however, does not provide information about specific emotions from direct recording of neocortical activities. The neocortex instead gives us "general-purpose 'computational space.'" "We must descend to the subcortical realm for more robust signals." The specific basic emotion coming from below and modified up the line gives us emotional and motivational urgency of a specific basic kind. It gives us conscious subjective feelings, Panksepp insistently maintains.

Affective neuroscience has revealed that there are seven basic emotional systems that are homologous in all mammals—that is, they are found in all mammals. For example, organs as different as a bat's wing, a seal's flipper, a cat's paw, and a human's hand are considered homologous because they "have a common underlying structure of bones and muscles."[66] That means the same neural architecture is retained across species, the most ancient systems underlying more recent additions as one goes up the evolutionary ladder. Panksepp comments that "the primal affective mechanisms exist in some of the most ancient regions of the brain, where evolutionary homologies are striking," and that "the basic neurochemistries for emotional feelings . . . are essentially the same in all mammals." The homologous presence of the emotion systems in both animals and humans means that the basic systems that give us humans the subjectivity of emotional experience also give animals felt experiences of emotion—that is, consciousness. Panksepp remarked to me one time that it is a "no-brainer" for any layperson (non-neuroscientist or philosopher) who has ever had a pet that animals have feelings, but it has been a hard sell for many behavioral neuroscientists, who have wished to remain agnostic on whether animals have felt feelings. (As I sit here writing, my cat Litchky is purring on my lap. And when I guiltily dislodge her in order to be better able to type, she voices what I can only perceive as a complaint.) Panksepp points out that it is a stretch to imagine the homologous affective systems that clearly produce both action and felt emotions in human beings would produce only actions and no felt emotions in animals, as the behaviorists (and some logical positivist and still Cartesian philosophers) claim. Felt emotions, the source of consciousness, are shared by all mammals, including humans; the neocortex is the source of self-consciousness or self-awareness, however, Panksepp says.

There are a number of other cogent and persuasive arguments Panksepp cites that also weigh in on the felt character of emotions in animals. Another source of evidence comes from studies of addiction in animals. It would be hard to explain, for example, why rats display intense seeking behavior for opiates if they did not find the ingestion of them pleasurable. Even the behaviorists' very notions of "reinforcement" and "reward" make sense, Panksepp points out, only if there is a pleasurable emotional feeling that makes a reward subjectively reinforcing for the animal. He further remarks that all kinds of research in science, not only the study of emotions in animals, who cannot tell us what they are feeling, involve indirect evidence for phenomena. Much evidence for quantum mechanics or even gravity, to take two examples that Panksepp cites, cannot be investigated directly. He quips, "Would physicists be searching for Higgs bosons if they did not value indirect measurement procedures?" Panksepp has been at the forefront of the effort to show that the neuroscientific evidence is on the side of the presence of feelings (not just behavior) in animals, and he has argued strongly for the ethical treatment of animals as a consequence. For if animals feel pleasure, they also feel pain, and the suffering caused in research labs to animals has a moral claim upon us. Hence Panksepp parted from behavioral neuroscience, carving out the field of affective neuroscience in order to understand human emotions, which investigates animal "MindBrains" (as he puts it) with homologous emotional structures to our own.

The seven basic or unconditioned emotional systems Panksepp has identified in all mammals include (1) seeking/expectancy, (2) rage/anger, (3) fear/anxiety, (4) lust/sexuality, (5) care/nurturance, (6) panic/separation, and (7) play/joy. The seven function not only discretely but also, and for the most part, in combinations. These seven basic emotions provide motivating power and flavor, while all kinds of cognition and experience modify these basic affects. "Each system has abundant descending and ascending components that work together in a coordinated fashion to generate various instinctual emotional behaviors as well as the raw feelings normally associated with those behaviors." The raw feelings are "ancestral memories (instincts) that promote survival," and they operate by "anticipat[ing] the kinds of survival needs that all organisms require to navigate the world." The seeking/expectancy system, for example, is "a very broad action system" and functions to promote the exploration of the world. The system "is designed to promote learning," but it has no particular aim; it does not seek anything in particular, Panksepp says, but is a "vast general-purpose appetitive motivational system." It "engenders enthusiastic affective-energy" that has at times been called "libido" in the psychoanalytic tradition and "euphoria" within modern drug addiction research; "it engenders curiosity, interest, enthusiasm, and cravings." The seeking/expectancy system is the motivational system, the source of agency, par excellence: "This dopamine-energized network helps coordinate MindBrain functions from anticipatory eagerness to feelings of purpose and persistence, all the while promoting planning, foresight and dreams through its multiple normal manifestations." The seeking/expectancy system "constitutes us as animated, energized, and expectant explorers of the world."[67] Crucially, this system to some degree "participates in the tasks of every other primary emotion . . . as well as everything we do." Panksepp told me I was correct that Spinoza's idea of the conatus has been borne out by the evidence for the seeking system, and it is indeed the deepest source of the self's selfiness.

Together these seven systems, and most important, the driving, seeking/expectancy system, constitute "a primary-process core-self," Panksepp argues. For a self needs be "anchored in the bodily emotional action coherences that characterize all organisms." "There is abundant room for 'core-self' type concepts in the subcortical emotional terrain of animals that engenders an 'organismic coherence' . . . that makes us all creatures of the world, as opposed to mere information processing automata," Panksepp concludes. This work "is showing how the higher rarefied subjectivities of self may find constant tether to our very specific animal identity."[68] At bottom, we are agents and derive our agency as mammals—albeit highly sophisticated, cognized ones—from a seeking self, a feeling of self, a conatus, that seeks to survive and flourish and explore the world. As this seeking self, we maintain our survival by fearing danger and experiencing rage at violations; we seek out the means of survival and objects of lust for reproduction; we seek nurturance and seek to provide care; we panic at the loss of care; and we seek play in order to practice our skills, learn to be social, and get rewarded for all the rest. This origin in the action motor structures and processes of the brain is one crucial source of a deep self and of the underlying biological grounding of selfiness. This origin makes the "archaic level of self . . . not cognitive."[69] Panksepp is arguing that the self emerges as an *emotional* system in order to further survival, and that it is derived from evolutionarily early layers of the mammalian brain.[70] "The self first emerges in the precognitive ability of most organisms to operate from an egocentric point of view."[71] It is the "point of view of survival" of each of us and our biological urge is to preserve, protect, and further this center. The philosopher Stephen Asma, remarking on the overall significance of Panksepp's research, suggests that "Panksepp is finally delivering on the Darwinian promissory note: a subject, an 'I,' that is truly born out of the struggle for survival."[72] Spinoza anticipated this discovery by more than three hundred years.

Unconscious Deep Selfiness

In an essay exploring the implications of Jaak Panksepp's discoveries in affective neuroscience for understanding what a self is and how agency and moral agency function, philosopher Paul Sheldon Davies argues that "Panksepp's exploratory neural archeology . . . gives us . . . our most powerful general framework for future studies of the human self." Davies regards Panksepp as "one of today's most audacious naturalists of the mammalian mind; his discoveries and innovations in affective neuroscience are changing our knowledge of the minds of all known living things." He points out that "the failure to factor our animal history prominently into a theory of the self is a virtual guarantee of intellectual bankruptcy." Panksepp's "discoveries reveal something of profound importance about the architecture of the human brain—that it comprises affective systems with inherited systemic functions shared across all mammals, that there are indeed ancestral affective voices in all mammalian minds." Moreover, "this is not just an idle just-so story, nor a theoretical expectation, but something we know from controlled experimental brain stimulations across different species."

Davies proposes that we can draw two important and startling conclusions about the human self from Panksepp's research. The first, Davies says, is that Panksepp has vastly complicated the picture of motivation, so we need to rethink in a very deep and wholesale way "what kind of agents we are." We need to relinquish the Western cultural confidence in ourselves as rational decision makers who have a capacity for self-conscious self-mastery, for it is rare that we do what we do for the reasons we claim. So, first of all, Panksepp's discoveries ought to shake our confidence in our self-knowledge, in the complete and direct availability to ourselves of our actual deep motives, since the basic emotional systems that he has discovered in all mammals are not directly or completely available to self-conscious awareness. The second conclusion that can be drawn, Davies says, is the thesis that "Panksepp's substantive discoveries force us still further toward a skeptical view of our capacities." For "his methods and discoveries give rise to a form of skepticism concerning agency." They "undermine substantive portions of our traditional humanistic view of ourselves." Davies argues that what Panksepp's research has shown is that we cannot definitively or completely know, by looking inside ourselves introspectively, what our real motives and the causes of our actions are. "For any action we perform," he says, "we cannot justifiably claim to know, from the first person perspective, whether our prior choices are among the actual causes of our action." This is the case because our "non-primary capacities are causally dependent on the workings of primary systems," and hence "it is a methodological mistake to theorize about any psychological capacity from a perspective that is top-down." So "our experiences as agents, in sum, are an unreliable source of knowledge for most or all the capacities in our psychological repertoire." In other words, what we can know about the thoughts in our conscious mind is limited to data from our neocortical capacities, and those do not tell us much, or much that is reliable, about the basic motivational systems that are largely driving, or contributing to, our actions. "The problem is," Davies concludes, "at least some of the substantive things we think or feel about our capacities are demonstrably false." So "nothing in mammalian psychology makes sense except in terms of shared primary process affective systems." We cannot give reliable *reasons* for our actions and we can know only partially through direct introspection the deep self constituted by our basic (primary-process) mammalian motivations.[73]

Hence we do not have the full self-availability entailed by the notion of self-control that we designate as free will or rational action. So Panksepp's research, Davies argues, is challenging the cultural commonplace that shapes how we interpret our inner subjective experience, exposing it as both false and provincially Western rather than as a fact of universal inner experience. Davies proposes that our Western senses of self and agency are in fact cultural inheritances subject to correction rather than universal truths of interior experience that all human beings share, as most philosophers mistakenly maintain. He insists that "we must recalibrate many of our traditional expectations concerning the nature of our own capacities and cultivate the very different expectation that our humanistic understanding is incomplete, misleading, or, in some instances, flat-out mistaken."[74] In order to understand our agency, the self in itself, we must take account of the mammalian basic emotions that are our deepest motivational systems, and these are unconscious—which does not mean that they are nonconscious or unavailable to be rendered self-conscious. We also know from Panksepp's account of basic mammalian emotional systems that at bottom our emotions are action systems that promote our survival, and hence our basic agency is self-preserving and self-serving, that is, selfy. Yet since these systems function partly beneath our self-conscious awareness, in order to come to know ourselves more precisely as agents—why we do the specific things we do—we must gather knowledge of ourselves, our motives, and our capacity for agency and moral agency, indirectly as well as introspectively. For why we do what we do can be discerned only with the inclusion of indirect evidence—looking at ourselves and drawing conclusions about our emotions and motives as if looking at a third person—as well as direct introspective evidence. For even when it comes to looking at ourselves, at our own actions and motives, indirect evidence fills in a picture of ourselves not available to us otherwise.

Stephen Asma, reflecting generally on Panksepp's body of research and discoveries about mammalian basic emotions, points out that "Panksepp's archaic self—with its primary consciousness—rescues the body and feelings from the long philosophical tradition that characterized them as purely unconscious [i.e., nonconscious] machinery." He adds, "Panksepp's approach

suggests that consciousness is not superadded to otherwise functioning survival machines, nor can consciousness be abstracted out of the physiochemi-
cal system."[75] Panksepp's position and the weight of the evidence he educes
to support it pose a vital challenge to many philosophers' and scientists'
prevailing claims about the confinement of all consciousness to human beings engaged in symbolic and logical thought and the like. The philosopher
Thomas Metzinger, for example, proposes that when we act, the self is
"translucent"—which means that we act as an agent, but the sources of our
agency are fundamentally invisible to us. On the other hand, at other times
the self is what he calls "opaque."[76] By this he means that we can have access to ourselves in certain kinds of cases and ways as the (conscious) framer
of the content of our experience. Similarly, philosophers Daniel Dennett
and David Chalmers, important thinkers about the relation of biology to
consciousness, continue to make a sharp distinction between conscious
and unconscious, narrowing the conception of "mind" to encompass "neocortical computation" and the like—that is, high-level abstract thinking of
various kinds, which, following Panksepp, I have subsumed under the rubric
"self-conscious." Chalmers holds that "fully functioning animals with intact
brains and bodies could be zombies—'all is dark inside with nobody
home.'"[77] Dennett poses a hypothetical question, "What is it like to notice,
while sound asleep, that your left arm has become twisted into a position
in which it is putting too undue strain on your left shoulder?" and answers
with the claim that it is "like nothing: it is not part of your experience."[78]
This denial of the possibility of subjective unconscious experience (that is, of
Panksepp's conscious but not self-conscious experience) holds, he says, even
though you have moved to release your shoulder and become more comfortable. So for Dennett, as for Chalmers, "whatever . . . problem-solving is going on at these biological levels, is not a part of our mental lives at all," Asma
remarks. Panksepp's discoveries challenge Dennett's and Chalmers's claim
that unconscious (Panksepp's conscious but not self-conscious) levels of
MindBrain are not at all subjectively experienced. Asma goes on to argue
persuasively that Panksepp's approach to the self through the investigation
of basic mammalian action-initiating emotional systems ought to lead us to
conclude that "one of the implications of our 'mammalian agency' approach
to the self may be that subjectivity is never purely opaque nor transparent,
but somewhere in between."[79] A further advantage is that "Panksepp's approach offers the tantalizing possibility that we can get into the muddy so-
called dynamic unconscious,"[80] that is, the conscious feelings deeper down

than self-conscious awareness—and it also provides evidence for the validity of pursuing that possibility.

The Survival Value of (Partial) Unawareness

In the 1980s the science writer Daniel Goleman, who's also a clinical psychologist, wrote *Vital Lies, Simple Truths: The Psychology of Self-Deception* to present to the general reader a revised model of human behavior sparked by the information-processing revolution of the mid- to late twentieth century. He described his project in the book as extending cognitive science's information-processing model of the mind to the understanding of personality, group dynamics, and social reality. Goleman speculates about what evolutionary survival advantage the self-servingness and hence (implicitly and necessarily) self-deceptiveness of our beliefs might have, besides making life more fun and less scary. After all, being unable to face reality squarely doesn't really sound like a good evolutionary strategy for the survival of the species! Goleman says that the upshot of the whole thing is that our brain is masking pain at the cost of awareness: research on how pain operates has revealed that opioids (among which are endorphins, often called "the brain's own morphine") lessen pain by creating blind spots, a mechanism that functions at every level of behavior, he says, from the psychological to the social.

Goleman begins with animals. The response to danger or injury involves a series of neurophysiological changes in which the brain signals the hypothalamus to release a substance called cortico-releasing factor (CRF), which travels to the pituitary gland and among other things triggers the release of opioids, particularly endorphins. The brain is set up so that the relief of pain is wired into perception itself. Endorphins inhibit attention. Reading this in Goleman's book, I was dumbstruck by how counterintuitive, even ridiculous, this seems. Why would it be of survival advantage to an animal to become *inattentive* at the moment of danger? Why wouldn't it instead become hypervigilant? But as Goleman describes what happens, I began to see the survival value of it. The blocking of pain allows an animal to cease to be concerned with the injury or threat, and as a result it's able to focus on fleeing or fighting. It is not distracted, slowed down, or frozen by the pain or threat. Endorphins make escape and other saving actions possible in the face of immediate danger or pain. There are lots of stories of people doing almost superhuman things under threat—picking up pianos and other enormously heavy objects, running at extraordinary speeds, carrying children for tens of miles, and so on. It's the numbing of pain and the high that make all these extraordinary self-saving actions far more possible.

Yet the mechanism does not consist only in the numbing provided by endorphins. In the first split second another chemical, adrenocorticotropic hormone (ACTH), floods the brain; unlike the opioids to follow, it heightens awareness and hence also the momentary sensation of pain. ACTH and the endorphins are split off from the same molecule, Goleman tells us, and they have opposite effects. It's all about timing: first hyperawareness, then the setting in of obliviousness to pain and terror to enable a quick decision and response. Goleman points out that the neural network pathways that connect pain with attention, and relief of pain with inattention, are "permanently fixed in the brain." We are all wired this way because of the evolutionary advantage it has bestowed of being able to put into effect a life-saving plan of action—fight or flight—that is unclouded by pain or panic. Rats, for example, exhibit a period of pain diminishment after fighting for territorial dominance. The endorphin system has been found in the most ancient parts of the brains of lower animals as well as in those up the line. Even leeches, it turns out, have opiate receptors. It makes sense that not having one's attention drawn to the wound as a result of pain or to panic at the threat of danger makes it possible for attention to be focused on putting a survival strategy into effect. Tuning out pain at that moment, after vigilantly assessing the dangers of the situation, is a good thing.

This bodily mechanism that promotes self-protection and survival in the face of danger spreads upward into our human responses to *psychological* threat and pain as well. Danger comes to mean not only the lion and the bear but also the accountant and the chemistry test. Psychological dangers— threats to our psyches, our symbolic and cultural and biographical selves— replace or are added to physical dangers as sources of the vigilance-numbing flight-or-fight sequence.[81] The brain protects the mind as self from mental pain and fearful thoughts in the way the mind protects the body as self from marauding elephants. What seems to be happening is that the mind protects its symbolic self, its mental processes and states, beliefs, relationships, and social and cultural identities as if they were body parts.[82] Psychological "denial," Goleman proposes, "is the psychological analogue of the endorphin attentional time-out. . . . Denial, in its many forms, is an analgesic, too." When painful negative attributions to the self are made, self-awareness tends to be withdrawn from them. So the protection of one's sense of self, the psychological and social self, is what promotes patterns of self-deception, which are as basic to the mind as the attention-pain system is to the body.

We big-brained creatures do have resources to outsmart our self-deceptiveness, but those big brains also mean that we have better ways of maintaining self-deception through elaborate rationalizations. While the mechanisms that evolution has developed certainly keep us alive in the face of immediate physical danger, the accompanying mechanism of selective and partial personal and social and cultural blindness seems to pose a challenge to our ethical lives, to our relationships, and even to our long-term survival as a species. Wouldn't it have been better if nature had designed us to feel the pain but had also given us better strategies for responding to the problem at hand?

Some Insights About Coming to Self-Awareness from Studies of Multiple Personalities

If being self-aware is difficult for a single self, how much more fraught might it be for someone with more than one personality, more than one self? Nevertheless, there are some insights to be gained about how we can overcome self-deceptive numbing and unawareness from studies of "switchers," or those who have full-blown multiple personalities (now referred to technically as "dissociated ego states") and also from those who have lesser forms of the same disorder.

Martha Stout, a clinical psychologist, has written a moving account of case studies from her practice with patients who have had varying degrees of impairment due to dissociation as a result of trauma.[83] Stout points out that during trauma a dissociative response is "extremely adaptive" and has "a survival function."[84] The hallmark of dissociation has been defined as "the ability of the human mind to adaptationally limit its self-reflective capacity."[85] Dissociation numbs the pain and terror of trauma, and with it comes a loss of emotion and a loss of feeling in the moment, so that the present feels less real (a phenomenon that psychologists call *de-realization*). When the world feels less real and emotions are muted, the feeling of self is also weakened. In an episode of dissociation the traumatized person is not aware of the trigger but only of the sense that she is suddenly far away in a dream world. She may feel that she has "departed from her 'self'" and begun to act "without self-awareness."[86] In the most extreme traumas the self can even become divided into different selves in order to isolate and compartmentalize the pain and terror in alternative selves that have little or no self-reflective capacity. Dissociation induces a sense of "going through the motions" rather than really being there, of departing from one's self. We recall the numbness that Holocaust victims experienced, the feeling that many had that what they were going through didn't feel real or that they were watching it play out as if from a distance or in a movie. That phenomenon of de-realization is an aspect of dissociation. Stout defines dissociation as the "separation of emotion from thought and action," and it results from the withdrawal of attention, of conscious awareness and of the sense of "me-ness" from the traumatic experience. Dissociation, she further suggests, induces a trance-like state and can be experienced as a sort of hazy dream world. Dissociation serves a "selfy" purpose, the purpose of one's own survival and coping, paradoxically, by getting rid of the full sense of self in conscious awareness.

Trauma induces the secretion of neurohormones that produce a fight-or-flight response. But in the long run the stress response becomes problematic, for it plays havoc with brain regions that integrate cognition and emotion, particularly the amygdala and the hippocampus, Stout explains. The hippocampus, which is involved in long-term memory, receives sensory information that has been evaluated for emotional significance in the amygdala via the thalamus. In response to the level of emotional intensity, the hippocampus is activated to organize the new inputs and integrate them into existing information about similar inputs. The consolidation of memories takes place according to the emotional priority given them by the amygdala. But when trauma occurs, the hippocampus, rather than prioritizing the emotionally overcharged experience and working to integrate it posthaste, instead exhibits *decreased* activation and does not effectively organize or integrate the new experience. Portions of traumatic memories are stored as isolated images and bodily sensations rather than as meaningful experiences and memories that are integrated into larger systems of memories. In addition, trauma may at times disable Broca's area of the brain, an area that translates experiences into language. So traumatic events tend to remain both unintegrated and fragmented, and hence inchoate. Stout calls them "chaotic fragments . . . sealed off from modulation by subsequent experience." They are "wordless, placeless, and eternal," she says.

In later life, however, these fragments come to be easily triggered by more complete, integrated memories.[87] The traumatized person may feel that she is not really present.[88] This episodic reactivation of dissociation is maladaptive to a normal life, gutting the realness of experience and relationships. In the worst cases of dissociation, whole days or even weeks or months can be lost to awareness and memory. In "switchers" with multiple personalities, the dominant "I" can be lost for a time while another personality, another identity, takes over, one whose consciousness and memories may or may not be subsequently known by the dominant one. But in either case, the dominant "I" sheds the feeling of "me" during the switch.

The kind of "going through the motions" that we all feel at one time or another, or even the experience of being distracted or being totally absorbed in a movie or play or game, Stout believes, are mild forms of the dissociation that occurs as a response to the terror of trauma. When we dissociate, "we go someplace else," so to speak. "Plainly stated," Stout says,

> it is the case that under certain circumstances, ranging from pleasant or unpleasant distraction to fascination to fear to pain to horror, a human being can be psychologically absent from his or her own direct experience. . . . The part of consciousness that we nearly always conceive of as the "self" can be not there for a few moments, for a few hours, and in heinous circumstances for much longer.[89]

In response to severe trauma, however, a person can get stuck in a deeper dissociative state, a mode of nonfeeling, of de-realization, of absence of self-conscious self from the present moment. In general we recall only those things that were in our self-conscious awareness. So, for example, those who have had traumas in childhood may have almost no memories at all before a certain adult age. For switchers, the self lacks a normal biographical arc. Dissociation in these cases is an involuntary flight from the truth of a past reality too painful and terrifying to acknowledge and feel.

The self that is missing is what psychologists call the "observing ego," and what I have sometimes designated self-conscious awareness, following Panksepp and harking back to the philosopher G.W.F. Hegel. Freud's "unconscious" roughly corresponds to Panksepp's "conscious," and Freud's "conscious" corresponds to Panksepp's "self-conscious." I use both sets of terms, but the context makes the meaning clear. In multiple personalities, Stout says, usually only the dominant personality has an observing ego, while the others do not. A lesser trauma can result in demi-fugue, a state in which the world seems unreal, seen as if through the wrong end of a telescope. Only the contents of one's own mind seem real; one's own body (and hence one's emotions) seems distant, as do other people, even friends and spouses. In this state people feel as if they are missing out on life. The mental mechanisms that give a person awareness of what's going on in the body, including access to one's own emotions, have been dissociated.

Stout identifies six degrees of detachment from self: (1) *Brief phasing out.* This can happen in situations of performance anxiety, such as public speaking or at your own wedding. These are short periods of feeling that you're watching yourself from a distance and create small gaps in our memory. Think of times you arrive home after work but have a sense that you don't remember the details of driving home. You wonder a little, "How did I get here?" (2) *Habitual dissociative reactions.* This is the person who gets the label "space cadet," who habitually is distracted and phases out, loses the gist of a conversation, daydreams, or is often detached from emotional issues. (3) *Dissociation from feeling states.* This is a state of feeling nothing much at all, although one's intellectual capacities remain intact. Stout describes it as a "cordoning off" of emotion, either of particular emotions or of all emotions. (4) *Intrusion of dissociated ego states.* This is the intrusion of an extra personality, the classic multiple personality. (5) *Demi-fugue.* This is a "generalized detachment from self and others" in a cognitive and emotional fogginess. It also engenders a feeling of the separation between body and mind, an inability to feel the body, to experience pain or illness. (6) *Fugue.* This is a total blackout. It results in dissociative amnesia, in which days and sometimes weeks are lost.

All of this gives us some new insights into what a feeling of self entails and how it becomes at times attenuated and even for periods can disappear in normal people. Under threat or stress, this loss of the feeling of self is in the interest of a broader protection of the self and can even be paramount to its survival. The loss of conscious self is an unconscious strategy of an unconscious force of selfiness—in Panksepp's terms, conscious but not self-conscious—of the organism preserving itself and furthering its survival. It's like a lizard shedding its tail in order to survive. An organism clearly does not need conscious self-awareness to protect itself and enact its own survival, its selfiness. Our self-conscious feeling and awareness of self offer an emotional, embodied, and cognitive grasp of the moment (and past and context, etc.) along with a reflective sense of knowing that we know, being aware that we are aware. That's the self-observing ego. When it disappears or is weakened, the sense of self in the present can become temporarily eclipsed and even disappear or become fragmented—with the resulting gaps in our memory, since we remember only what we have (self-) conscious awareness of. But even the mere loss of self-conscious self-awareness that we feel when we get really absorbed in an activity or a daydream involves un(self-)conscious selfy processes fully taking over the present.

The loss of self-feeling, of me-ness, of the self-observing ego, has implications for our sense of agency. The sense of me-ness is necessary in order to have a sense of ownership of one's actions as one performs them, since one has to be (self-)conscious of what one is doing and remember it. It turns out, however, that it is also possible to claim ownership (and responsibility) retroactively, in a self-aware conscious present, for actions that were at the time un(self-)consciously performed and in some instances not even remembered. Martha Stout notes, "Just as it is more than possible to operate in the world with a single, constant identity while evincing no notion of accountability at all, it is possible also to have a deeply bewildering identity disorder that exists simultaneously with a committed sense of responsibility." Stout describes a patient of hers whom she calls Garrett, a house painter, who had been terribly abused in childhood, including being falsely blamed for having kicked his little brother to death. Garrett had a number of intrusive separate personalities whom he had named. His dominant self nevertheless maintained a strong sense of personal responsibility for the un(self-)conscious behavior of its alternative personalities or ego states. Stout has enormous admiration for Garrett's courageous commitment to moral responsibility. And she comments with him in mind that "perhaps nothing defines unified personhood so solidly as the courage of strong commitment to personal responsibility. Garrett could easily have claimed to be *non compos mentis*, with gigantic believability and teams of medical research to back him up. But he never made that claim." Stout presents a conversation she had with Garrett on the topic of responsibility. I found his words profoundly moving and inspiring, as clearly she did, too. Here's an excerpt. Martha Stout is asking the questions and Garrett is answering them:

Question: Do all your personalities remember the same life, have the same memories?

Answer: No, not even close. I don't think James remembers much of anything. He'll always be a child. Gordon remembers the fights, and the really bad times. Willie remembers Sundays, mostly and a few other times. And me, well, I'm not really sure what I remember.

Question: Do your personalities know one another?

Answer: Some of them do. James and Big James know each other, and they know Gordon. . . . But Willie doesn't know anyone except me, of course. And Abe is really a lone wolf. . . . And I think there may be more of them, ones that even I don't know.

Question: Do they have conversations with each other in your head?

Answer: Constantly, constantly, constantly.

. . .

Question: Could one of them, say, buy a car, or have a friend or lover the others disliked?

Answer: Sure. Sometimes I get strange messages on my answering machine from people I don't know. I have CDs that I hate. I have no idea how I got them. Sure.

Question: What if one of your personalities did something really bad—committed a crime, perhaps—who is responsible in that case?

Answer: I am.

Question: Wait a minute. I thought you didn't have a sense of *I.*

Answer: Doesn't matter. I am.[90]

Stout has found that whether or not a person with multiple personalities claims ownership of, and hence takes responsibility for, the actions of all of his or her alternative selves is a strong predictor, and possibly the only predictor, of whether he or she is capable of recovery and will in fact recover. Garrett took responsibility for his un(self-)conscious selves, for alternative egos, that didn't even feel like "me" to him. How much more possible is it, then, for normal people like most of us, who do not have dissociated ego identities, to come to own and hence take responsibility for our un(self-)conscious, denied, and disavowed thoughts, feelings, and actions, all of which are expressions of our self and of its selfiness? Indeed, I will argue, such a step is necessary for moral agency.

Garrett is like a contemporary Oedipus, acknowledging as his own and taking retroactive moral responsibility for actions done of which he was ignorant. I'll end this account with Martha Stout's own reflections from her clinical experience on the source of the capacity for responsibility in the ownership of the full self:

People who are compelled and organized by a sense of responsibility for their actions tend to recover. . . . [T]he difference [between those who are likely to recover and those who are not] is that of tenaciously assuming personal responsibility for one's actions, and therefore taking on personal risk, versus placing the highest valuation upon personal safety, both physical and emotional which often precludes the acknowledgement of responsibility. . . .[91]

[W]e cannot simultaneously protect ourselves and experience life fully. . . .
Maybe there is no salvation for any of us outside of the meaning system provided by personal responsibility, despite all the daunting risks. Perhaps this is
why we so doggedly look for examples of accountability in our role models, our
parents, our leaders.[92]

As Stout puts it, owning up depends upon an "increased self-observation
[that] exercises the observing ego, the part of the self that will be able to
view dissociation as a currently unnecessary limit upon one's freedom."[93]

The philosopher and psychoanalyst Jonathan Lear calls the owning of disavowed aspects of the self, and hence the taking of responsibility after the fact
for the discovered givennesses of the self, its tendencies and actions, which one
could never have anticipated, understood, or even been aware of at the time,
"the other Oedipus complex." As he puts it, why should Oedipus accept responsibility for his acts, since both his actions and his fate were ordained by
the gods before he was born? This question implies our own: why should each
of us assume responsibility for our own actions and our own fate if they are
determined, if not by the gods, then by who our parents were, what world
and situation we were born into, and who we became as a result of our early
experience, our genetic inheritance, and on and on? This is Lear's answer:

> Now, Oedipus, in accepting responsibility for his acts, is claiming that the
> truth that ultimately matters is, as he says of his blinding, that "I have done it
> with my own hand." Whatever the gods ordained, Oedipus says, the fact is
> *that I did it.* Oedipus is in effect claiming a part of nature for himself. . . . In
> accepting responsibility, Oedipus is making an altogether more elemental
> claim: "these acts are *my acts.*" Oedipus thus constitutes himself as an agent, a
> locus of activity. *"Where it was, there I am."*
>   It is this claim which reverberates at the deepest levels of our souls.[94]

Lear concludes: "If I am to integrate the 'it' into my life, I must actively take
it into my soul."[95] Responsibility, a claim of the moral agency of the self,
consists in owning up to the factuality of the self; the self is something that
in a sense has happened to me, and I discover it as if it were outside me. It is
not something I invent, originate, can completely control, or even have full
self-conscious awareness of. Nevertheless, it is even more deeply "me" than
the actions I take toward my goals and ideals (or away from them), actions
about which I have some self-awareness. This given "me" is that by which I

am constituted. It is the "me" I find, and I resign myself to accepting it. In so doing, Lear says, I become transformed from being passively acted upon into a morally responsible agent.

Yet in embracing the enhancement of self-awareness, we are confronted with the vast problem of the un(self-)consciousness of most of our cognitive and affective mental life and even the nonconsciousness of some of them. The cognitive neuroscientist Michael Gazzaniga has proposed that 98 percent of our cognition is unconscious, or "implicit," and very little of it actually reaches full self-conscious awareness. So, much of selfiness, the urge to preserve oneself both in us normal folks and, as we have seen so dramatically, in severely dissociated people, operates beneath self-awareness. It even includes at times the loss of the conscious feeling of self and a disavowal of and dissociation from self-presence, from the self-observing self (or observing ego) with the concomitant loss of the feeling of agency and, ominously and sadly, of a sense of moral responsibility both in that present and in retrospect. Nevertheless, enhanced self-awareness of some aspects of what is usually or partly un(self-)conscious and also of what once was un(self-)conscious is also possible.

Conscious Awareness and the Feeling of Agency

Psychologist Daniel M. Wegner draws two conclusions that are crucial for rethinking moral agency. The first is that there is no such thing as a free originating will, so actions and agency ought to be explained in terms of their causes, and so ought the (somewhat misleading) feeling of self-origination that accompanies them.[96] The second is that self-conscious mental processes are accompanied by a sense of self-agency, whereas un(self-)conscious mental processes do not have an accompanying sense of self. Self-conscious awareness feels selfy, but un(self-)conscious thinking and feelings do not, for "automatic actions do not support inferences of agency during the action . . . they do not feel willed as they unfold," he says.[97] Nevertheless, when un(self-)conscious mental processes come into conscious self-awareness, they can acquire a sense of self-ownership, selfiness. "When automatic processes do happen to announce their resulting actions and thoughts to consciousness," Wegner remarks, "they may then be eligible to give rise to a sense of agency. But controlled processes do this every time."[98] By "controlled processes," he means those in which we (self-)consciously envision an aim and make that aim come about; they are actions we believe we control by our conscious thoughts and intentions. Their hallmark is stability and predictability.

Wegner wants to decisively challenge the notion that there is a little self inside us, a homunculus, who controls our actions merely by making decisions that cannot be further causally accounted for by citing prior causes constitutive of that exercise of will.[99] He calls the notion of free will a kind of non-explanation, which presupposes a God-like control over our own behavior, a God-like self.

> Another way to explain a homunculus is simply to say that it has free will and can determine its own behavior. This means the homunculus causes things merely by deciding, without any prior causes leading to these decisions, and thus renders it an explanatory entity of the first order. Such an explanatory entity may explain lots of things, but nothing explains it. This is the same kind of explanation as saying that God has caused an event. . . . Just as we cannot tell what God is going to do, we cannot predict what a free-willing homunculus is likely to do either. There cannot be a science of this.[100]

Wegner remarks that there is still a tendency among scientists to read (self-)conscious and controlled mental processes as the real human person, whereas the un(self-)conscious processes are thought of as a kind of robot within, "little smarter than a bar code reader." That we have that feeling of aim and fulfillment from those controlled processes tends to engender in us the illusion of a homunculus inside us—someone directing the action. The Blakeslees address the false homunculus presupposition, too, and offer a clearly stated alternative view of the self:

> The illusion of the self isn't that there is no such thing as you. Nor does the illusion of free will mean that you cannot make choices. Instead, the illusion is that the self and free will are not really what they seem to be from your, the "end user's," perspective. The illusion of free will is that free will has infinite scope, rather than being a flexible set of feedback loops between higher-order maps and emotional and memory storage systems in the brain. The illusion of the self is that the self is a kernel, rather than a distributed, emergent system.[101]

While no one comes out and claims the reality of a homunculus, it's an unspoken assumption, Wegner says. It involves the presupposition that

"controlled processes are . . . conscious, moral, responsible, subtle, wise, reflective, and willful" while the "automatic processes may be unconscious,
unintentional, primal, and simple-minded—as well as impulsive, selfish,
and prejudiced to boot." Even today only the "conscious self is still accorded
full status as a human agent."[102] This is to marvel at the conscious self rather
than to explain it, he suggests, and leaves "controlled processes . . . seem[ing]
less than genuine, reflecting unpredictable human choices rather than scientifically respectable causes."[103] Wegner sets out to demystify some aspects
of controlled processes—the ones we feel we control by conscious
awareness—by explaining causally how they arise and why they feel self-
originated in a way that belies further causal origins and explanation.

The Because actions that are (self-)conscious involve a preview loop that turns
out to accord with the results as we observe them after the fact, we attribute
to those actions a feeling of self-ownership—and also, mistakenly, often a
claim that they are freely willed. In fact, we are merely inferring after the
fact that we originated them. "The self," Wegner proposes, "can be understood as a system that arises from the experience of authorship, and is developed over time by a set of controlled processes that manage memories and
anticipations of authorship experiences. We become agents by experiencing
what we do, and this experience then informs the processes that determine
what we will do next."[104] But some of the most astonishing human feats
of skill and thought often do not feel self-originated because they are
un(self-)conscious; composing a symphony, writing a poem, having a great
discovery or insight, or breaking an athletic record can feel this way. We feel
we *are* the composing or the writing or the running; we do not stand outside
the process directing it, but rather it directs us, as it were. We are delighted
to claim these achievements ex post facto, Wegner says, but we did not feel
that we *willed* them in the doing; rather, they somehow happened to us.
"The authors of skilled actions often report feeling like spectators who happen to have particularly good seats to view the action."[105]

The sense of self-ownership of a (self-)conscious action (and the illusion of
free will) play out in the following way, according to Wegner: "Why does it
feel as though we are doing things? The experience of consciously willing
our actions seems to arise primarily when we believe our thoughts have
caused our actions. This happens when we have thoughts that occur just before the actions, when these thoughts are consistent with the actions, and
when other potential causes of the actions are not present."[106] The thought
must occur just before the action in order for us to make the inference that

we have engaged in a voluntary action, and the action must turn out to be what we expect. Equally important and paradoxical, however, is the necessity of the absence of a causal explanation in order for us to feel that we caused something to happen. *We attribute actions solely to the self, to our thoughts— and hence we claim that they are willed—only when we do not know the causes.* Wegner comments that "without a perception that one's own thought is the exclusive cause of one's action, it is possible to lose authorship entirely and attribute it even to an unlikely outside agent." He suggests that a feeling that one is not the complete cause may be what happens in trances, in spirit possession, in speaking in tongues, even perhaps in Ouija board writing and the like. A sense of not having full causal agency may explain the results of the Milgram experiments and those like it, Wegner says, insofar as Milgram pointed to an experience that his experimental subjects articulated of an "agentic shift," "a feeling that agency has been transferred away from oneself."[107]

Wegner developed experiments that tested and confirmed that the sense of ownership of one's actions arises as a result of an external inference, rather than privileged introspective access to ourselves as exclusive causal agents, as we assume. He describes an experiment in which experimental subjects looked in a mirror at their own reflection as someone else hidden behind them extended arms on each side of them as if they were the experimental subjects' arms. One of the hands snapped a rubber band on the wrist of the other. Only those experimental subjects who heard previews of the upcoming hand movements and then saw them had a skin conductance response and also reported that they experienced the hand movements as their own and as being controlled by themselves along with an emotional feeling of ownership of the hands. Those who heard no previews or heard previews inconsistent with the subsequent actions had no such heightened conductance, nor a feeling of ownership. Wegner concludes that the feeling of mental causation is thus "an inference we draw from the juxtaposition of our thought and action, not a direct [internal] perception of causal agency . . . it is merely an estimate of the causal influence of our thoughts on our actions, not a direct readout of such influence." It's a rough, useful estimate rather than a true reading. The inference from conscious self-aware thoughts to observable actions is the source of a feeling of authorship and ownership of our actions. So the feelingthat our thoughts are the *exclusive* causal origin of our actions, that there is a virtual agent, a free will, a homunculus inside who is performing the actions, is an illusion. The illusion is an artifact of the psychological mechanisms—mechanisms that are nonconscious—underlying the sense of ownership, of agency. Wegner sums it up: "The way the mind seems to its owner is the owner's best guess at its methods of operation, not a revealed truth."[108]

The uncanny conclusion we come to is that introspection is open to doubt and to external critique. We do not have the privileged access to or control over much of our internal workings that we presume. We cannot overestimate the implications for ethics of what Wegner has exposed about our sense of agency: of how we come to own our actions, which ones feel selfy to us and which do not, and also how and why the selfiness we feel in actions has degrees of intensity. He has pointed out that the mechanisms of the feeling of agency tend toward two crucial errors: they lead us to claim too much personal, individual ownership of our (self-)conscious actions and too little of our un(self-)conscious actions. The latter type of action feels as if it happens to us from the outside and is not really our own. Moreover, into this latter category of actions that we feel are not really our own fall all those actions whose agency does not feel as if it originates exclusively in our own (self-)conscious thoughts; this includes joint actions, group actions, and actions carried out under the authority of another or others. This has ominous implications for both individual and social moral responsibility. Furthermore, our overweening sense of ownership of our (self-)conscious actions—even if we don't also attribute to them the illusion that we completely originated them and hence "freely willed" them—is also pernicious, for it shows no mercy to ourselves or others, while letting off the hook all the other contributing causal factors—social, biological, situational, cultural, tribal, hierarchical, and so on.

Looking Outward to Discover the Self

> I no more wrote than read that book which is the self I am, half-hidden as it is from one and all.
>
> —DELMORE SCHWARTZ

Timothy D. Wilson, a professor of psychology at the University of Virginia, aims to give a global picture of what the unconscious does, how and why it came about, and what we can do about some of its downsides by increasing our self-knowledge. The unconscious isn't *one* thing, and it is certainly not one receptacle or one mode of operation; rather, it is a variety of capacities and operations that take place unconsciously. Many of these are inaccessible to consciousness, that is, nonconscious, possibly because they evolved prior to the evolution of consciousness, Wilson suggests. But the basic reason that many processes are unconscious is that (self-)conscious awareness is quite limited, whereas the un(self-)conscious mind has a far greater capacity to hold and work with information. It is largely a matter of efficiency, Wilson says. "Just as the architecture of the mind prevents low-level processing (e.g., perceptual processes) from reaching consciousness, so are many high-order psychological processes and states inaccessible." Having all this thinking and feeling and judgment occurring outside of (self-)awareness enables a great deal of parallel processing about the environment and other things to take place while a person is thinking and doing something else entirely. All this efficiency has a cost, however, in that much of it is and remains inaccessible to us. As a result, we cannot for the most part know ourselves directly (*pace* Descartes), and in many cases the only way to come to greater knowledge of our unconscious selves, our hidden minds, is to observe our own behavior as if from the outside, and also to attend to people's reactions to us and be open to their assessments of us. The new scientific grasp of the un(self-)conscious ought to make us much less sanguine about our powers of introspection. In many respects we may know even less about ourselves than an outside observer does. *To come to greater knowledge of ourselves, we will have to look outward to see inward.* And we will have to direct a skeptical eye at what seem to be the assurances of introspection. We are indeed, as Timothy Wilson quips, "strangers to ourselves."

> A picture has emerged of a set of pervasive, adaptive, sophisticated mental processes that occur largely out of view. . . . There is more agreement than ever before about the importance of nonconscious [i.e., un(self-)conscious, Panksepp's "conscious"] thinking, feeling, and motivation.
>
> The gulf between research psychologists and psychoanalysts has . . . narrowed considerably as scientific psychology has turned its attention to the study of the unconscious. This gap has not been bridged completely, however, and it is clear that the modern, adaptive unconscious is not the same as the psychoanalytic one.[109]

James S. Uleman defines the current view of the unconscious as including "internal qualities of mind that affect conscious thought and behavior, without being conscious themselves."[110] Uleman says that in the last twenty years the multidisciplinary scientific picture of the range and content of unconscious processes has changed the landscape considerably. "Unconscious processes seem to be capable of doing many things that were, not so long ago, thought of as requiring mental resources and conscious processes," he writes. "These range from complex information processing through behavior to goal pursuit and self-regulation." Wilson proposes that the unconscious and the conscious are two information-processing systems that function in different ways and have different evolutionary histories.

So what does the un(self-)conscious include? First of all, one category encompasses thoughts and memories that are not being recalled at the moment but can be brought to mind when we want them; Freud called that the preconscious. But more interesting is the recent growing understanding that many sophisticated and high-level modes of thinking also take place un(self-) consciously. These modes include "sizing up the world, warning of danger, setting goals, and initiating action," understanding and using language, perception generally, recognizing faces, quickly evaluating the environment and things and events in it as good or bad, and also aspects of the self (such as personality traits and tendencies).[111]

The systems that produce our perception, language capabilities, and motor capacities are unconscious. Wilson refers to these as "low-level" unconscious mental processes. These are largely nonconscious. We even have a sixth sense, proprioception, which gives us the position of our bodies and our limbs. We are not consciously aware of how we know where our feet are right now or how we got that sensory information, but without it none of us could stand up or move around. "An important role of the nonconscious mind [which, for Wilson, includes both nonconscious and un(self-)conscious processes] is to organize and interpret the information we take in through our senses," Wilson says.[112] Scientists, for example, have discovered that our senses take in and send to the brain for unconscious processing about eleven million pieces of information per second, whereas conscious awareness can process a mere forty pieces of information per second! An important operation that the unconscious mind performs with this vast amount of data is the detection of patterns. There are also high-level unconscious mental processes, which include "our ability to think, reason, ponder, create, feel, and decide," Wilson writes.[113] High-level unconscious processes select, interpret, and evaluate data, set goals, and initiate action. Learning can be, and is, both with and without conscious awareness. We learn complex patterns in the environment unconsciously, too. We certainly learn language without a conscious awareness of how we do it and what we are doing—although we often try to learn a foreign language in a quite conscious manner, sometimes to good effect and sometimes not. Learning a second language self-consciously may be something like learning to drive or to knit: first there is much self-conscious awareness of what we are doing and how we do it, and then the performance becomes habituated, automatic, and un(self-)conscious, only calling us back to self-conscious awareness when, say, a car stops short in front of us or we drop a stitch. It is simply not the case, as it used to be supposed, that lower-level processes are unconscious and that higher-level conscious processes, such as thinking and reason and deciding, take over to use the data and direct the capacities made available to it through the unconscious. That model of the mind has been proven wrong. Both lower and higher processes are at times unconscious.

An important job of the unconscious is to select what reaches consciousness. That is "selective attention," which is an unconscious filter that allows only certain things to reach conscious awareness. For example, I know that I have a keen eye for handmade pottery, homemade ice cream, and cats. My unconscious radar is always sorting for these Heidi-salient categories and picking them up in the environment, and I snap to awareness out of any reverie or conversation or daydream when they appear in my environment. And then there are somatic markers, the emotional charges we build up through the associations of events and people and things with our memories of their emotional valences in our past experiences. "A case may be made," Wilson continues, "that the most important function of the adaptive unconscious is to generate [the] feelings" that drive quick decision making. The adaptive unconscious scans the environment, detecting patterns, and signals a good or bad evaluation, an almost instantaneous thumbs-up or thumbs-down. It is also, according to the neuroscientist Joseph LeDoux, a "danger detector." One job of conscious thought, on the other hand, is to look more carefully at the environment and give a more detailed analysis of it than the initial quick, unconscious, and often affective one that is so prone to error and prejudice, and to anticipate, mentally simulate future effects, and plan for tomorrow and next year.[114]

The adaptive unconscious is also a spin doctor, interpreting information and people's behavior outside of awareness in a way that masks subliminal influences and that comports with both their view of the world and their view of themselves. Wilson and a colleague have dubbed this the "psychological immune system." We keep coming back to the self-protective unconscious, and now we have a name for it. Wilson writes:

People's judgments and interpretations are often guided by . . . the desire to view the world in the way that gives them the most pleasure—what can be called the "feel-good" criterion. . . . People go to great lengths to view the world in a way that maintains a sense of well-being. We are masterly spin doctors, rationalizers, and justifiers of threatening information. . . . Just as we possess a potent physical immune system that protects us from threats to our physical well-being, so do we possess a potent psychological immune system that protects us from threats to our psychological well-being.[115]

Wilson says that we differ somewhat from person to person and from culture to culture in what drives our unconscious pleasurable distortions in our thinking and beliefs. "What makes us feel good," Wilson goes on, "depends on our personalities, our culture, and our level of self-esteem." But thinking and belief un(self-)consciously shaped by the desire for thoughts pleasing to the self is universal. Wilson concludes that the conflict between thinking that grasps reality and thinking that makes the self feel good "is one of the major battlegrounds of the self." The self-deception that we are all prone to as a result has some advantages in keeping us optimistic and happy with ourselves. It also has some obvious disadvantages. The psychological immune system is like many other unconscious mental systems, Wilson says, in that it is a trade-off between accuracy and efficiency. The need for efficiency often trumps the need for accuracy and, as a consequence, produces illusions and deceptions that are not in themselves adaptive but rather unfortunate by-products of adaptations. We have optical illusions; we overcategorize people and develop irrational prejudices; we are overly committed to our unconscious categories and fail or are quite slow to assimilate information that contradicts and challenges them. We have little internal access to our unconscious processing and even heartily deny the errors and weaknesses our unconscious is prone to. Nor do we have much direct introspective insight into our motives and personalities.

"Many of people's chronic dispositions, traits, and temperaments are part of the adaptive unconscious," Wilson points out, "to which they have no direct access." So we all tend to make up theories about ourselves from what our parents have told us, from our cultural environment, and generally from our surroundings. People's stories about themselves often have very little correspondence to their actual unconscious personalities and talents.[116] One of the most recent and sophisticated theories of personality considers personality to be a combination of "cognitive and affective variables that determine how people construe . . . and evaluat[e] different situations." People then act on their characteristic interpretations. It is these patterns of emotionally charged interpretation and evaluation that are now thought to constitute a "cognitive and affective personality system." This system is fundamentally unconscious; it is implicit and procedural. People's self-reports, by contrast, capture almost exclusively their conscious self-construals, their consciously constructed selves, and "there is increasing evidence that people's constructed self bears little correspondence to their nonconscious self." For the unconscious self, its dispositions and personality, cannot be accessed directly but is amenable only to a third-person perspective. We can come to know ourselves better only by becoming good observers of our own behavior as seen from the outside, as others do.[117] In fact, studies have shown that others make better predictions about how we will behave than we do. So, contrary to our deepest convictions and expectations, we tend to know others better than we know ourselves, and other people know *our* unconscious better than we know our own. In predicting our own behavior we rely only on our self-reports and self-constructions, but others see what we do in a more complete and empirical way. When it comes to ourselves, each of us tends to focus on our internal self-constructed view of ourselves, built up from our self-conscious thoughts and our self-protective rationalizations and confabulations.[118] Other people are less inclined to see us through the rose-colored selfy glasses through which we generally see ourselves.[119] This is a startling finding and a downside to the unconscious that we are loath to accept. "Maybe it's true of others," we say to ourselves, "but certainly not of me!"

Furthermore, our unconscious motives, unlike our (self-)conscious ones, are generally opaque to us. Wilson suggests that the line between what we know and what we don't know of our own minds cuts between the conscious and the adaptive unconscious. "To the extent that people's responses are caused by the adaptive unconscious, they do not have privileged access to the causes and may infer them," and hence they confabulate reasons for doing what they do. Furthermore, many "judgments, emotions, thoughts, and behaviors are produced by the adaptive unconscious" and hence are unknowable directly through introspection. "But to the extent that people's responses are caused by the conscious self, they [do] have privileged access."[120] People not only have little access to their motives but can also be wrong about their emotional feelings. Our access to our feelings, contrary to our Cartesian presuppositions, is *not* infallible. "People can be wrong," Wilson says, "when they report a feeling."

We can also have emotions that we are not aware of at all, Wilson says. "The adaptive unconscious," he writes, "can have its own beliefs and feelings" that we are not aware of since it operates "independently of consciousness."[121] The neuroscientist Joseph LeDoux has identified two emotional pathways in the mammalian brain, which he calls the high road and the low road. The low road is fast and dirty and unconscious, with little information processing. The high road, by contrast, is slower and goes through the cortex. It involves more cognitive nuance, detail, and thoughtful conscious analysis. The low road, LeDoux says, is a kind of early warning system, alerting us to danger, whereas the high road looks again and lets us assess situations more slowly and completely—whether what looks like a snake in our path (which we have already unconsciously responded to emotionally, with our heart rapidly beating) is instead a stick, for example. Sometimes the two systems work together well, but at other times the two systems are out of sync. Prejudice against an ethnic or religious or sexual group is a great example of a context in which we find that phenomenon frequently played out. We think we are not prejudiced, but if we look at ourselves as if from the outside, we can often observe our own biased behavior.[122] It takes actually observing ourselves as if from the outside as we act (as if we were observing someone else), to see how we ourselves really feel. We indirectly gather how we feel by drawing implications from our behavior rather than by direct introspection. Wilson reports that the research averaged across a number of studies has even shown that "there seems to be no net advantage to having privileged information about ourselves, [for] the amount of accuracy obtained by people about the causes of their responses is nearly identical with the amount of accuracy obtained by strangers."[123]

The best way of looking inward is not through introspection; instead, as Wilson puts it, "we must look outward to know ourselves." In particular, people's self-theories, to which we are all inordinately attached, interfere with and obscure our actual selves. Others' appraisals of us and our own often disagree; moreover, others' appraisals may capture us far better than our own. So in this case it is really our un(self-)conscious self (reflected back to us by others) and our (self-)conscious self (our fabricated self-stories) that are at odds. If we want to become aware of our unconscious self, the best place to begin is by becoming good observers of our own behavior and inferring from it—as others do—what our feelings must be. Wilson proposes that the realization that we have to look outward in this way to see inward is "truly a radical proposition."[124]

But what is the advantage of knowing our unconscious? Why bother? An answer may come from some studies done of students who were tested at the beginning of the semester to determine to what degree their unconscious motives and goals coincided with their (self-)conscious motives and goals. Those whose unconscious and conscious were more internally consistent and aligned had a greater sense of well-being at the end of the semester than those whose motives and goals showed a greater disparity between the conscious and the unconscious.[125] Such self-consistency also had the advantage of enabling those who had it to predict their feelings in future situations far more accurately than those who did not, and as a result they could work toward realistically achieving goals that would make them happy. Increased self-awareness and consequent self-ownership offer us the possibility of enhanced control over our actions and lives; they mark the birth of greater agency and a greater sense of responsibility.

The Fundamental Openness of the Core Self: Brain Maps of Bodily Encounters with the Environment

The primary affects and basic emotional systems that Jaak Panksepp has identified in all mammals produce the root sense of self because, as the philosopher Stephen Asma puts it, "they have to be 'owned' by the organism to work properly."[126] Ownership of even unconscious aspects of the self, it turns out, can extend outward to include our relationships to others and to the environment. This is one conclusion that we can draw from research conducted by Antonio Damasio. Both Panksepp and Damasio are avowed Spinozists, embracing a monism according to which mind is embodied and body enminded—Spinoza's "dual aspect monism"—so that consciousness and self emerge from deep body levels, and both reject the Cartesian notion of consciousness as a disembodied spectator. And while both Panksepp and Damasio hold that deep and evolutionarily ancient layers of the brain produce rudimentary consciousness along with a basic sense of self, for Panksepp "subjectivity resides first in the biological realm of action," whereas for Damasio it emerges from the brain's sensory structures. Both cite the self-mapping of bodily states and engagements produced by our neurons as crucial to the emergence of the feeling of self and consciousness. Yet while Panksepp focuses on "motor-action coordinates that create a primal representation of the body"—that is, what he calls the core "SELF," an acronym for "simple ego-type life form"[127]—Damasio focuses instead on the affective feeling of the neural self-mapping of sensory states, which capture the presence of the self in relationship to a given object. In its simplest form this relationship of self to object occurs as "the kind of image that constitutes a feeling . . . the presence of you is the feeling of what happens when your being is modified by the acts of apprehending something."[128] Internal states "occur naturally along a range whose poles are pain and pleasure," and that pleasure or pain is then attributed to the external (and sometimes internal) object or event. This is the way that our internal states "become unwitting nonverbal signifiers of the goodness or badness of situations relative to the organism's inherent set of [survival] values."[129] They define an embodied personal point of view, an interested perspective from the standpoint of the body.[130] Damasio now locates the most primitive and basic level of self even below Panksepp's emotion action systems in the brain stem. In his most recent book, *Self Comes to Mind*, Damasio puts forth the position that it is the painful and pleasurable body-states produced low in the brain stem that are the deepest origins of a feeling of self. Only subsequently, later in the evolutionary inheritance, do these self-feelings get taken up into the slightly higher motor action systems—where Panksepp sees them as originating— and eventually (from the standpoint of evolution) into the nested hierarchy of neural networks, where autobiographical linguistic memory develops.[131] The exact origin of the most basic feeling of self, whether it turns out to be Damasio's or Panksepp's or some other, cannot be determined as yet from the evidence. Nevertheless, the importance of both Damasio's and Panksepp's data is that they provide much evidence that the feeling of self—the origin of self, the urge to self—is in very ancient animal parts of the brain.

The evidence provided by both Panksepp and Damasio lends itself to a reading of the self as unitary (although not a homunculus). To say that the self is unitary at the deepest level and not merely integrated only at the high level of language and autobiography is to challenge not just the Cartesian self as a "substance," that is, a primary metaphysical unit, but also Hume's view that the self is not unitary at all but instead is an ongoing subject of momentary raw feelings, or Kant's view that these disconnected, discrete feelings get organized and unified (what philosophers call "the binding problem") as a self only at the upper reaches of our cognitive capacities and functions.[132] For what Panksepp has surely shown is that the self of the basic emotion action systems has a unitary orientation toward its own survival, which animates the egoistic (selfy) point of view it expresses in all its basic emotions. It is an orientation toward the overall survival and furthering of this singular organism (for Panksepp), or a more sensory self-orientation (for Damasio), that survives at the increasingly cognized and linguistic higher levels. Damasio's sensory origin of self has a further important and fascinating implication: that the sense of self emergent in sensation opens each of us to the world, for it is in its encounters with the environment in which our sense of self is born. The self thus comes about in relation and external engagement. The self as a self is thus fundamentally relational and environmentally engaged. It opens outward as well as securing at the same time inward stability and survival.

According to Damasio, feelings, including emotions, are the feelings of our own body. They are "the direct perception of a specific landscape: that of the body," he says.[133] Feelings offer us momentary glimpses into body states, into salient parts of the body landscape.[134] They provide us with "a glimpse of the organism in full biological swing, a reflection of the mechanisms of life itself as they go about their business."[135] Damasio says that an "emotion [is] a reflection of a *self*. It shows who one is" at the deepest level of our being, down to our core, our self-organization, and literally our viscera. He writes, "Feelings let us *mind the body*. . . . They let us mind the body 'live,' when they give us perceptual images of the body, or 'by rebroadcast,' when they give us recalled images of the body state[s]."[136] This kind of feeling expresses the "global ownership" of self, the sense of "mine-ness" we feel about our bodies. This feeling of holistic ownership can be contrasted with the localized feelings we have of parts of our body—of a sore hand or an itchy knee. It is what we all experience as our identification with our whole body, our self-location, our first-person perspective. It is a sense of being a holistic entity, a body located in particular space and time.[137] The neural self-mapping, Damasio says, is carried out by "a collection of brain devices whose main job is the automated management of the organism's life." The information needed to accomplish this task is provided by "neural maps which signal, moment by moment, the state of the entire organism."[138] Damasio proposes that consciousness brings to the organism the capacity to connect inner regulation with the processing of images. It thus enables regulation to be fine-tuned to the precise details of the environment.[139] "It places images in the organism's perspective by referring those images to an integrated representation of the organism, and in so doing allows the manipulation of the images to the organism's advantage."[140] At the farthest reaches, even our experiences and our stories about ourselves, and finally our social and other identities, are mapped onto more basic maps of the feeling self, Damasio conjectures.

The "I That Is We" and the "We That Is I"

The crucial point I want to emphasize is that the global feeling of self, because it includes in its composite images the history of our interactions with, responses to, and manipulations of our environment (in other words, our emotional associations with the objects, situations, and people in our environment) is an extended self. This self extends not only back into the past and forward into the future but also outward into the world; the world is taken in as self-related and self-constituting. As Damasio puts it, "The presence of you is the feeling of what happens when your being is modified by the acts of apprehending something."[141] It is this extended self—the self that grasps itself as interacting with, responding to, and encompassing aspects of the environment—that, I believe, makes ethics possible. Selfness turns out to be the record of organism-environment engagements introjected into self-maps, and hence those engagements are what the self is and what is felt as self. It is also projected into both past and future. So a self cannot be self-enclosed; it is never an internal program playing out, but instead a dynamic history of interactions, the self-maps, projected into the future. A self is never capable of definition apart from its environment and interactive history. A self is only a self in and as a product of action and interaction. "Consciousness," Damasio writes, "from its basic levels to its most complex, is the unified mental pattern that brings together the object and the self."[142] Blanke and Metzinger make a related claim in their paper on how the feeling of self comes about. They argue (following Metzinger's earlier work on self models)[143] that it is only when self and object are brought together that a strong point of view of the self comes about:

A strong 1PP [first person perspective] appears when the system as a whole is internally represented as *directed* at an object component, for example a perceptual object, an action goal as internally simulated, or perhaps the body as a whole. A strong 1PP is exactly what makes consciousness *subjective*: the fact that a system not only represents itself as a self but also as *"a self in the act of knowing."* . . . It co-represents the representational relation during the ongoing process of representation.[144]

Damasio in his research on self-maps develops this point. On one hand, feeling and emotion have as their primary object the body, so thinking itself is about the body. As Damasio puts it, "The mind had to be first about the body, or it could not have been."[145] Yet at the same time it is about the world.

> Our very organism and not some absolute external reality is used as the ground reference for the constructions we make of the world around us and for the construction of the ever-present sense of subjectivity that is part and parcel of our experiences. . . . On the basis of the ground reference that the body continuously provides, the mind [must be first about the body but] can then be about many other things, real and imaginary.[146]

Feelings based in the body and extended to the fully cultural and biographical self connect us to the world. As the late philosopher of the emotions Robert Solomon put it, emotions are our "engagements with the world" and give us "our basic orientation to the world and to one another." Damasio defines a feeling as the relation between self and other, self and world:

> If an emotion is a collection of changes in body state connected to particular mental images that have activated a specific brain system, *the essence of feeling an emotion is the experience of such changes in juxtaposition to the mental images that initiated the cycle.* In other words, a feeling depends on the juxtaposition of an image of the body proper to an image of something else, such as the visual image of a face or the auditory image of a melody.[147]

For Damasio, what is represented in consciousness is neither the "basic self structure" nor the object but rather always and necessarily "*the interaction of the two.*"[148] To be a self is to be in relation with objects and others and to track those relations as self.

So we are indeed selfy, but that selfiness comes to include others and our worlds, as well as our own un(self-)conscious thoughts and feelings. As Damasio proposes, we come to have internal subjective experience (both un[self-] conscious subjective experience and [self-]conscious awareness) neither in the self-mapping nor in the mapping of objects but instead just when we know that our own state has been changed by an object such that the object has become salient and affectively charged—which is to say it is has become an emotional object for us and placed within memory.[149] So subjectivity in its very nature is a feeling of neither self nor object but of the dynamic relations of self and objects, self and environment, and the emotional memories of those relations. We internally express and track those interactions and produce them interiorly as self-related. "Because the sense of that body landscape," Damasio writes, "is juxtaposed in time to the perception or recollection of something else that is not part of the body—a face, a melody, an aroma—feelings end up being 'qualifiers' to that something else."[150] *Feelings connect the external world to internal body states.* Our internal states thus "become unwitting nonverbal signifiers of the goodness or badness of situations relative to the organism's inherent set of [survival] values."[151] They define an embodied personal point of view, an interested perspective from the standpoint of our body.[152] Subjective experience, at its most fundamental, is the story of our body's experience in the world, Damasio proposes. We are selves open to the world, engaged with the world, constructed in response to the world. Our selfiness is our felt survival-promoting perspectives on our engagements with the environment. This person is good; that person is bad; this situation is both good and bad . . . such evaluations are written into our neural pathways and evoke and modify Panksepp's primary emotions; they are felt images, according to Damasio, yet not of discrete objects outside the self but instead of the composite of self and object with its survival-based feeling tone and urge. Selfiness opens to self-in-world and world-in-self. That is the topic of the next chapter.

9

The Self Beyond Itself: The "We That Is I" and the "I That Is We"

The mass [or collective mind] seems to function at an archaic [psychological] level. . . . The person has abdicated conscience to the external world.

—JONATHAN LEAR

Compassion is selective and often ultimately self-serving. . . . Man defending the honor or welfare of his ethnic group is man defending himself.

—EDWARD O. WILSON

The "I That Is We"

I continue here to present a range of neurobiological and other evidence for the claim that the boundaries of the self extend way beyond the scope of the body and that our investment in others and in the world is finally what ethics is all about. Along with self-mapping, the discovery of the self beyond itself, invested in and spanning its worlds both social and natural, is what we must now realize is the source—or a major contributory source—of ethics. I argue here that it is this sense of self as spanning mind-body and world that is the origin and nature of our moral investment and agency in the world. We locate our basic biological sense of self-preservation and self-furthering in a self distributed beyond our skin into our environment, natural and hu-

"The I that is We and the We that is I" is a phrase coined by the philosopher G.W.F Hegel in the early nineteenth century.

366

man. This is why we care about the world and why it is the arena of our moral concern and of our ideals.[1] In this chapter I review some of the mounting evidence that the scope of the self as moral agent, of who is performing a given moral action, can be distributed beyond the individual to groups, and even extend at times to whole contexts. The scope of the self as actor, its agency, can in certain circumstances be laid at the feet of social-cultural-historical systems, spanning time and place and even generations.

The self has turned out to be permeable and relational (as well as self-promoting, self-protecting, and self-furthering) rather than closed, discrete, and playing out its own internal program upon the world stage. Yet we have not yet seen *how* permeable, *how* open it can be and usually is. Body maps extend the "me" to include the hammer I use when I nail the picture on the wall or the car when I'm driving. Research reveals that the feeling I have that Tessie, my metallic light blue Acura, is an extension of my body when I drive in fact reflects the neural reality, for my body maps are extended to include the car's proportions and motions within the bounds of my self that I feel and control. They are mapped within the "body mandala," as the Blakeslees put it. There is a "tool-body unification" or extension. There is also a certain amount of space surrounding our body, "peripersonal space," like a bubble around us, that is included in our neural self-maps.[2] The expansion to include tools and other objects that we use to do things and carry out our aims, as well as all kinds of biographical and cultural and familial information, is now referred to in neuroscience as the extended self. We take so much in.

The philosopher Andy Clark, in his 2008 book *Supersizing the Mind*, has written about how the mind spills over into the world.[3] When he argues that the mind is extended, he means that "at least some aspects of human cognition . . . [are] realized by the ongoing work of the body and/or the extraorganismic environment," so that the "physical mechanisms of the mind . . . are not all in the head" or in our central nervous system.[4] When we use a computer or a pad of paper, a calculator or our address book, our mind has distributed memory and even operations outside itself. This view of the matter radically complicates and reconfigures the nature of the relationship between mind and world, the neurophilosopher David Chalmers points out in his foreword to the book.[5] Clark says that we have "a fundamentally misconceived vision . . . that depicts us as 'locked-in' agents—as beings whose minds and physical abilities are fixed quantities apt (at best) for mere support and scaffolding by their best tools and technologies." He proposes instead that our "minds and bodies are essentially open to episodes of deep and transformative restructuring in which new equipment (both physical and 'mental') can become quite literally incorporated into the thinking and acting systems that we identify as our minds and bodies."[6] As an example, Clark mentions a robot arm, an arm that extends one's reach and gets mapped into self-maps whose scope now includes its reach. Clark's analysis supports what Antonio Damasio proposed about the incorporation of the extended biographical and cultural aspects of the self in a third level of self-mapping: becoming and being a self, on one hand, and responding to and incorporating multifaceted contexts, on the other, are one and the same ongoing process. Yet there is more. For we not only discover the world within us but also discover ourselves in the world, identifying ourselves with parts of it. The psychoanalysts call this projective identification. Philosophers since Spinoza have referred to the group mind. Psychologists have studied mass psychology. And neurophilosophers have begun to explore the phenomenon of distributed agency, a subject of action that is larger than the individual. The distributed self leaks out of its boundaries of the skin and can even feel itself somewhere else entirely outside the body.

The "I That Is We": Co-Consciousness

Experimental psychologist Philippe Rochat designed and implemented a series of research studies on infants and children to investigate how and when they develop of a sense self. He has proposed that, developmentally, the levels of consciousness culminate in what he calls co-consciousness. Rochat defines co-consciousness as "knowledge of [one's own] knowing," the most complete form of self-reflexivity because it "is not just individual but collective." This "metaknowledge is embodied in the group," he says, and "will survive individuals." It consists in "shared representations" resulting from "our *co-conscious* experience." It is the melding or integration of first- and third-person perspectives on the self, Rochat says.[7] If Rochat's observations are correct, then the self is ultimately extended not only biographically and culturally and to its tools and projects but also, finally, into a "we"—a sharing of selves and ultimately a shared world. The self as fully self-aware includes perspectives that were initially "other": third-person perspectives now taken in as one's own have relocated the self outside itself and in another. In acknowledging them, owning them, and integrating them with my first-person perspective, I create a shared self, a self that is both mine and yours. The self in a sense "others" itself by casting itself into the world and finding itself there in the eyes of another. It returns to itself as including the other *as integrated within the I*. I discover myself in others' eyes. I have a socially defined self resulting in a shared reality, a shared world.

Relational Selves: Now We Are Two

> The observation that the self cannot exist for long in a psychological vacuum is a finding that is not easily accepted. . . . This discovery heralds the end of another cherished illusion of Western man, namely the illusory goal of independence, self-sufficiency, and free autonomy. . . .
>
> Thus self psychology strikes deeply at a politico-religious value system in which the self-made individual is the ideal. We now have a deeper appreciation of our inescapable enbeddedness in our environment.
>
> —ERNEST S. WOLF

At any given time the self is constructed by a relation to another person. "People have as many selves as they have significant relationships."[8] These selves emerge from the early relationships with parents, siblings, extended family members, and others who have had an impact on one's life, and they profoundly affect motivation and emotions. Unconsciously perceived similarity triggers the relational patterns to take hold, without our conscious awareness or even the ability to control the process. So each of us has "an overall repertoire of relational selves—aspects of the self tied to a significant other." Our sense of selfiness is distributed into the relationship and allowed to be co-defined with the other. The other and the relationship are necessary for the feeling of "me," of this "me." So what we have here is a notion of how our feelings about our self are co-constructed with other people and, via them, stretch outward to the world at large. We perceive the world as resembling one or another of these models of relationship time and again, thereby reawakening the "me" of that relation, of that context, and its attachments and aversions. Our selfiness has an ongoing vulnerability to the world's feedback, and so it has a stake in the world. The evidence also appears to support claims of the psychoanalytic theory of self psychology. Developed by psychoanalyst Heinz Kohut, self psychology presciently hypothesized "selfobjects," external significant persons, caregivers, that are not separate in the mind of the infant but are experienced as part of the self, functioning as "self-machinery" to "complete" the self. Kohut held that it is through "expanding its self-experience to include the whole surround" that an infant comes to be able to (1) feel itself confirmed as a self, (2) pursue its ambitions, and (3) pursue its ideals.[9] We find the regulation, definition, and control of our selfiness, of our self-regard, outside ourselves in another—and our attachment to the world is thereby set in motion.[10] In primitive versions (and some pathological ones) we merge with the group or the parent. Yet Kohut suggests that a mature form of discovering ourselves or even relocating ourselves in the world is the deflection of the sense of self away from body and mind and its projection into our creative works—art, music, writing. "The self-esteem regulation that relates to one's own body and one's own person is surrendered to the work," he writes. "The self" in these cases, Kohut comments, "is now in the work."[11]

The Neurobiology of Out-of-Body Experiences

My investigation contains hundreds of pages dealing with the psychology of the self—yet it never assigns an inflexible meaning to the term self, it never explains how the essence of the self should be defined. The self . . . is, like all reality—physical reality . . . or psychological reality . . . —not knowable in itself.

—HEINZ KOHUT

How radically our sense of self can overshoot the boundaries of the skin has been demonstrated by Olaf Blanke and Thomas Metzinger's research on out-of-body experiences. Out-of-body illusions locate the feeling of self, of having a point of view, as coming from an illusory body outside one's own actual physical body. Sometimes the illusion is that the sense of the location of one's body comes from two places: both from one's actual body and at the same time, or alternately, from the illusory body. These experiences also involve a feeling of being disembodied and seeing one's body as if from the outside. The sense of where our body is and where its feelings come from can be projected onto parts of the environment and felt as coming from else-where, beyond our own skin.[12] Earlier research into the neurological underpinnings of the global sense of self and self-ownership (the feeling we have of the body as a whole and that it is under our control) led the two to try to reproduce in the laboratory the out-of-body experiences that have been observed in people who suffer from certain neurological injuries. They found that the subjective states of the feeling of the body as outside itself and located somewhere else can be induced in experimental subjects in the laboratory. The scientific explanation of how these subjective experiences come about has something to tell us about the nature and boundaries of the self: about what a self is and what it is not, and what the feeling of self actually can tell us about ourselves—and what it can't.

Blanke and Metzinger's experiments exposed subjects to "conflicting multisensory bodily cues by means of mirrors, video technology or simple virtual reality devices." These manipulated the sense of self-identity, locus of experience, and point of view so that they were experienced as displaced outside of the subject's own body, thus reproducing the types of illusions brought on by specific neurological injuries. For example, subjects were hooked up to a video camera and their body image projected in front of them. When they saw themselves stroked on the virtual back as they were being stroked on their real back, they experienced themselves as located in the illusory body projected in front of them.[13]

Some out-of-body and displaced-body-part experiences can be attributed to injuries to the body-mapping capacities. The Blakeslees describe a patient who had a growing brain tumor and had lost the sense that she had a right arm and then a right leg. She felt as if they had flown off somewhere unless she looked at them, moved them, or had a heavy object touching them. Another patient, the victim of a stroke, didn't have sensations in her left hand and claimed that when looking at her niece's hand, she experienced her niece's hand as her own. Her body map of her left hand, the Blakeslees point out, had been projected onto her niece's body.[14]

In trance and trancelike experiences, and in Buddhist meditation, there is a weakening of the self-in-body feeling, which becomes diffused into the environment or beyond the skin. Research into this Buddhist and trance experience of loss of self is now showing what seems to be an attenuation of self-maps, the Blakeslees suggest.

> When people enter deep meditation or trance, they say that their bodies and minds expand into space. Body awareness fades, and they are left with a unitary

yet diffused and nonlocalized sense of themselves. Along with it come feelings of joy, clarity, and empathy. When Buddhist lamas meditate in brain scanners, activity in their parietal lobes plummets. It can't be a coincidence that the dissolution of the bodily self accompanies the shutting down of the body and space maps that create it.[15]

Blanke and Metzinger's experiments on out-of-body states and the Blakeslees' accounts of certain impairments of stroke victims and of altered states of consciousness have revealed something important: the sense of self is not a simple direct readout of a homunculus inside us, as we tend to think. Our internal feelings are anything but indubitable, as Descartes insisted. They are hardly even reliable. Instead, our feelings of self, self-location, and the like are flexible, complex, and indirect outputs of various coordinated neurological systems. That we have a sense of self-identity and self-location and a personal point of view does not mean that the self that is pinpointed in these ways is a concrete thing whose internal state we experience and report on in any direct way. Our lack of accurate self-access is glaring.

The nature and scope of the self is turning out to be flexible and permeable, and even capable of being relocated. That the feeling of self can be extended beyond itself (and sometimes even detached and displaced to someplace beyond one's own body) does *not* suggest the immortality of the soul, but instead quite the opposite: these discoveries lead us to question the "thinginess" of an interior self. The sense of global self arises, we now see, from various distributed neurological operations and systems that do not directly reveal an interior reality, but instead construct it. In other words, the self is not an interior person to which our inner consciousness gives us simple access. Instead, the feeling of self is a mental capacity that can be projected inward or even outward onto the world. It is a feeling of ownership, of selfiness, and it is malleable and expandable. That feeling can be attached to relations we enact in the world, claiming all kinds of things as our own and as who we are, seeing ourselves in worlds beyond our mere bodies and taking into ourselves identities we find ourselves enacting in the world. We mistake the feeling of self as our interior, a soul-thing, a solid bounded essence that is our true self that we alone know and disclose to the world. But that is a false picture. We are more like verbs than nouns; we make parts of the world feel like self, and we fill our feeling of self with our engagements in the world.

The Neurochemistry of "We"

Let me be clear. In company with most of my fellow neuroscientists, I believe strongly that the brain does not have a signaling circuit dedicated to ethics. . . . What the brain has is a mechanism to make use of circuits that are already there, in order to disable the self-preference that is akin to our instinct for survival.

—DONALD W. PFAFF

From the uncanny out-of-body experiments of Blanke and Metzinger I turn here to some discoveries about the neurochemistry of the self-other boundary discussed by Donald W. Pfaff.[16] Pfaff shows that the feeling of self and where we draw the line between self and other, self and world, depends upon particular neurochemicals. These chemicals that produce the self-other divide, he says, are sometimes turned off, and when they are we experience others as if they were ourselves. The neurochemistry that produces the feeling of self as extending into others, he argues, underlies ethics. He says the chemicals that turn off the feeling of a self-other boundary operate in fear and in love, in empathy and in aggression, and in other situations. These chemicals create a sense of shared experience and even of merger. Pfaff believes that this mechanism produces ethical action by creating empathic responses to others. It works by inducing a kind of "blurring and forgetting of information."[17] He theorizes that before people take any action, they represent to themselves mentally the action and its effects (a feed-forward loop). The would-be actor foresees the consequences to the recipient of the action. The next step, Pfaff argues, is the potentially ethical step: the actor can blur the distinction between self and other both cognitively and emotionally and experience the action he or she is about to perform as if he or she were about to receive it and not just commit it. Pfaff gives the example of a woman, Ms. Abbott, who is about to stab a certain man, Mr. Besser, in the stomach. If "instead of seeing the consequences of her act for Mr. Besser, with gruesome effects to his guts and blood, she loses *the mental and emotional difference* between his blood and guts and her own," her final step, the decision of whether or not to knife the man, may be to hold back. She feels a "blurring of identity—a loss of individuality—the attacker temporarily puts herself in the other person's place." Ms. Abbott has an anticipatory sense of "shared fear" and as a result refrains from knifing Mr. Besser.[18]

What Pfaff proposes is happening is "a reduction in the operational efficiency of neural circuits that discriminate between self and not-self," thereby blurring the difference between self and other. This "reduced image processing take[s] place in the cerebral cortex." There is a reduction in the neural circuits that discriminate between self and other. All kinds of self-mapping that represent to ourselves our position in space, what is happening in our musculoskeletal frame, our possible movements, our sense of being touched in a particular spot, and so on—generally, our moment-by-moment monitoring of ourselves—if toned down in any of various ways would create a blurring of self and other, "allow[ing] us to run our sense of self together with our sense of another human being."[19] A general excitation of emotion merges the identities of two people by making the neuron assemblies that mark particular identities fire together. So several identities would be signaled at once. This is what Pfaff proposes is happening in ethical behavior. When the self-other distinction is shut off between, "say, myself and the target of my intended action," the images merge. "The result is empathic behavior that obeys the Golden Rule." The merging of images within me of myself and the person at whom my actions or feelings are directed also entails a flow of information within my brain from my more evolutionarily recent cerebral cortex to "the more primitive forebrain, connected with emotions and drives."

Since the discovery of mirror neurons (discussed in detail later), the plausibility of the merger of the sense of self and other has been given a big boost. For it is the same neural circuitry that is operative both when we perform an action or feel an emotion and also when we see someone else doing or feeling it. For example, "some of the same brain circuits for pain become activated whether the person is receiving the painful stimulus or observing another person receiving it." Only the intensity differs, as does the exact location of the activation within the circuitry.[20] We simulate others' actions and feelings internally as if they were our own—and that's the basis for recognizing what they are doing or feeling. This constant simulation of others within the self, taking place by recruiting our own mechanisms of perception, emotion, and action, makes the potential loss of the distinction between self and other an easy move. So Pfaff brings to bear evidence from the neurobiology labs that our feeling of self need not, and at times is not, confined to within our skin. We can distribute or project our sense of self and our selfiness—our self-protection and self-furthering—into parts of the world and into other people. Emotions, and the beliefs that underlie and drive them, can be contagious. We can identify with others and can use that ability to empathize—or not. We have, Pfaff tells us, "shared fates, shared fears."[21] Not surprisingly, we also can have shared love.

In social attachment generally, in sexual and parental love, and even in friendship, we are infused with neurochemical mechanisms that erase the borders between self and other. The hormone oxytocin, for example (which is involved in parenting behavior), "appears to contribute to the blurring of the 'me-you' distinction in a variety of social situations." Parental love and especially motherly love "take the blurring of identity between two living beings to new heights," he says. The upshot of all this is that the various mechanisms that induce sexual and parental attachment can be harnessed and recruited for all kinds of prosocial behaviors and attachments.[22] A case in point is oxytocin, which in evolution began as a way to induce maternal care but is the major hormone of sociability in human beings. Oxytocin is widely operative in both male and female adults and especially in their bonding and increases feelings of trust, dampening the fear activation of the amygdala.[23]

Pfaff argues that there is a basic underlying neural mechanism operating in all the kinds of sociality that are now being investigated in neuroscience labs: it is the diminishment of the recognition of distinct identity. It is a surprising finding, he says.

> Throughout the range of all possible social interactions, neuroscientists are beginning to piece together the influences on neural networks that regulate sex, social recognition, and sociability. . . . Wonderfully, . . . we do not need to posit increased levels of performance in these neural mechanisms, but instead can be confident that once we recognize another as a friendly, nonthreatening presence, then it is a *decreased* social recognition that leads someone to obey the Golden Rule. *A person forgets the difference between himself and the other.*[24]

A failure to blend self and other can now be seen as one of the component mechanisms of violence and aggression.[25] But of course there is also the bonding of shared aggression in groups. "The violence committed by an individual may actually reflect some kind of bonding with other members of the group," Pfaff says. "And bonding is a brain question." He goes on to illustrate: "In gangs, aggressive acts might be both an issue of loyalty and a requirement for membership." Again, the me-you, us-them boundary is pushed outward from the discrete self. Even in terrorism, Pfaff remarks, the explanation may reside more in the individual's group identification than in his actual personal proneness to violence. This is an important and pertinent finding in this age of terrorist attack on the United States and elsewhere. Groups can ensure loyalty through violence and members may also be violent in response to their merger with the group, perceiving threats to the group as threats to themselves. "Even some terrorists," Pfaff points out, "might be seen as normal people responding to threats against their cause or their group."[26] In sum, Pfaff's theory is that altruism amounts to the sharing of identities, the lowering of the feeling of the boundary between self and other, the feeling of self (and selfiness) as it extends or is distributed into the other. That's what Pfaff thinks is behind the action of a man, a Mr. Aubrey, who jumped onto the subway tracks in front of a train to rescue a stranger who had fallen in front of the train. At that moment, Pfaff says,

> Mr. Aubrey's brain must have instantly achieved an identity between his self-image and the image of the victim who fell in front of the subway train. This identification did not occur by some complex highly intellectual act—it came about by *losing* information, that is, by blurring the distinction between the two images. In addition, Mr. Aubrey was demonstrating the kind of prosocial caring feeling that (I hypothesize) normally develops from parental or familiar love.[27]

I think it would be better to jettison the term *altruism* and replace it with a better description of what's going on: shared selves, a selfiness that loses the boundaries of the skin and is extended and distributed to others. The case of this Mr. Aubrey may seems like quite the anomaly, but actually we live in a world in which people give up their lives for the larger community all the time: that's war. We tend to think of war as a case of violence to be explained. But it's just as much or more a case of self-sacrifice for the group, and hence the full merger of a self into the group. The young soldier willing to die, when we think about it, is an uncannier phenomenon in need of explanation than group outrage and violence. Both are products of the erasure of the self-other boundary or its displacement beyond the skin into us versus them. Mr. Aubrey is merely a surprising case of a phenomenon so familiar that we tend to fail to think of it as in need of explanation.

Mounting data from the neurosciences show that evil is rooted in the failure to see others as ourselves. In the case of genocides such as the Holocaust, it is a collective failure of society-wide proportions. The self-other boundary beyond the skin is of central importance to a rethinking of ethics, to a deeper understanding of both good and evil. And adding neural plasticity into the mix (which makes any belief and any norm possible) gives us the true depth and scope of the problem. But this is still not the whole story. There is yet more to learn about how and why selves merge, about the sharing of feelings and actions, attitudes and beliefs, leaders and ideals.

Overlapping and Shared Self Maps

When monkeys' personal space is invaded and threatening objects approach the face or body, neuroscientists have discovered that special "flinch cells" fire. It's the neural system that keeps you from bumping into things and from falling off a cliff, its discoverer, Michael Graziano of Princeton University, remarks. So it is likely that there is an opposite mechanism that allows objects and people to come close, be within one's personal space, and engage in joint action. The Blakeslees dub these speculative entities "hug cells."[28] Our self-mapping can extend to another's peripersonal space, coming to include theirs and theirs ours. We unconsciously take account of our space bubble all the time when we move, reach for things, dance, do martial arts. We know where our hands can move and where there are objects in the way; where our feet are stationed on the floor and where the table leg intrudes. We not only feel where our body parts are but also what the space immediately enveloping them is like—and we feel that we own it as much as we feel that we own the space of our body parts and the actual space taken up by them. This personal space around each of us can merge with others' personal space. This is what the Blakeslees refer to as "blended personal space," a "we-centric" space. They call this an "envelope" that encompasses blended selves. Great examples of shared space, they say, would be holding your child on your lap, riding a horse, or making love. They all induce in us a sense of shared space. "It is likely, but not yet proved," they say, "that your brain contains spatial mapping cells that specialize in 'affiliative behavior,' which is a clinical term for cooperation and intimacy."

Researchers Greg J. Stephens, Lauren J. Silbert, and Uri Hasson used fMRI to study the spatiotemporal activity of the brains of people engaged in conversation. When the research subjects engaged in "natural communication," the spatiotemporal brain activity of the listener was coordinated or "coupled" with the spatiotemporal activity of the speaker. When they failed to communicate, the "coupling vanished." They found that generally the listener's brain activity follows that of the speaker with a small delay, but sometimes the listener's brain activity would anticipate the speaker's. Using a quantitative measure of story comprehension, they also showed that the greater the anticipatory listener-speaker coupling, the greater the understanding. When the same speakers and listeners were not engaged in comprehensible conversation (the speaker started speaking in a language that the other did not understand), the coupling of brain activity patterns ceased. The more people communicate and understand each other, the more their brains simulate each other's brain activities, and the more they meld in the moment, researchers concluded.[29] So neither our minds nor our bodies are as discrete as we think they are. We are connected to each other *on the inside*. Our very souls or selves—our bodies and minds—are bound up with those of others, as the Bible says of the love of Jonathan and David. We are anything but Cartesian subjects, solipsistically isolated in our internal feelings, thoughts, actions, and even bodies.

Mirror Neurons: Shared Action, Emotion, and Understanding

> Mirror neurons will do for psychology what DNA did for biology.
>
> —V.S. RAMACHANDRAN

The discovery of mirror neurons begins with a delightful story. Giacomo Rizzolatti and the folks in his lab at the University of Parma were investigating how the neurons in the premotor cortex, which is located in the frontal lobe of the brain, operate in guiding movements, in grasping, in bringing some food to the mouth, and the like. They had recruited a monkey, into whose brain they implanted electrodes, in order to monitor where and how the planning and carrying out of goal-directed movements occurred. One day in the summer of 1991, a graduate student walked into the neuroscience lab where the monkey was sitting in its special chair, waiting. It was just after lunch and the student was holding an ice cream cone that he brought up to his mouth and licked. And lo and behold, the parts of the monkey's brain devoted to hand-to-mouth motor movements became active. But the monkey's own paw had not moved! As the monkey had watched the student bring the ice cream cone to his mouth to take a lick, its own motor map had simulated the same movement—but without moving! The neuroscientists demonstrated that the same simulation in the monkey occurred when they ate peanuts. Both when the monkey itself picked up peanuts and when it watched a researcher picking up peanuts but didn't itself move its own arm, the same group of cells fired. The neuroscientists demonstrated the phenomenon again and again. The brain's "mirror system" automatically produces an internal replica of the perceived action as a kind of internal reenactment of both the action and its goal. A series of neuron recording experiments carried out in the 1990s showed "a particular set of neurons, activated during the execution of purposeful, goal-related hand actions, such as grasping, holding or manipulating objects, discharge also when the monkey [merely] observes similar hand actions performed by another individual."[30] This was the discovery of an "action/observation/execution matching system."[31] Scientists invented the term *mirror neurons* to describe this set of brain cells.

The mirror system enables not only joint action and agency but also a form of immediate understanding of other people's actions, goals, and emotions, too. For we recognize both actions and emotions through the mirror system—so we are able to imitate them from the inside out.[32] The human system "includes a rich repertoire of body actions" and operates at the preverbal level.[33] It functions via automatic "direct matching" rather than via thinking and analysis.[34] Giacomo Rizzolatti comments on his own discovery:

> At the cortical level the motor system is not just involved with single movements but with actions. Think abut it; the same is just as true for humans [as monkeys]: we very rarely move our arms, hands and mouth without a goal; there is usually an object to be reached, grasped, or bitten into.
>
> These acts, insofar as they are goal-directed and not merely movements, provide the basis for our experience of our surroundings and endow objects with the immediate meaning they hold for us. The rigid divide between perceptive, motor, and cognitive processes is to a great extent artificial; not only does perception appear to be embedded in the dynamics of action, . . . but the acting brain is also and above all a brain that understands.[35]

The motor schema of the observer enacts, in an "as if" pattern, the motor schema of the actor. Neuroscientist Vittorio Gallese proposes "that this link [between the observed agent and the observer] is constituted by the embodiment of the intended goal, shared by the agent and the observer." The observer's motor system "resonates" with the system of the actor, and that creates a kind of empathy and, sometimes, contagious behavior.[36] Gallese claims that action is primarily "relational."[37] So fundamentally we act together from joint agency rather than individually. The social scope of agency, the group as actor, is what's normal, and individual agency is rarer and more of an achievement—the direct opposite of what we, at least in the West, tend to assume and take for granted.

An advanced kind of mirroring is involved in understanding without imitating, and it often occurs when actually engaging the action wouldn't be useful.[38] The interior experience of understanding without imitation may be due to "super mirror neurons," one of whose major roles may be to inhibit actual imitation by the "classical mirror neurons," neuroscientist Marco Iacoboni suggests.[39] That is to say, we reproduce the *motivation* of the acting other within ourselves. In his analysis of the research, Gallese proposes that the mirror system is a vital contributor to empathy and social cognition. And there is now a great deal of evidence "suggesting a strong link between mirror neurons (or some general form of neural mirroring) and empathy."[40] The German word for empathy, *Einfühlung*, "feeling with," is apt, for it was invented to describe aesthetic experience linking the observer with the work of art through a kind of internal simulation. Iacoboni describes how the empathic recognition of others' emotions works: we unconsciously mimic or imitate internally another's facial expression, and it is through that inner experience that we recognize what emotion that other person is having.

Iacoboni devised experiments to investigate if and how links between the mirror system and the emotion system are forged in the brain, and if they come to work together in empathy. He set up an experiment in which he monitored the brain activity of volunteers as they either watched or imitated pictures of faces exhibiting fear, sadness, anger, happiness, surprise, and disgust—the basic or primary emotions. Three brain areas—mirror neurons, the limbic system (controlling emotion), and the insula, a region that connects the other two—did in fact show activity that indicated their connection when the volunteers observed emotions and even more activity when they imitated those emotions. Because of the involvement of the limbic system, we can feel the emotions that others feel. Iacoboni reproduced the experiment to investigate the empathic sharing of pain. "Mirror neurons," Iacoboni says, "typically fire for actions, not pain." (Another type of cell in the cingulate cortex fires for pain; it does not have an action component.) Since "mirroring of emotions is mediated by action simulation," the viewing by volunteers of filmed painful scenes (a needle going through a hand) elicited an inhibited motor response simulating the withdrawal of the hand from the needle, which is to say a mirror response; the greater the mirror response, the greater the empathy the observer felt. The conclusion was that "our brain produces a full simulation—even the motor component—of the observed painful experiences of other people." Rather than a private experience, pain is a shared experience.[41] Studies of children's capacity for empathy correlated the scores of children from behavioral tests of empathy and interpersonal competence with brain activity measured by fMRI. The more empathic a child, the more his or her mirror neurons would fire while watching people expressing emotions. Social competence was also correlated with high empathy scores and high mirror neuron activity.

Vittorio Gallese believes that we ought to extend the concept of "empathy" to explain all the behaviors that enable us to establish a meaningful link between ourselves and others.[42] In its basic form, he suggests, empathy means that "the other is experienced as another being like oneself through an appreciation of similarity."[43] Most important, Gallese believes that this mechanism expresses itself not only in mirrored actions but also in shared emotions, body schemas, or maps.[44] Gallese remarks that "there is preliminary evidence that the same neural structures that are active during sensations and emotions are active also when the same sensations and emotions are to be detected in others." So it is likely "that a whole range of different "mirror matching mechanisms" may be present in our brain in addition to the action simulation architecture first discovered. In fact, he concludes, mirroring "is likely a basic organizational feature of our brain."[45]

So even when we don't explicitly act together in concert, we engage in automatic, unconscious, and unstoppable social behavior because our emotions and motivations are shared. It takes an extra inhibitory neural process to stop joint action and render it merely into joint perception, let alone individual autonomous action. Mirror neurons, along with the other neural evidence I have cited—self-maps, the extension and movability of the self-other boundary, and the like—provide further evidence that free will, even in the limited sense of actions stemming from autonomous individual reasoning, is fraught and highly unlikely. As Marco Iacoboni puts it:

According to free-speech theorists, we are all rational, autonomous, and conscious decision makers. However, the data . . . ranging from the unconscious forms of imitation observed while people interact socially to the neurobiological mechanisms of mirroring that have their key neural elements in mirror neurons . . . suggest a level of uncontrolled biological automaticity that may undermine the classical view of autonomous decision making that is at the basis of free will.[46]

We are internally connected and wired together—for good and for bad. When people observe a behavior in others, the intensity of their internal mirror simulation is stronger if they have engaged in the behavior before. This is benignly and even beneficially true, for example, of dancers watching dance but ominously true of smokers watching others smoke, experiments have shown.[47] Iacoboni hypothesizes that the level of activity in a person's mirror system in a given context is an index of his or her "identification" and "affiliation with other people" in that context. Mirror neurons, he continues, "seem to create some form of 'intimacy' between self and other" that may also be relevant to the sense of belonging to or being affiliated with a specific social group whose members, we feel, are more similar to us than other people. The more people like each other, the more they imitate each other, studies have shown. Yet research also has consistently shown that "there is a much stronger [neural] discharge for actions of the self than for actions of others." Mirror neurons fire for self and other but more strongly for self. We *do* know when we act, Iacoboni says. He hypothesizes that "mirror neurons in the infant brain are formed by the interactions between self and other." The baby smiles, the parent smiles; next time the parent or somebody else smiles, "the neural activity associated with the motor plan for smiling is evoked in the baby's brain, *simulating* a smile." We use the same cells in developing a sense of self because they reflect back to us our own behavior. "In other people, we see ourselves with mirror neurons." In self-recognition there are two selves: a perceiving self and a perceived self. The mirror neurons map the perceived one (in a picture, for example) onto the perceiver self. The perceived self is perceived as the "other" and is mapped onto (and simulated as) self—but it is already another self. So the other-self (in the picture) awakens a motor repertoire already belonging to the self. The firing of neurons for self is doubled, and the mirror activity for self is stronger than the mirror activity for other.[48] This explains, too, why similarity and familiarity create higher mirror activity than the less similar and less familiar. Selfness obtains even—and perhaps especially—in mirror neurons and in the sociality, empathy, and groupiness they spark and maintain. The selfier the other, the more it moves us.

The evidence from mirror neurons on shared emotions drives home the idea that personal individuality is an unusual achievement. Sociality is our immediate, default state. Moreover, we humans, not monkeys or apes, are the imitators par excellence. An unexpected and perhaps almost uncanny discovery about mirror neurons is that when it comes to goal-directed actions, other primates are better than we are at reproducing a goal (like finding a way to pick up a raisin), while we tend not merely to strive for the goal but also imitate all steps of a procedure, even the ones that are not contributory to the goal. Yet what this enables, I think, is the extraordinary retention and transmission of culture and tradition. In the end, the monkey is ever reinventing how to pick up a raisin, and we are amassing knowledge (albeit retaining the less than optimally functional as well as the functional) and hence in the end reclaiming our inheritance and moving on. Just think about what complete imitation must contribute to the acquisition and transmission of language—and of culture as well. Language and culture enable the joint agency of groups not just to be in the moment but to span time and place and even generation.

Our neural mirror systems ensure that we act more often than not as collaborative agents, as members of a social group, rather than independently and individually. This social or relational root of human acting helps to set the scope of our responsibility as individuals. We are not less responsible because we act jointly; instead, we are responsible individually for *all* our actions, both individual and collective. Unfortunately, the nature of our un(self-) conscious processes, as we saw in the previous chapter, tends to make us blind to and hence disavow responsibility for our joint agency within the group. Consequently, the central moral problem of human beings is not how to get individuals to care about each other and about the common welfare (and to use their alleged free will to choose to do the right thing), but rather the human tendency to fanatic belonging and loyalty to the group (whether family, tribe, political party, church, ideological comrades, etc.) and its unfortunate concomitant, the disavowal of personal responsibility for the group's actions. Empathy and joint action, while sounding wonderfully moral, are a double-edged sword. All of this, down to its neurobiological basis, Spinoza intuited and anticipated. He knew all too well, in his own case and in his own flesh, what the basic moral failing of human beings was, and he hypothesized its origins. He also proposed the outlines of the moral solution— one solution for the social and political management of society as a whole and another for the few who could overcome social slavery, joint agency, merger in the group, and transform themselves toward both personal freedom and also universal love and the love of nature.

The Secret of Success: Context, Context, Context

> No one was used to thinking of health in terms of *community*. Wolf and Bruhn
> had to convince the medical establishment to think about health and heart
> attacks in an entirely new way: they had to get them to realize that they
> wouldn't be able to understand why someone was healthy if all they did was
> think about an individual's personal choices or actions in isolation. They had to
> look *beyond* the individual. They had to understand the culture he or she was
> part of, and who their friends and families were, and what town their families
> came from.
>
> —MALCOLM GLADWELL

> No one—not rock stars, not professional athletes, not software billionaires,
> and not even geniuses—ever makes it alone.
>
> —MALCOLM GLADWELL

Against so much evidence—and against our own experience, if we think
about it clearly—we nevertheless tend to attribute our choices, actions, and
decisions to ourselves alone. This is especially true of our successes. We see
ourselves as having pulled ourselves up by our bootstraps, and therefore anyone else could have—and should have—done the same. We do not think of
ourselves primarily as lucky but rather as talented, smart, hardworking. We
may be all those, but we also owe a great deal—much more than most of us
believe—to our social and other contexts. Our successes owe far more to
others than we assume—and so, too, do our own and others' failures. Some
humility and gratitude are in order here, plus compassion for those who have
not been so lucky.

In 2008 the science writer Malcolm Gladwell published a book, *Outliers*,
about people who had achieved outstanding success. Gladwell uses well-
known examples of extraordinarily successful people, from Bill Gates to the
Beatles, from winning sports teams to Nobel Prize winners in medicine, and
changes the way we think about them. Ordinarily we think of the extraordinarily successful and even the highly successful as geniuses, people quite
unlike ourselves, who are blessed with some very unusual and special God-
given talent, whether it be musical, athletic, mathematical, intelligence, or
other. Gladwell says that almost all of us (and especially Americans) sub-
scribe to a myth that extraordinary success is due to a special endowment of
genius of one form or another. It is this myth that he sets out to shatter in

Outliers. In case after case, Gladwell exposes the (necessary and sufficient) conditions of success as a confluence of unusually enabling (in other words, lucky) circumstances, on one hand, and outstanding personal devotion to the achievement of a specific goal (practice, practice, practice), on the other. It turns out that the folks who are at the top of their fields have a lot in common with other folks at the top of entirely different fields. These star achievers, irrespective of area of success, have something in common. Gladwell makes the case again and again for what this common factor is: a combination of a highly specific fortuitous context and roughly ten thousand hours of devoted hard work. So success is first and foremost a matter of context— often very specific and even quite temporary contextual factors. Second, it is a matter of a great deal of relevant hard work. More than anything else, he shows again and again, it is the constellation of particular features of a context that makes it possible for an individual to become successful at something that that context, at that specific time and place, fosters in some extraordinary way. So it is not individuals alone who act but instead entire contexts that are operative in success stories.

Not surprisingly, it turns out that in success-producing contexts it is often a number of people—not just one lone supposed genius—who achieve outstanding success of the particular kind that that context fosters. So it is not the lone superachiever who is Gladwell's outlier; the real outlier is a particular extraordinary context that provides the necessary conditions for outstanding success of a very specific kind, again and again.

In *Outliers* Gladwell uses case after case to show that we cannot understand why certain people become stars just by dissecting them in every possible way as individuals. "Personal explanations of success," Gladwell argues, "don't work." Individuals' success is not due principally to what they "are like," but instead it is fundamentally a result of "where they are *from.*" Gladwell says that he wrote the book to convince us that explaining outstanding success by looking at people's special personal qualities, talents, personalities, or intelligence simply doesn't work. "People," he says, "don't rise from nothing." And they don't do it all by themselves, even if they look as if they did—and even if they, and we, would prefer to think it's all about Horatio Algers, self-made successes. "But in fact," Gladwell goes on, these outstanding successes "are *invariably* the beneficiaries of hidden advantages and extraordinary opportunities and cultural legacies."[49] His book, he says, using an apt simile, "is not a book about tall trees. It's a book about forests." It's about how it's the forest that makes the particular tallest trees in it possible. The tallest oak tree in a forest gets to be the tallest, ecologists point out, not because it had the genetically best acorn of all but rather because it grew in a spot where it got an unusually large amount of sunlight, the soil was particularly rich, it was not readily accessible to lumberjacks, it was not eaten by deer or rabbits early on, and the like. That's the kind of ecological explanation that accounts for outstanding success in human endeavors as well, Gladwell argues, and he offers some fascinating examples—examples that defy our expectations.

I'll give a few of Gladwell's examples of famous people whose remarkable successes, when studied closely, turn out to be case studies of extremely special and sometimes unique opportunities that were taken and used to fullest advantage. And isn't having a particular talent, one ripe for the unusual context, perhaps the luckiest advantage of all? Gladwell makes the case that in the end even talent is far more a matter of gaining expertise than of possessing some extraordinary innate property. Take the Beatles, for example, or Bill Gates. Gladwell shows that in both instances their phenomenal successes arose from "a combination of ability, opportunity, and utterly arbitrary advantage." Gladwell says that when you look at them closely, "what truly distinguishes their histories is not their extraordinary talent but their extraordinary opportunities."[50] The Beatles had a lucky invitation to play in Hamburg, Germany, when they were still a struggling high school band. That came about because a German impresario who had gone to London to look for bands to invite to Hamburg had happened to meet an entrepreneur from Liverpool, so he invited bands from Liverpool to Hamburg instead. In addition, in that period the only venue for live bands in Hamburg was strip clubs, and in those clubs the bands had to play not just for a couple of hours at a stretch but for eight hours at a time. So the Beatles, because of this unusual confluence of circumstances, were forced to play for hundreds of hours. Gladwell quotes John Lennon on their time in Hamburg:

> We got better and got more confident. We couldn't help it with all the experience playing all night long. It was handy them being foreign. We had to try even harder, put our hearts and soul into it . . .
>
> In Liverpool, we'd only ever done one-hour sessions, and we just used to do our best numbers. . . . In Hamburg, we had to play for eight hours, so we really had to find a new way of playing.[51]

The Beatles also recall that they often played seven days a week. Gladwell remarks that in about a year and a half of playing in Hamburg, the Beatles played a total of around 270 nights, and by 1964, when they had their burst of extraordinary success, they had already done approximately twelve hundred live performances—an astoundingly large number. They were extremely seasoned onstage. They came to Hamburg quite ordinary and left quite extraordinary. Gladwell concludes that "the Hamburg crucible is one of the things that set the Beatles apart." Indeed, Beatles biographer Philip Norman comments that Hamburg "was the making of them."[52] Gladwell does not at all deny the Beatles' musical gifts, especially the outstanding songwriting talents of Lennon and McCartney. "But what truly distinguishes their histories," he remarks, "is not their extraordinary talent but their extraordinary opportunities."

This is the case for Bill Gates as well. The standard story of Bill Gates goes something like this: "Brilliant, young math whiz discovers computer programming. Drops out of Harvard. Starts a little computer company called Microsoft with his friends. Through sheer brilliance and ambition and guts builds it into the giant of the software world."[53] In fact, Bill Gates came from a well-to-do and well-connected family in Seattle. He went to an elite private school where, by chance, a mothers' fund-raising committee one year just happened to use some of the funds they had raised to put a computer into a basement room in the school. The school had started a computer club, an unusual thing for the late 1960s. And the computer wasn't the punch-card kind but a link to the mainframe computer at the University of Washington. As an eighth grader Bill Gates had a link to a time-share system that enabled him to do real-time programming, Gladwell points out—an incredible, and incredibly unusual, opportunity. Not only that, the mother of another kid at the school was a founder of a company associated with the University of Washington that leased computer time-shares to companies. She suggested to the school that perhaps their computer club might want to test out some of the software her company was developing, and she would give them free computer time on the weekends at her company in exchange. The club started hanging around the computer lab at the University of Washington, and another lucky break gave the club's members an opportunity to have free computer time in exchange for testing out software. Gladwell estimates that over a seven-month stretch in 1971 Bill Gates, then an eleventh grader, averaged eight hours a day, seven days a week, on a mainframe computer. Gates says he was obsessed. He and a friend would even sneak out of the house in the middle of the night and go up to the University of Washington to use some computers—"steal some computer time," he himself put it—that were hooked up twenty-four hours a day but were less intensively used in the middle of the night.[54] Then in his senior year in high school the technology company TRW needed computer programmers familiar with the mainframe that Gates and his high school computer club had been working on for years, and turned to the club to help them out. He spent his senior year at TRW doing the specialized programming in which he had already developed a rare expertise. And his supervisor at TRW was a whiz and taught him a tremendous amount. Gates reflects on his own opportunities, commenting that he "had a better exposure to software development at a young age than . . . anyone did in that period of time, and all because of an incredibly lucky series of events."[55]

In 1975, the people who became the movers and shakers in the computer industry, like Bill Gates, had to be not too old and already settled down, nor too young and still in high school. So they had to be in their early to mid-twenties. Bill Gates was twenty; Paul Allen, who founded Microsoft with Gates, was twenty-two; the third in charge at Microsoft, Steve Ballmer, was nineteen. Steve Jobs, co-founder of Apple, was twenty, and all four founders of Sun Microsystems were between the ages of nineteen and twenty in 1975.[56] There was a window of incredible opportunity for those who also had had the luck to be prepared to take advantage of it.

The same story holds for the great captains of industry of the nineteenth century. There was a nine-year window in the 1860s and 1870s when those who were just the right age could take advantage of the great industrialization of the American economy and the rise of the railroads. Of the seventy-five richest people in the whole history of the world, fourteen of them were Americans born during the nine-year period between 1831 and 1840.

Gladwell concludes:

> There are very clear patterns here. . . . We pretend that success is exclusively a matter of individual merit. But . . . these are stories, instead, about people who were given a special opportunity to work really hard and seized it, and who happened to come of age at a time when that extraordinary effort was rewarded by the rest of society. Their success was not just of their own making. It was a product of the world in which they grew up.[57]

High IQ: It's About Culture and Context

It is not only the outstandingly successful who owe so much to context. Intelligence generally is also far more about context and culture, family and heritage, than any inborn raw brainpower. The cultural component has been shown to be the most important factor in IQ scores, not inherited intelligence. Richard E. Nisbett's exhaustive study *Intelligence and How to Get It: Why Schools and Cultures Count* leaves us in no doubt of that.[58] Nisbett points out that studies have shown that Americans of East Asian background have a slightly lower IQ than Americans at large. Yet their achievements far outstrip not only their own IQs but those of other Americans. "Asian intellectual accomplishment," Nisbett concludes, "is due more to sweat than to exceptional gray matter."[59] Nisbett is particularly concerned to dispel the myth that lower-income groups are underachievers due to genetically inherited lower IQ. Instead a careful review of the evidence from multiple studies shows that the IQ gap is the *result* of entrenched poverty and its accompanying social and cultural deficits rather than its cause. Nisbett argues that the evidence clearly exposes that "blacks and other ethnic groups have lower IQ and achievement for reasons that are entirely environmental. Most of the environmental factors relate to historical disadvantages but some have to do with social practices that can be changed," he says. Moreover, "some cultural groups have distinct intellectual advantages. . . . These include people with East Asian origins and Ashkenazi Jews."[60]

I'll end this gesture at the vital importance of context with an account of the success of the Jewish Americans who hailed from European backgrounds. We Jews do not owe the preponderance of our relative success to being smarter, it turns out. In a twist of black humor, we are *luckier*—our relative success is due primarily to cultural and other contextual factors. How successful have Jews been in America? Nisbett points out that Jews of European descent are overrepresented, compared to their minuscule population in the world, by a factor of 50 to 1 in the percentage of Nobel Peace Prizes and by 200 to 1 in Nobel Prizes for economics. If we count as Jews those who have at least one Jewish parent, American Jews have received 40 percent of all Nobel Prizes in science awarded to Americans. If we count only those who have two Jewish parents, the figure falls to 27 percent. Yet Jews are only a little over 1 percent of the American population. Other science and math awards display similar percentages of Jewish winners. More than 30 percent of students in the Ivy League are Jewish, and the same per-centage applies to faculty at elite American colleges and universities. Su-preme Court law clerks are also Jewish by about the same percentage. Jews have been extraordinarily successful in all kinds of endeavors, including business, in which intelligence makes a big difference. While average Jewish IQ is the highest of any American ethnic group—it's a little less than one standard deviation above the white American average, which means it aver-ages about 110 to 115—Nisbett points out that that factor cannot alone ac-count for the extraordinary success of American Jews, for the record of success outstrips the IQ advantage by leaps and bounds.[61] Given the average IQ of American Jews, Nisbett says, the number of Jews in the genius range, with an IQ of 140 or over, would only be six times that of the standard American population. But the overrepresentation of Jews as winners of No-bel Prizes is at least 15 to 1 for those with two Jewish parents. In the Ivy League and as professors and law clerks, the IQ difference would predict an overrepresentation of 4 to 1, but the actual overrepresentation is again 15 to 1. What these statistics show is that Jews are substantially overachieving their IQs. Sephardic Jews, who are Jews of North Africa and Asia, have IQs that are on average the same as Americans at large. Yet they were the great achievers under Islam between the years 1150 and 1300 C.E., when 15 per-cent of all scientists were Jews. Nevertheless, Jews and Jewish achievement are not in any way unique, Nisbett points out, for regional differences in in-tellectual accomplishments, both worldwide and also within the United States, display far greater differences than those between European Jews and other groups—consider the vast disparity in accomplishments in sci-ence, philosophy, and the arts occurring between, for example, Texas and the Northeast. "The magnitude of the differences in intellectual accomplish-ment between Jews and non-Jews in the West," Nisbett concludes, "pales beside all these national, ethnic, and regional differences." Nevertheless, the Jewish difference is still in need of explanation. Like Confucians, Jews have a strong emphasis on education; also similar to those in the Confucian milieu, Jews have very strong family ties, and family expectations of the in-dividual are demanding and hard to resist. Achievements are seen to re-dound to the whole family and even the whole community. In sum, Nisbett says, "Jews place a high value on achievement, period." And the emphasis on achievement is not only in academics and intellectual and cultural endeavors but in business and sports as well. Nevertheless, this explanation is less re-search than anecdote. What is clear from the hard evidence, however, is that

Jews achieve far more than their somewhat higher IQ averages would pre-dict. So the difference is a result of environmental, contextual factors—whatever they may be.[62]

Going Out of Our Minds: Thinking Through Embodiment, Embeddedness, Extendedness, and Enactment

> Humans are collective thinkers, who rarely solve problems without input from the distributed cognitive systems of culture.
>
> —MERLIN DONALD

> Where do you stop, and where does the rest of the world begin? There is no reason to suppose that the critical boundary is found in our brains or our skin.
>
> —ALVA NOË

I have been approaching the discovery of the self beyond itself and in the world from a number of angles. The evidence is building for the *extension* of our selves into our tools and computers and pencils, into robot arms and cell phones and cars, and for the *distribution* of our sense of self into shared environments and contexts, from culture and family to nation, from school and neighborhood to generation and church. Here we find the source of a sense of distributed agency: it can be the group, rather than the individual, who is performing an action or making a decision. We have reviewed evidence from widely different quarters—from psychological studies of infants in their development of co-consciousness, a shared world and a self co-constructed by self and environment; from studies of our un(self-)conscious thinking and feeling, which reveal that each of us has a number of implicit working self-concepts rooted in two-person repertoires that arise from our relation-ships with significant others (mother, father, siblings, and the like) in early childhood and triggered ever anew by the environment; from the surpris-ing neurobiology of out-of-body experiences, which reveals that we can dis-cover our feeling of self outside of our bodies and lodged in parts of the environment; from the neurochemistry of the self-other boundary, a bound-ary that breaks down and enables the other to feel like self in empathy and love but also in shared anger and fear; from the discovery of mirror neurons, which directly cause homologous brain cells to fire in mere observers of an action, creating a shared experience of actor and observer from the inside;

and from the sociological analysis and meta-analysis of success and intelligence, whose findings identify the major causes of outstanding individual achievement as environmental, social, and cultural, rather than individual or genetic. Taken together, the evidence ought to begin to change where we look when we we're searching for ethics. We should begin to look not inside the individual, as we have assumed, but rather outside, in the environment. Some philosophers and other theorists have begun to do just that.

There is a growing movement to rethink thinking, and the mind more generally, as embodied. Embodiment means that the mind is not a brain in a vat. The days when thinking is likened to a computer program are coming to an end. As the philosopher Alva Noë puts it, the standard view, not only in philosophy but in neuroscience, has been that "we are brains in vats on life support. Our skulls are the vats and our bodies the life-support system."[63] But that standard view is turning out to be wrong. One of the things that the old view assumed was that it made no difference whether thinking takes place in an embodied person, in a machine, or somewhere else. A good analogy to the way thinking was supposed to work was that it was like seeing a movie in a theater, on TV, or on your computer—and it wouldn't matter much except for the scale and the clarity. The movie was the movie, and it was just different technology bringing it to you. But that's not the way the mind is turning out to work. Instead, and contrary to decades of the dominance of the standard "movie" account in cognitive science, the ways that the brain is biologically, neurologically, and ecologically constructed are coming to be appreciated as supremely relevant to the content of the mind. The mind is not a computer running a discrete genetic or other kind of internally constructed program that would be the same on any type of hardware. That computer or media metaphor, a metaphor that has driven a great deal of research, is simply misguided when it comes to human thinking. For the body shapes the mind—and the content of the mind—in crucial respects rather than merely underlying it. Second, the new conception of the mind is that it is not only embodied but also embedded in its environment: in its contexts and situations and histories and communities of all kinds, social and cultural and linguistic and natural. As Noë puts it:

> The limitations of the computer model of the mind are the limitations of any approach to mind that restricts itself to the internal states of individuals.[64]
>
> The content of experience—what we experience—is the world; in the world's absence we are deprived of content.[65]

Finally, in addition to the embodiment and embeddedness theses, there is the extendedness thesis. This is the claim that the mind is not confined to the skull. It means that "the boundaries of cognition extend beyond the boundaries of individual organisms."[66] Instead, both what's in the mind and who's doing the thinking and acting are "boundary crossing" and "world involving."[67]

Our thinking and our acting are not separated, which is how we tend to think of them—as cognitive reflection on, and an internal picture (representation) of, a world that is separate from ourselves and upon which we take independent action. Instead, perception and cognition depend upon and are crucially constructed by the way we *interact* with the world. Cognition is now being shown to involve the sensorimotor brain, that is, "motor capacities, abilities, and habits."[68] This may occur both "online," so to speak, and "offline." What that means is that the bodily involvement or "embodiment" in question may consist at times only in the involvement of the sensorimotor areas of the brain but not necessarily the body proper; that is, it need not involve current perceptual or action input or output. The relevant point for us here is that action and cognition are, in some not yet fully delineated or completely understood ways, bundled together, causally interdependent, rather than discrete and independent processes. Our perception is interdependent with bodily motor relationships—which is to say perception is an interaction and what is perceived *is* the interaction, rather than a self-removed grasp of the external environment per se.

The theory of how perception is shaped by how we interact with things was first put forth in the 1960s by psychologist James Jerome Gibson. Gibson theorized that we—and animals as well—do not perceive objects, or the environment more generally, objectively in terms of the shape and volume of objects. Instead, we perceive the environment and objects in terms of how we envision how we can interact with them; we see not objects per se but rather "affordances that make possible and facilitate certain actions." The Blakeslees explain what Gibson meant by "affordances" by suggesting that "handles afford grasping. Stairs afford stepping. Knobs afford turning. Hammers afford smashing." We perceive the world, according to Gibson, "through an automatic filter of affordances."[69] As the Blakeslees put it:

> Your perception of a scene is not just the sum of its geometry, spatial relations, light, shadow, and color. Perception streams not just through your eyes, ears, nose, and skin, but is automatically processed through your body mandala to

render your perceptions in terms of their affordances. That is generally true of primates, whose body mandalas have grown so rich with hand and arm and fine manipulation mapping, and even more so for you, a human animal.[70]

If thinking and even our basic perceptions of the world around us are not separate and separable from acting, then moral agency cannot be, as we tend to assume, about understanding and assessing situations from a removed perspective and then rationally and independently choosing the right action. For on the new model, all three are bound together—perhaps in the way that emotions and cognition have been found to be bound together in neural packages and pathways; think of mirror neurons and how acting out scenarios within us is the basis of understanding others' actions. Perception, emotion, action, cognition, empathic understanding of others—all seem to be integrated. What we believed to be discrete and bounded mental processes that "we" then in some sense preside over from above and bring together are turning out to be more intimately bound together from the start and all the way up the line. There is no independent "we" or "I" outside of these bundled perceptual, conceptual, affective, and enactive processes, no "I" who stands above them as if they belong to someone else or are distant parts of the world and looks down upon them and then decides or feels or chooses or acts. As Noë puts it, "Scientists seem to represent us as if we were strangers in a strange land. . . . Our relation to the world is not that of an interpreter. . . . Our relation to the world is not that of a creator. The world is bigger than we are; what we are able to do is to be open to it."[71]

Openness to the World: Cognitive Externalism and Distributed Agency

It's not what is inside the head that is important, it's what the head is inside of.
—JAMES JEROME GIBSON

Human decision making is most commonly a culturally determined process . . . When the individual "makes" a decision, that decision has usually been made within a wider framework of distributed cognition, and in many instances, it is fair to ask whether the decision was really made by the distributed cognitive cultural system itself, with the individual reduced to a subsidiary role. . . . Distributed systems are able to change where in the system each component that influences a certain decision is located.

—MERLIN DONALD

Openness to the world would seem to be our fundamental posture. We are of the world and in it, engaged in and engaging the environment and context. The misleading but dominant metaphor of "seeing" as our basic relation to the world obscures this reality. Seeing places us too much on the outside looking in. Let's replace sight with touch. If we think of ourselves as fundamentally touching and being touched, acting and being acted upon, and acting together with others, then we can grasp our fundamental openness. Each of us can come to be aware of ever larger contexts and environments in which we are embedded, as affecting and being affected. The local context we grasp is too narrow to contain or explain the scope of the openness of the self. The self we are and with which we can identify moves ever outward.[72] We discover our thinking, emotion, and action as a product of the group and of ever-wider contexts and environments. We come to know ourselves by discovering ourselves beyond ourselves. The self is *distributed*.

The brain interacts with the world in ways that influence our perception itself. Pivotal findings of neuroscience, particularly those of Jaak Panksepp about the crucial role of action in perception,[73] were anticipated by the philosopher Susan Hurley beginning in the 1990s, and especially in her first great breakthrough work, *Consciousness in Action*.[74] Her insights go a long way toward explaining and establishing that the boundary between self and world is not set by our skin. Instead, patterns of interaction are what thinking is all about. Hurley argues that action is distributed among mind, body, and world rather than being attributable to the individual alone. And her conjecture has turned out to have a great deal of neurobiological evidence in support of it.

Hurley suggests that our standard (and generally unconscious) assumption that perception and action operate according to an input-output model is simply incorrect. This means that we do not simply have sensory faculties that bring us raw data from the world (input), which we make sense of according to some internal genetic or other program, and then act upon (output). We falsely presume that the mind is bounded in a way that separates it (and us) from the world, so that input and output are distinct processes. As Alva Noë puts it, "We are not world representers. . . . Our worlds are not confined to what is inside us, memorized, represented." Instead, "we live in extended worlds" that are "reachable" rather than "depicted."[75] By trying to rid us of the presumption that we are "representers," Hurley is banishing our sense of ourselves as observers of the world rather than participants in it. Another way to put it is that we are not somehow separate from the world, our brains constructing a common intersubjective internal world by imposing standard patterns upon a chaos of disorganized perceptual data. In the standard Cartesian-Kantian story, these subjective and intersubjective constructions play out as films in our heads, and, as a result, they have a tenuous relation to the actual external world. We are locked in our heads. Hurley sets out to refute this standard view that the mind is an *internal* program playing out upon a world stage, which she argues is wrongheaded.

Part of the conceptual problem, she says, is that we tend to focus almost exclusively on the input-to-output direction, how the mind structures incoming percepts. Consequently, we tend to ignore the functions from output back to inputs, "the way environments, including linguistic environments, transform and reflect outputs from the human organism." In other words, the world we encounter is not an unstructured arena of chaotic raw data but in fact is (pre-)structured by human practices, linguistic meanings, institutions, histories, cultures, and nature itself. So both directions are just as complex, Hurley remarks; not only that, but they "are causally continuous." "To understand the mind's place in the world," she goes on, "we should study these complex dynamic processes as a system, not just the truncated internal portion of them." Our place in the world is "a complex dynamic feedback system [that] includes not just functions from input to output, but also feedback functions from output to input, some internal to the organism [that is, from internal data about the body's state back to the mind], others passing through the environment before returning." The upshot of this approach, Hurley tells us, is that there are no "sharp causal boundaries either between mind and world or between perception and action."

Hurley points out that we must not take at face value the notion that the mind provides a ubiquitous and knowable human structure for the raw environmental data coming in through the senses. That scenario, she says, is just as much based on unfounded faith—in this case, in the transparency and availability of the mind—as is the naive acceptance of the external world as being just as we perceive it. So the same kind of Cartesian-Kantian skepticism about the world ought to be directed at the mind itself. If we don't take the world as given, why should we so take the mind? Why should we assume that the mind is subjectively available to us as providing reliable information or accurate readouts of our internal mental states? And why should we assume that the ways the mind perceives external data are ubiquitous across human beings? Why should we take our internal mental states naively, as we have learned not to do with our external senses?

Hurley also challenges the notion that the contents of our mind and the structuring of the mind are independent of the world, and the world we take in is independent of the mind. Her book is an extended argument for externalism, the view that the self is in the world and that self and environment are related in and as interacting open systems. What that means is that there is nothing that is either pure self or pure environment. Both are always interactively constituted. Hurley proposes that

> the revolution that began with Kant's arguments about perceptual experience should be carried through to agency. Action is no more pure output than perception is pure input. The whole of the Input-Output Picture should be rejected, not just half of it.
>
> We get a new angle by making the ninety-degree shift: by making the focus of our scrutiny the perception/action cut rather than the mind/world cut.[76]

When we act, we create a relationship to the environmental context we inhabit, and that relationship both influences what we perceive, on one hand, and structures the mind's way of perceiving, on the other. People, for example, who live in remote locations and build only round buildings and hence have no experience of corners have difficulty seeing corners when they later encounter them because their faculty of sight in early life was not influenced and in part constituted by its interaction with right angles.

The self is not deeply hidden within, able to be discovered only by solipsistic introspection or surmised solely via indirect clues from others' behavior. Instead, "the self is in the open, where it seems to be," Hurley says. "It is a mistake to think that the processes in brains that make subjecthood and agenthood possible relocate subjecthood and agenthood internally. These processes make it possible for us familiar persons to be selves, embedded in the world, here where we seem to be. They don't replace us with other hidden selves."[77] That is, it is a mistake to confuse the vehicle with the content. The fact that we have neural architecture that makes possible a sense of self does not mean that the content of that self is a product of that architecture alone.[78] Nor are our minds completely determined by the world or totally passive to it, Hurley points out, for we can make mistakes about the way it is. So instead we can understand our internal neural architecture as making possible openness to the world and shaping by the world, along with its own shaping of the world. Person and world are relational, interactional, and also contextual.[79]

The themes of Hurley's revised approach to the mind—decentralization, self-organizing systems, context dependence, feedback, emergence—have resonance in research programs in connectionism, dynamic systems theory, and artificial life. So they are not only good philosophy but also where neuropsychology is leading us.[80] We might reflect that it's taken this long to begin to banish Augustinianisms—which severed the human psyche from the body and from nature, and hence from both desire (from cognition, understood as "will," from affect and emotion) and the world, natural and social—not just from Western cultural, philosophical, and of course religious presuppositions but also from the scientific ones that have driven research agendas. Hurley identified the problem but not its theological origins. She also pointed to its tyranny in ethical thinking when she wrote that "the Input-Output Picture that is here [in *Consciousness in Action*] criticized has had significant implicit influence in ethics." She then raises a question: "What are the implications for ethics, and for social and political philosophy, if this [input-output] view cannot be justified?"[81] She calls attention to the fact that the input-output view in ethics presupposes and bolsters the claim of free will, for it conceives us human beings as originating sources of causal chains—our minds impose an order upon the world. Hurley's own view, in contrast, gives a death blow to free will since it envisions human beings as contextually embedded in natural and social causal networks and webs.[82] Here we have come back full circle to Spinoza, who presciently anticipated the direction now being taken by cutting-edge contemporary philosophy of mind in response to the new brain sciences. And here we are rethinking agency and moral agency via systems theory, again with Spinoza as pioneer and guide, even though the formal discovery or invention of systems theory would not take place until more than three hundred years in the future.[83]

Spinoza anticipated externalism and rethought moral agency in terms of it. He of course didn't call it that, but he envisioned nature as a network, a system of causes at all levels from culture to physics. Each person, animal, or thing was a location in the system of networks, a location that defined the point in the overall system that makes each thing what it is and serves as its definition and identity, an "essence," a basic desire, that tries to maintain its integrity amidst change. A thing's essence was not a static thing or content or a quasi-genetic program, Spinoza believed, but instead a "ratio," which is to say it was itself also a system, an open system within open systems, and each system at every level strove to maintain its internal homeodynamic organization while being open to the larger systems, environments, that were its constitutive causes and to which it also contributed.

As a consequence, when it came to ethics, Spinoza conjectured that the move inward to discover the causes of how we each come to be this self and maintain ourself as this self, and the move outward toward identification with the world, turn out not to be discrete processes but instead interwoven. For there is an irony at work here that gets us out of our solipsism indirectly, perhaps even surreptitiously: the process of filling the self with its unconscious content, bringing its un(self-)conscious and even nonconscious causes to light, entails connecting the self to all the contexts, environments, worlds, people, situations, culture and history and biography, and event and memory of which it is composed. So the self, in becoming a self and in becoming itself, acknowledges and becomes a more internally coherent, self-organizing internalization of its immediate world—Spinoza's name for that dynamic well-functioning of self as a coherent system is *activity*—and then of its more distant environments. The paradox is that to be truly yourself *is* to be your world, and ultimately the universe that created you. That was Spinoza's insight, long before neural self-maps ever were a gleam in any scientist's eye. To be this self was to be this point in the universe, Spinoza thought, and it took the whole universe up till now to produce any given "me." So to attain what Spinoza regarded as a state of personal autonomy or "freedom"—to achieve the spiritual and moral psychological aim of the *Ethics*—was to come to understand and own as self all that has come to make up this (biologically, psychologically, socially, culturally, historically, biographically, cosmologically, quantum mechanically, etc.) situated and constructed self.

The world thus is systematically introduced into the self as causes of the self and hence as self—but in the doing, the self now flips and sees itself in terms of its world, in terms of those parts of the world that appear now as personally constitutive. There is no limit to that centrifugal force. We are in principle at home in the universe, and our freedom lies in making that real to ourselves. The environment is not foreign but constitutive. So the irony of autonomy is that its achievement comes to fruition only in the embrace of the environment and of those things within the environment in which one now sees oneself, and progressively more so to infinity. To see aspects of the environment as self rather than only as other is to feel the world not as merely an external limit to the self but as constitutive of the self and finally as harboring the possibility of further self-formation in it and also of it. To include the world within the self (self-mapping) is to open the self to the world, as Antonio Damasio realized, and also, eventually, to extrude the self into our constitutive environments. It is to come to accept the self and embrace the world. *It is to love the other as the self.* Yet to embrace the world or aspects of it without the arduous path of self-discovery is to magically (or "imaginatively," as Spinoza designated it) extrude and lose the self in the immediate environment, in a merger with the present situation and moment and world. A self that has not discovered its own unique environmental constitution is all too vulnerable to being filled by its immediate environment—or, if it rebels, by nothing much at all, or by chaotic impulses and wild shifts in identification and viewpoint. So the human moral danger is more often than not that of the fanaticism of the group mind. That marks the devolution of the systematicity of the self into its environment and the relinquishment to the group of its internal cohesive identity of its own "ratio" or "essence" or homeodynamic stability. That is the real moral danger, not the rare psychopath who has no stake in others or the world. Moreover, the danger of the individual merged into the group self and group mind plays itself out not only between external enemies but also, ominously, between subgroups hierarchically organized in a society.

Thus the group poses another danger as well: it is an internal danger rather than an external one. I am referring to the sacrifice of some component groups for the sake of others within a society. While subgroups are often designated as those to be sacrificed for the benefit of the whole group (the military, for example), it is also the case that certain subgroups are chosen to be sacrificed to other subgroups within the larger social group that is the society as a whole. The obvious example is slavery. The history of American slavery takes that form: the group as a whole includes both masters and slaves, but the hierarchical structure benefits only the elite group. We can think of the Holocaust in this way, too, for not only was it group against group (Germans against Jews) but also a hierarchical group's war against a subgroup (Jews were Germans, too, after all) that those in power systematically denigrated and legally disenfranchised. The Milgram experiments can be interpreted along these lines as well. The experiments were set up so that those designated as the legitimate authorities and the top subgroup in the hierarchical order of subgroups—that is, the psychological researchers—were (or, actually, appeared to be) in effect sacrificing the lowest subgroup in the group hierarchy, the (mock) experimental subjects, on the altar of science. The middlemen were the subgroup of unwitting volunteers. They came into the social system and group structure as the means between the hierarchical elite and the lowest subgroup. It was they who were willing to sacrifice the subgroup for the seemingly legitimate purposes defined by the hierarchical elite.[84] Seen from a wider perspective, the danger to individual members of the group, especially those who are part of subordinate subgroups, is as high as its danger to outside groups. Moreover, the danger from within is always present, whereas the dangers of war and genocide are more episodic. I am reminded of Spinoza's lament that human beings are far too prone to fight for their own slavery as if it were their salvation. His urgent cri de coeur was a plea to envision the moral route to human freedom. Malcolm Gladwell relates a news story that illustrates how common the danger to subgroups within the larger group can be even in normal peacetime situations. He reports that a series of Korean plane crashes occurred because junior pilots (a subordinate subgroup) were loath to criticize or even to call attention to the mistakes of senior pilots. The junior officers were disastrously engaged in what Gladwell says is technically called "mitigated speech," the kind of downplaying talk that people tend to use in deference to authority. As a result, hundreds of people died, including a number of pilots.[85] The normative authority conferred upon those in a high subgroup position in the social hierarchy is a social urge so overwhelming that it has a tendency to trump life itself. If we think that this is merely an ethnic or national issue, we recall that the Milgram experiments had comparable results across cultures.

This story has a good ending and an instructive one. Korean Air recognized cultural deference to authority as a problem and called in Delta Airlines' David Greenberg to help them. Research has shown that the United States is low on a comparative scale of nations' relative deference to authority. Greenberg says that what he did with the Korean Air personnel was that his team "'took them out of their culture and re-normed them.'"[86] The intervention was a success. What we should take away from this example is that intervention from the broader world and context can critique and revise internal group norms and the hierarchies among constituent subgroups of the larger group. It is just as much the internal environment that is in danger from a group as its external enemies—perhaps even more so, generally speaking. That was one lesson of the Stanford Prison Experiment, which also had its subgroup of internal victims rescued only because of an outside intervention. International human rights organizations attempt to intervene in this way across cultures. Intervention from a wider perspective and the world at large is a function of diversity and suggests how much it should be cherished, bolstered, and institutionalized. Diversity creates the possibility of groups, which are adaptive systems, opening themselves to ever-wider perspectives.

Systems Theory

> Human cognitive processes are inherently social, interactive, personal, biological, and neurological, which is to say that a variety of systems develop and depend on one another in complex ways.
>
> —WILLIAM J. CLANCEY

> The mind leaks out into the world, and cognitive activity is distributed across individuals and situations. This is not your grandmother's metaphysics of mind: this is a brave new world. Why should anyone believe it? . . . One part of the answer lies in the promise of dynamical systems theory . . . as an approach to modeling cognition. . . . Insofar as the mind is a dynamical system, it is natural to think of it as extending not just into the body but also into the world. The result is a radical challenge to traditional ways of thinking about the mind, Cartesian internalism in particular.
>
> —PHILIP ROBBINS AND MURAT AYDEDE

The insight that a person is an open system in relation to other open systems, natural and cultural, has begun to be rigorously articulated and theoretically worked out in the developing field of systems theory. Systems theory, which originated in the 1940s, derives its principles from physics, biology, and engineering, and was influenced by both cybernetics (the study of communication and control systems) and semantics (the study of meaning). Systems thinking seeks to encompass the viewpoints of the many different disciplinary approaches to the parts of the system being analyzed. So, "for example, when building a highway, one can consider it within a broader transportation system, an economic system, a city and regional plan, the environmental ecology, and so on." The different perspectives highlight different relationships, different parts, and different causal processes. The issue is how to understand a dynamic process at many levels of analysis and within a hierarchy of levels of analysis (the inorganic, the organic, the neurological, the symbolic, and the cultural, for example). So what *are* the identifiable features of complex systems? A complex system is defined as one "whose properties are not fully explained by linear interactions of component parts," that is, it is a system whose properties are not merely explainable by adding them together and in which small changes can produce large effects. The whole is more than the sum of the parts.[87] The systems perspective on human behavior is all about context; it takes into account "psychology, anthropology, sociology, ethology, biology, and neurology, and their specialized investigations of knowledge, language, and learning" in emphasizing the "contextual, dynamic, systemic, nonlocalized aspects of the mind, mental operations, identity, organizational behavior, and so on." The approach was well established in biology before it reached other disciplines. Biologists realized that in order to come to understand the sustenance, development, and evolution of life, an organism could not be isolated from its environment nor a cell from the organism. "Systems thinking, involving notions of dynamic and emergent interactions, was necessary to relate the interactions of inherited phenotype, environmental factors, and the effect of learning."[88] The overall approach of systems thinking to human cognition and behavior is as follows:

> An all-encompassing generalization is the perspective of complex systems. From an investigative standpoint, the one essential theoretical move is contextualization (perhaps stated as "antilocalization" . . .): we cannot locate meaning in the text, life in the cell, the person in the body, knowledge in the brain, a memory in a neuron. Rather, these are all active, dynamic processes, existing only in interactive behaviors of cultural, social, biological, and physical environment systems.

A self, according to this approach, is "self-organizing" and "unfolding" and always contextualized or "situated."[89]

Thinking about human behavior in terms of systems changes dramatically the way we conceive agency: what it means to act and even who is doing the acting. That is the conclusion of the computer scientist Merlin Donald, who argues that although decision making "seems to be a very private thing: individualized, personal, and confined to the brain," when it is looked at from system theory, we realize that culture is a major factor in how the brain self-organizes during development, both in its patterns of connectivity and in its large-scale functional architecture.[90] So it is the system that makes the decision: "The mechanisms in such decisions must be regarded as hybrid systems in which both brain and culture play a role."[91] Donald, of course, does not deny that decisions are made in individual brains. Nevertheless, he points out that "human brains . . . are closely interconnected with, and embedded in, the distributed networks of culture" that "define the decision-space."[92]

Neurobiological research is further elaborating how decision making can have a range of scopes, from a part of the brain to large groups of people. To discover who is actually acting in a given case, all the facts need to be taken into account and then analyzed from multiple standpoints, from the brain sciences to organizational behavior to culture and history, and so on. Only through this multidisciplinary approach can the attribution of agency and responsibility be accurately distributed across people and levels of organization and participation and authority. Christoph Engel and Wolf Singer raise the question of how many agents are involved in decision making, and offer this answer: "This perspective treats human decision makers as [possible] multiple agents [both internal and external]."[93]

Complex Adaptive Systems

Complex adaptive systems are quite different from most systems that have been studied scientifically. They exhibit coherence under change, via conditional action and anticipation, and they do so without central direction.

—JOHN H. HOLLAND

Complex adaptive systems [are] those that learn or evolve in the way that living systems do. A child learning a language, bacteria developing resistance to antibiotics, and the human scientific enterprise are all discussed as examples of complex adaptive systems.

—MURRAY GELL-MANN

The human person not only is an open system within others but also adapts. Complex adaptive systems are a special kind of system in which emergence and self-organization hold sway. In contrast, Murray Gell-Mann, Nobel laureate in physics and one of the founders of the science of complexity, points to galaxies and stars as examples of systems that are complex and evolve but which are nonadaptive.[94] The control in a complex adaptive system is decentralized and widely distributed, rather than being under some central control. The patterns of activity arise or emerge from the interactions of the agents rather than from some overall plan. Nevertheless, there is dynamic stability, identifiable patterns that are neither utterly chaotic nor substantially fixed. These patterns evolve, changing over time as the system itself changes and evolves. It is the individual "agents" in the system that, from their location and environment, develop adaptive behavior.[95] They exhibit the same patterns of the whole at various scales within the system. Learning is an important feature of complex adaptive systems even though there is no central consciousness involved. And they are highly resilient. Ever increasing diversity is an important feature of complex adaptive systems and crucial to their capacity to adapt and survive.

Diversity in Complex Adaptive Systems

The hallmark of complex adaptive systems is perpetual novelty, according to John Holland. Diversity arises from how this kind of system recycles its resources.[96] For example, if you have a network with an ore extractor, a steel producer, and a car manufacturer, and three-quarters of the steel from cars is recycled and goes back to the steel producer and then the manufacturer and comes out again as cars, then with the same input of steel, recycling creates more resources at each stage. A tropical rain forest illustrates the point perhaps even more dramatically, and from this second example we can see how diversity enters in. The soil in a tropical rain forest is quite poor because minerals are constantly leached out of it and into the rivers by heavy rains. So tropical rain forests are terrible places for introducing agriculture. Nevertheless, the rain forest is hugely rich both in numbers of living beings and in the diversity of species. Why and how is this so? Rain forests depend for their fecundity on the recycling of critical resources. The rain forest is not like a simple hierarchy in which the top predator consumes the resources. Instead, "cycle after cycle traps the resources so that they are used and re-used before they finally make their way into the river system." This system of recycling is so effective and creates such richness in the rain forest that ten thousand different insect species may inhabit a single tree! The recycling effect produces resources available to be used in new environmental niches, and these niches are filled by increasingly diverse species: "each new adaptation opens the possibility for further interactions and new niches."[97]

A complex adaptive system does not settle into locked-in patterns but keeps producing change and self-correction.

It is the particular niche that defines the kind of diversity that will arise. In evolution, what this results in is convergence. A particular niche produces species that exploit it in very specific ways, and if that species disappears, another species, one that can be entirely unrelated, appears that nevertheless has similar characteristics to the earlier one in the way that it fits into the niche. Holland gives as examples the prehistoric ichthyosaur and the modern porpoise, two species in no way related to each other yet similar in their habits (their prey, for example) and their form. That the niche determines the kind of diversity that arises to occupy it is a stellar example of what is meant by externalism: the context determines the individual.

To understand what any item in this kind of system is, we need to look at its location, function, and interactions, rather than primarily inside it. What you get inside a complex adaptive system is what the outside, the environment, has made it—through either the "fast" externalism of changes in behavior or the "slow" externalism of evolutionary adaptation. While the quintessential example is an ecosystem, two other examples of the context defining the niche are New York City, which has thousands of different kinds of wholesale and retail businesses, and the mammalian brain, with its patterns of neurons organized into hierarchies and regions.[98] The overall system and its dynamic relations define its component building blocks. The dynamism depends upon ongoing diversity.

Some Further Thoughts on Complex Adaptive Systems, Diversity, and Ethics

Each of us humans functions in many different ways as a complex adaptive system. . . . When you are investing in a financial market, you and all the other investors are individual complex adaptive systems participating in a collective entity that is evolving through the efforts of all the component parts to improve their positions or at least survive economically. Such collective entities can be complex adaptive systems themselves. So can organized collective entities such as business firms or tribes. Humanity as a whole is not yet very well organized, but it already functions to a considerable extent as a complex adaptive system.

—MURRAY GELL-MANN

Murray Gell-Mann comments that the "human race" can be thought of as a complex adaptive system engaged in "evolving ways of living in greater harmony with itself and with the other organisms that share the planet Earth." At different scales, "a society developing new customs" and even an individual "artist getting a creative idea" are complex adaptive systems both within and engaged with other complex adaptive systems.[99] How can thinking in terms of complex adaptive systems help us rethink moral agency and come up with ways to make societies and all kinds of groups function more ethically? Crucially, complex adaptive systems theory suggests that to change the person we need to look at the system. Interventions in context and environment, rather than in the brain or mind of the person (for example, through drugs or the training of the individual will), seem to be the place to start.

Diversity is crucial, too. We need to think about diversity and its role in complex adaptive systems to ensure their ongoing vitality and continuing evolution. In social systems, diversity plays out as the introduction of diverse people, practices, and points of view that challenge and disrupt the stable social system, sparking a more complex and inclusive reordering and reintegration. It is a trade-off between closure to variation and the resulting static internal coherence, on one hand, and on the other, openness so great that the system cannot accommodate the differences fast enough through internal systemic reintegration. Spinoza advocated a systems theory of ethics that was perpetually reorganizing at the brink or "edge" of chaos.[100] This means that he advocated a personal maximal openness to others and to the world while retaining the capacity for dynamic self-organization. So at best, in this view, each of us ought to cultivate an openness to others that doesn't overwhelm us but can be integrated into our sense of self and what we care about through understanding, the increased capacity for empathic identification, standing in the other's shoes, perspective taking, and openness to critique and self-critique. As a friend puts it, it's not just about tolerating differences; it's also about finding in oneself the capacity to enlarge one's empathic acceptance. This involves the ability to learn from others, both about them and about one's own self from another's perspective. So ongoing diversity is necessary in order to overcome personal self-deception, dogmatism, and denial— our most ubiquitous and corruptive moral dangers. Professor Pat Longstaff of the Newhouse School at Syracuse University points out that diversity enhances the resilience of systems, so that diverse groups and institutions are more likely than homogeneous ones to survive under threat or in times of adversity.

We should now rethink the problem of selfiness from a systems perspective: it is the attempt to maintain a narrow systematicity and coherence that won't allow in challenging data from others or even from the implicit un(self-)conscious meanings and intentions of our own actions. Selfiness tends toward the refusal to acknowledge that one is a part of larger cultural, social, and natural systems. It is the arrogance of the myth of self-creation, of free will. The overcoming of narrow selfiness of this kind in an expansive self-coherence that enlarges the self to include more of the world and others is a lofty ideal for the individual and a noble and difficult path. It is also a rare one, as Spinoza pointed out.[101]

On the societal level, more is called for. Protecting diversity, especially diverse opinions, is absolutely vital to the possibility of social moral agency, for it makes available critical perspectives that foster both whistle-blowing and creative thinking and solutions. Whistle-blowing and creative thinking enable social systems to adapt to changing environments instead of stagnating. Just think of oil companies or car companies that fail to adapt to new energy sources, or of large corporations that corrupt scientists and government agencies to the point where they deny global climate change. Our human survival, as well as our moral integrity, lies in the balance. Groups and individuals alike are the subjects of moral integrity or corruption. We're all systems in need of the critique and larger perspectives that can come from the outside. We all need to foster moral life at the edge of chaos. We cannot allow self-systems and group-self-systems to push for a selfy (that is, self-serving and hypocritical) coherence that eliminates—and so allows us to deny—the challenge of external perspectives and critiques at the price of our moral integrity, the betterment of the world, the health of the natural environment, and even our survival as a species.

Evolutionary Cooperation and Group Selection

True, humanity never runs out of claims of what sets it apart, but it is a rare uniqueness claim that holds up for over a decade. This is why we don't hear anymore that only humans make tools, imitate, think ahead, have culture, are self-aware, or adopt another's point of view. If we consider our species without letting ourselves be blinded by the technical advances of the last few millennia, we see a creature of flesh and blood with a brain that, albeit three times larger than a chimpanzee's, doesn't contain any new parts. Even our vaunted prefrontal cortex turns out to be of typical size: recent neuron-counting techniques classify the human brain as a linearly scaled-up monkey brain. . . . No one doubts the superiority of our intellect, but we have no basic wants or needs that are not also present in our close relatives. I interact on a daily basis with monkeys and apes, which just like us strive for power, enjoy sex, want security and affection, kill over territory, and value trust and cooperation. Yes, we use cell phones and fly airplanes, but our psychological make-up remains that of a social primate.

—FRANS DE WAAL

We are social primates. We are defined by our niche and by our evolutionary inheritance. We are different in degree but not in kind from other primates in our brain and behavior, capacities and needs, and desires and tendencies. Aristotle was right: human beings are social animals. Cooperation is the norm, and it occurs on a grand scale in human beings. Human beings are "groupy" to an astounding degree, as well as selfy. From an evolutionary biological standpoint as well as from a systems theory standpoint, the group, not the individual, is often the significant actor. New thinking about evolution is now focusing not just on competition among individual members of a species as the driver of selective pressures, but crucially on group-against-group competition within species and between species. Some biologists even think cooperation is as important and basic an evolutionary principle as competition.

Evolutionary biologists Martin A. Nowak, Corina E. Tarnita, and Edward O. Wilson argue that natural selection for kinship is proving to be unfounded.[102] What that means is that altruistic behavior in living beings (specifically, in insects of various kinds) is not proving to be tied to kinship. Instead, behavior benefiting the group, rather than the singular individual, is far more widespread and the scope broader than mere kinship relations. Altruism is not driven by an urge to pass on the genes of related individuals. Since the 1950s the standard theory has been that close kin are most likely the ones who give up their lives altruistically so that those who share their genes are able to survive and reproduce. The sparse evidence available at that time seemed to prove the case. Yet with much more evidence now accumulating, what was known early on has turned out to represent not the norm but the exception. It is now known that most eusocial (what is commonly called "altruistic") behavior in insects, for example, takes place not between kin, nor is it to the particular genetic advantage of kin. There is even evidence that genetic variation is being favored, in contrast to genetic likeness: in the case of certain types of ants, for example, it is colony-level selection that obtains, and that favors genetic variability rather than similarity.[103] The authors continue: "Other selection forces working against the binding role of close pedigree kinship are the disruptive impact of nepotism within colonies, and the overall negative effects associated with inbreeding. Most of these countervailing forces act through group selection or, for eusocial insects in particular, through between colony selection."[104] So there is increasing evidence that evolutionary competition is not just between individuals but also between groups within colonies and between colonies. Moreover, the authors note that "in some cases, social behaviour has been causally linked through all the levels of biological organization from molecule to ecosystem."[105]

Having defeated kinship selection as the primary way that cooperation occurs, the authors put forth their own "alternative theory of eusocial evolution."[106] Animal groups assemble in all kinds of ways, the authors point out, from local nest and food sites to families staying together to flocks following the leader or even just local proximity. "What counts then," they say, "is the cohesion and persistence of the group."[107] According to Martin Nowak, director of the Program for Evolutionary Dynamics at Harvard University, cooperation is as vital and ubiquitously operative an evolutionary principle as competition. Individuals themselves, after all, are the result of cooperation among cells, among organs, at every level. Evolution favored cooperation in order to create organisms.

Moving On

The upshot of the evidence presented in this chapter and throughout the book is that sociality trumps individuality in human beings as a species. It is an idiosyncrasy of the particular religio-cultural trajectory of the Latin West that prosocial behavior, moral agency, came to be defined in terms of the individual standing beyond belonging—willing and choosing and deciding from a locus of self that could free itself from determination by group, context, world, and natural endowment and act. An alternative understanding of what it means and takes to be moral seems to have come on the horizon: a conception of ethics in terms of shaping, improving, and enlarging a fundamental human natural sociality. Sociality is the default position. On this view, individuality—freedom in some new form—is a difficult and heady achievement and one accomplished only through the embrace of broader belonging and wider critical perspectives.

10

What Is Ethics? How Does Moral Agency Work?

> But a limited causality is no longer a causality at all, as our wonderful Spinoza recognized with all incision, probably as the first one. And the animistic interpretations of the religions of nature are in principle not annulled by monopolisation. With such walls [between oneself and others] we can only attain a certain self-deception, but our moral efforts are not furthered by them. On the contrary.
>
> —ALBERT EINSTEIN

How can we rethink what it means and what it takes to be ethical and morally responsible, and get others to be ethical and responsible, without recourse to a notion of free will? How can we rethink both moral agency and moral responsibility within a framework that acknowledges that human beings are fully biological organisms, that they are within the world and not beyond it or above it, and that the mind, particularly consciousness, is real and as much a product of nature and embedded within causal networks and systems as the body? These have been the overriding questions that this book has raised and attempted to approach from multiple angles—from history, from comparative cultural analysis, from social psychology, from the new brain sciences, from evolutionary biology, from the theory of complex adaptive systems, and from the philosophy of Spinoza.

After having exposed free will as both a Western cultural myth and an implausible description of moral agency, we moved on to think through how moral action might actually work, delving into the Greek philosophical tradition of moral philosophy and psychology as it was extended and developed

412

within the Alexandrian orbit of *falsafa*. We undertook a broad overview of findings in social psychology, other branches of psychology, and the new brain sciences, and even glanced at complex adaptive systems theory and at the evolutionary biology of self-organization and cooperation as well as at the competitive struggle for survival, in order to gain a rough composite picture of how the mind operates in affective motivation and decision making.

We learned about the relations among cognition, emotion, and action from several perspectives, illuminating their interpenetration. We jettisoned any explanation of moral agency that depended upon the discreteness, the independence from one another, of both cognition (thought) and affect (emotion), and also self and environment.[1] We came to reject the standard view that reason could function independently of emotion, of context (cultural, social, and political), and of present situation to produce choices that intervene in our world from a separate quarter via a (mythical) free will. Cognition, emotion, and context were rethought as intertwined, and action, including moral decision making, was seen as emerging from that whole, a whole that included both self and world.

Relinquishing the myth of free will along with recognizing the vastness of the brain's neuroplasticity—in keeping with Jaak Panksepp's arguments against the discreteness or modularity of brain mechanisms directly playing out in isolated ways at the highest levels of human behavior—suggested a revised account of action. We learned that archaic, affectively laden functions trickle up into neural pathways, informing them and also being modified by them. Hence neural pathways were seen as emotionally rich with experiences and responses, interactions that incise into the brain patterns of unique histories of responsive engagement with the world, amounting to who each of us is. We come to act and enact who and where we are and who and where we have been, with all the associated emotional meanings and colorings and contextual interpretations, and also our present situation and its embedded meanings, interpretations, and incentives. Instead of a free will that is called upon to act beyond and outside its own constitution, I argued that the evidence suggests we act as who we have become, where we locate ourselves in the world, what we belong to at the present moment. This self beyond itself, one that has included its relations to its immediate worlds within itself as itself, is the default position. Being in and of the world is who we necessarily are. We discover and rediscover ourselves in the world.

Nature and nurture, present situation and social belonging and political contexts, and the emotional feelings of pain and pleasure associated with all these were understood to have come together to produce this "me," acting in this moment. None of this is rigidly set; despite the largely stable neural paths of connection that have been incised in us early on, our neural networks retain varying degrees of openness to change and also to expansion throughout life. Even memories are constantly revised and reconstructed rather than set. In regard to ethics, what may be most pertinent are the neural pathways that map the self according to its painful and pleasurable engagements with the world, going deep down even into those very basic, survival-related homeodynamic mechanisms that Antonio Damasio has called a "core self" and that Jaak Panskepp has located in a most ancient basic seeking system. Spinoza rooted a transformative ethics, an achievement of enhanced moral agency, in just such a core homeodynamic and desiring mechanism and its capacity for integrating into ourselves systemically our engagements in the world, anticipating contemporary neurobiology and neuropsychology.

How a sense of self arises, the ways experience writes itself upon the structures and functions of the brain, the brain's ongoing openness to change and restructuring throughout life, the affective embodied and situated character of perceiving and thinking—all suggest that transformation throughout life seems increasingly possible and even plausible. The brain's neuroplasticity, which is of a scale we know to be nearly immeasurably vast, can be retrained and hence restructured by specific interventions that are now being discovered. The degree of openness to neuroplastic change has been shown to be enhanced through meditation and other practices.[2] Moreover, greater understanding can now be understood as the modification of pathways of meaning (interpretation) and their associated emotions, which are the triggers from which actions follow.

Ethics in a Spinozist Key

Some ideas of Spinoza are still valid after the roughly 350 years that have elapsed. . . . Most of Spinoza's general approach to conceptualizing the world can still be utilized given a slight change in terminology. . . . Spinoza's approach . . . may be capable of shedding a new light on the ancient problem of the relationship between human beings and the rest of nature. The theory of evolutionary systems is a prime candidate for a conceptualization that might be useful in order to concretely develop this new insight. . . . Ludwig von Bertalanffy[, who] . . . also started from a holistic perspective . . . thought that the laws of biological systems such as those governing growth and adaptation of living beings might well be applicable to the human psyche, to social institutions, and to the global ecosphere altogether. He thought the natural laws of organization would govern systems on all levels of existence. And the search for these laws he called "General Systems Theory." . . . And this is clearly where Bertalanffy meets Spinoza.

—RAINER ZIMMERMANN

At the dawn of modernity, Spinoza, anticipating a number of important discoveries in the recent brain sciences, drew upon his protobiological, embodied account of the mind in order to rethink moral agency without free will. Spinoza's account of moral agency, which he called ethics, may provide the best working model available at the present time for trying to develop an account of how and why we are ethical, why and when we are not, and how to get people to be more ethical. Spinoza's philosophy perhaps represents the best starting point for trying to integrate the evidence emerging from the new brain sciences and other relevant disciplines into a composite view of the basic moral brain, the optimal route to its development, and the implications of such a view for how social, legal, political, and other institutions and practices might be redesigned.

Spinoza's principles—of the nonreductive identity of mind and body; of the affectivity of all thinking; of the mind as the very activity of understanding and desiring (rather than a Cartesian bounded thing that thinks and has ideas), expressive of one's engagements with the environment as self-enhancing or self-diminishing; of the un(self-)conscious character of our motivation; of the rejection of free will; of the biological urge to organic ever-expansive self-organization (mental as well as physical) within environments ranging from the local to those as large as the universe, a path he designated as freedom, to name some central theses—provide a fruitful starting point for making sense of and synthesizing a lot of the new thinking about how the mind produces action and, in particular, moral action, while at the same time recasting human nature within its environments, both social-cultural and natural.[3]

Spinoza regarded the search for self-understanding—within the broadest theoretical grasp of all the contexts and environments that contribute to making each of us who we are, and finally within the infinite context of the universe itself—as being at the root of the possibility of moral transformation. That path was held to be transformative of motivation and agency because it is knowledge about the self, about the very desire for organic survival (which Spinoza called the conatus) at the core of the person. Hence that rigorous knowledge about the world, because of its self-relatedness, was also held to be emotionally rich.[4] Spinoza believed that he had made a psychological discovery about how a general feeling of joy and a motivation of compassion toward all things could emerge from a path of coming to understand oneself in terms of the bodies of explanation of the sciences, explanations that thereby became part of what we would call today one's self-concept. That path he also understood as the route to freedom: freedom from the tyranny of social-hierarchically defined group and subgroup identities, from their narrowly and locally shaped desires and ideals, from local prejudices and attitudes—that is, from all the local incentives that capture and often enslave the mind and heart as much for self-destructive ends as for fanatic violent ones.

All Roads Lead to the Self Beyond Itself

The teaching of Spinoza, has nothing in common with the sickly atomistic conception of immortality, a conception which dissolves the unity of life. . . .
In the teachings of Spinoza . . . the individual is not treated as a separate entity, but as a part of the whole. According to Spinoza, eternity does not begin with our death, but always exists, is always present even as God himself.

—MOSES HESS

As I wrote in the previous chapter, we locate our basic biological sense of self-preservation and self-furthering in a self distributed beyond our skin into our environments, natural and human. This is why we care about the world and why it is the arena of our moral concern and of our ideals. We are fundamentally social and world-related. Aristotle was right: man is a social animal. That sociality, that world-relatedness, that constitutes each of us, can become enlarged through broader self-understanding and also more systemically internally integrated and coherent. It was Spinoza's brainstorm that the moral path was not from selfish isolated individualism to altruism via a free will, but instead from a diffuse localism in which self and environment were merged to an expansive identification with worlds and environments further and further afield yet understood as having combined to produce this unique "me." Thus the progressive inclusion of one's environmental causes, social and natural, in one's self-understanding organizes the self toward greater coherence and individuality, a view from this very point in the universe. Spinoza envisioned this as a path to freedom, a way to overcome passive definition by one's narrow and provincial social and cultural location, becoming free of the seduction and tyranny of its rewards and punishments. To expand the self's boundaries infinitely outward into an ever-wider world was to embrace freedom.[5]

It was this transformation toward an enlarged and coherent self—a self extended by its constitution by its relations with the world, distributed into its environments, and systemically integrated—that Spinoza envisioned as capable of true moral agency. Such a self was not a mere passive reflection of the immediate local environment (the basic human condition, the default position) but rather a desiring-focused point, expansively opening outward, systemically integrating its environments into self and self into its worlds. Spinoza regarded that progressive openness to the world as the source of our freedom. The paradoxical nature of this vision is that freedom is to be achieved through greater belonging, not by taking oneself out of the world but by becoming ever more deeply, broadly, and coherently of it and engaged in it.

A Systems Theory of Moral Transformation

We can come to a better understanding of Spinoza's ideas about moral transformation, his ethics, in terms of systems theory.[6] Becoming an increasingly open, complex, internally organized, "coherent" (that's Spinoza's term) systemic mind and self—consciously identified with, belonging to, and nested within larger and larger complex adaptive systems—offers a way (perhaps the only way) to free oneself from submergence in one's initial narrow, insular, provincial, and often fanatical immediate world, a world that all too often was willing to sacrifice individuals, especially members of subordinate subgroups, to others within it or outside it.[7] The moral problem is that people generally are passive reflections of their immediate local environments. Individuality is an achievement that, paradoxically, is to be accomplished by greater and broader yet unique belonging to a larger world and universe. That the self is fundamentally beyond itself means that nearly all of us are moral agents in the broader (and default) sense. Yet that does not protect us from group evil, internal and external; that is our great human weakness. The moral transformation toward greater social and natural environmental scope, broader identification, and internal self-organization brings about a stronger individuality as well, an independence from the immediate context, along with an ever-expansive and self-correcting yet solid point of view (and desire) from here. That freedom, in Spinoza's estimation, was the heady and rarefied achievement of the very few.

The recognition of ourselves as environmentally constituted is, paradoxically, the source of our capacity to transform and direct our basic biological desire for survival and self-furthering toward a more complexly internalized dynamic. We find the universe within ourselves and ourselves within it: we are the "we that is I" and the "I that is we." We are *of* the world, not beyond it. But we can occupy this unique point of conjunction of self and world with integrity and with a freedom that enlarges the self and its desires by inclusion of ever more of the world, and even the universe, as our vital personal concern. My systems theory of moral agency, my account of the self beyond itself, has attempted to translate Spinoza's moral philosophy into a contemporary idiom and revise it in the light of significant discoveries and insights from the new sciences of the mind and brain, and through greater cultural and historical understanding and our own cultural self-understanding.[8]

Why be ethical? For freedom's sake; for the heady joy of a self expanding itself beyond the narrow and provincial local environment; for a sense of living on in the wide universe of which one is a tiny, local expression; and for an enhanced sense of agency (activity) in a dangerous, unpredictable, and ephemeral existence—those were Spinoza's answers. All these are personal answers, goals that rest with the individual, with his or her desires. They appeal to no external or universal norm or value, to no sense of justice or responsibility, not even to conscience. But they do appeal to love of others, love of the world, and joy in them, through uncovering and forging connections to the world. They appeal to a transformed desire to be free, to be whole, to be at peace, to be with others and make the best of the world. Opening oneself to being more broadly acted upon by the world in order to discover oneself within it—surely as a basis for acting more broadly within it—is a paradoxical route to freedom.

A Final Word on Moral Responsibility

Those who believe in free will, both historically and even today, generally claim that without free will there could be no moral responsibility. In my many conversations about what my book was about, the response of perhaps most of the folks I was talking with, from the most rigorous philosopher to the acquaintance I bumped into standing in line at the grocery store, was usually along the lines of "Don't we need free will so that wrongdoers can be held responsible for their actions?" My answer is no, we don't, and here's why. One of the surprising findings of the new brain sciences is that the self-conscious sense of self—selfiness, as I have dubbed it—is simply attached to things that we are self-consciously aware of doing and being, even as un(self-)conscious selfiness is of course still driving the actions. It does not depend on a sense of free will. We have all been embarrassed by a faux pas, whether an unfortunate slip of the tongue or a clumsy accident on a first date. None of these things has anything to do with free will, but we feel intensely that these experiences are our own and that we are responsible for them. Moreover, the new brain sciences tell us that this sense of self-ownership—of personal responsibility—is weakened in key instances: for example, in un(self-)conscious actions, in actions where some other person holds authority, and also, especially, in group actions.

The sense of ownership of our actions and selves, research has revealed, can be enhanced by bringing actions of the kinds just mentioned into self-conscious awareness. Un(self-)conscious and implicit regions of neural self-maps, especially those in which there is an emotional charge in or near awareness, can be brought into awareness and hence brought under a sense of "I." We do not yet know the extent to which our implicit self-mapped relations to the environment can be uncovered.

Even ex post facto we can claim actions as our own and come to feel their selfiness, which is our sense of responsibility for them. Recall Martha Stout's patient Garrett, who had multiple personalities of which he was at times unaware but took responsibility for what any and all of them did. In addition, I argued that moral responsibility should be broadened to capture the real scope of the actual agent of actions, and that this scope could sometimes span (be distributed in) individuals and even societies, decades and even eras. It is a matter of future debate and planning to think through how our political, social, and legal institutions could better accommodate the real scope of agency rather than be tied to the myth of individual free will.

A revolutionary change in attitude follows from the conception of moral agency that I have developed here in this book. If free will is relinquished, we come to recognize that what must be *must indeed be*, and that what must have been *could not have been otherwise*. We can come to accept ourselves and others for who we and they are. Compassion (as well as reasonable self-protection) would then replace punitive moral outrage as the resulting emotion and motivation. Moreover, a transformation in our moral viewpoint toward the self beyond itself, a self embracing and discovering itself in larger and larger social and natural worlds, would be revolutionary in a different way: it would lead us toward an ecological and universal perspective. We cannot all get there through philosophy, through a Spinozist education of desire. But all of us can be set on a new path through changes in institutional structures that foster broader social diversity and critical perspectives on entrenched norms, legitimacies, and hierarchies. We can popularize ideas that support a new moral vision of freedom and joy for all individuals. Our default localism and provincialism can be socially, politically, and environmentally challenged and broadened, and scientifically informed, to bring out the best in us and also point the way to greater freedom for all.

Notes

1. Searching for Ethics

1. All the names of places and people used here are pseudonyms.

2. All quotations are as remembered.

3. This "pledge" (including its designation) is slightly disguised to obscure the identity of the school.

4. Words modified to disguise the identity of the school.

5. Cecily Kaiser, *If You're Angry and You Know It*, ill. Cary Pillo, Scholastic Reader Level 2 (New York: Cartwheel Books, 2005).

6. B. Edward McClellan, *Moral Education in America: Schools and the Shaping of Character from Colonial Times to the Present* (New York: Teachers College Press, 1999). Although McClellan's sympathies run to the moral character education movement, his history is basically fair to all sides of the debate and succeeds at a reasonably neutral presentation.

7. Ibid., 89–90.

8. Ibid., 90.

9. Ibid.

10. Ibid., 91.

11. "The Endowment," Lilly Endowment website, www.lillyendowment.org/theendowment.html.

12. "Guidelines & Procedures," Lilly Endowment website, www.lillyendowment.org/guidelines.html.

13. Thomas Lickona, personal communication.

14. Robert W. Howard, Marvin W. Berkowitz, and Esther F. Schaeffer, "The Politics of Character Education," *Educational Policy* 18, no. 1 (January–March 2004): 207.

15. McClellan, *Moral Education*, 89, 91.

16. Ibid., 91.

17. Ibid.

18. Sanford N. McDonnell, foreword to *Building Character in Schools: Practical Ways to Bring Moral Instruction to Life*, by Kevin Ryan and Karen E. Bohlin (San Francisco: Jossey-Bass, 2003).

19. Ibid.

20. Ibid., 75–78.

21. William J. Bennett, ed. with commentary, *The Book of Virtues: A Treasury of Great Moral Stories* (New York: Touchstone, 1993).

22. William J. Bennett, *The Broken Hearth: Reversing the Moral Collapse of the American Family* (New York: Doubleday, 2001).

23. Quoted from the frontispiece of Bennett, *Broken Hearth*.

24. Bennett, *Broken Hearth*, 170–71.

25. Ibid., 3.

26. Ibid., 4.

27. Ibid., 5.

28. Ibid., 38.

29. James Q. Wilson, *On Character*, expanded ed. (Washington, DC: American Enterprise Institute, 1995), 15.

30. William Kilpatrick, *Why Johnny Can't Tell Right from Wrong: And What We Can Do About It* (New York: Simon & Schuster, 1992). While the message is similar to that of others in the genre, there is perhaps a slight difference in emphasis. Bennett, for example, demonizes liberals as well as the poor and blacks, saying, "In the 1960s and '70s, and for the first time, truly significant numbers of Americans took the liberationist critique of society seriously and, cheered on by intellectuals and social scientists, acted upon it" (Bennett, *Broken Hearth*, 23).

31. Kilpatrick, *Why Johnny*, 225.

32. Ibid., 13. The *Wikipedia* article "Whole Language" (en.wikipedia.org/wiki/Whole_language#Contrasts_with_phonics) offers accounts of both methods, their histories, and the controversy. The beginning summary is helpful: "Whole language describes a literacy instructional philosophy which emphasizes that children should focus on meaning and moderates skill instruction. It can be contrasted with phonics-based methods of teaching reading and writing which emphasize instruction for reading and spelling. It has drawn criticism by those who advocate 'back to basics' pedagogy." The article goes on to describe the influence of research in linguistics, philosophy, and the neuroscience of cognition and learning on the development of the holistic approach to reading instruction as well as an account of the phonics method of teaching, the history of the controversy, and the current state of practice.

33. Kilpatrick, *Why Johnny*, 15.

34. Ibid., 14.

35. Ibid., 14–15.

36. Ibid., 15.

37. Ibid.

38. Ibid., 15–16.

39. Ibid., 15.

40. Ibid., 226–28.

41. For a fascinating analysis of the role of the unconscious or implicit conceptions of authoritarian versus more egalitarian models of family life in driving conservative versus liberal political presuppositions, see George Lakoff, *Moral Politics: How Liberals and Conservatives Think*, 2nd ed. (Chicago: University of Chicago Press, 2002). Lakoff's study is an application of recent findings in cognitive science about the operation of implicit categories and cognitive schemas to the nature of thought and especially ethical thinking.

42. Howard, Berkowitz, and Schaeffer, "The Politics of Character Education," 189.

43. From the Center for the 4th and 5th Rs website, www2.cortland.edu/centers/character, and "Assessment Instrumentals," Center for the 4th and 5th Rs website, www2.cortland.edu/centers/character/assessment-instruments.dot: "The Center for the 4th and 5th Rs (Respect and Responsibility) . . . [p]romotes a comprehensive approach to character education that has been widely adopted in the field. This approach uses every phase of school life—the teacher's example, the content of the curriculum, the instructional process, the rigor of academic standards, the handling of rules and discipline, and the school's intellectual and moral climate—as opportunities for character development."

44. Kevin Ryan, founder of the Boston University Center for the Advancement of Ethics and Character, is another.

45. Thomas Lickona, *Educating for Character: How Our Schools Can Teach Respect and Responsibility* (New York: Bantam Books, 1991), 3.

46. Ibid., 3–4.

47. Ibid., 5.

48. Thomas Lickona, *Character Matters: How to Help Our Children Develop Good Judgment, Integrity, and Other Essential Virtues* (New York: Simon & Schuster, 2004), xxiii.

49. Lickona, *Educating for Character*, 4.

50. Ibid.

51. Ibid., 5.

52. Ibid.

53. Lickona, *Character Matters*, xxvi.

54. Thomas Lickona, personal communications, fall 2006; Thomas Lickona, preface to *Moral Development and Behavior: Theory, Research, and Social Issues* (New

York: Holt, Rinehart and Winston, 1976); Lickona, *Character Matters;* Lickona, *Educating for Character;* Thomas Lickona, "Educating for Character: A Comprehensive Approach," in *The Construction of Children's Character: Ninety-Sixth Yearbook of the National Society for the Study of Education, Part II,* ed. Alex Molnar and Kenneth J. Rehage (Chicago: NSSE, 1997), 45–62; "Thomas Lickona Presents Character Education: Restoring Respect and Responsibility in our Schools," videorecording, National Professional Resources, Inc., 1996. The threefold formula seems to be standard and recurs in other character educators' definitions of character and descriptions of moral agency; see, for example, Ryan and Bohlin, *Building Character in Schools,* 5: "Good character is about *knowing* the good, *loving* the good, and *doing* the good." And Sanford N. McDonnell's foreword to Ryan and Bohlin's *Building Character,* xi, summarizes the authors' position as follows: "When you build character you must address the cognitive, the emotional, and the behavioral—the head, the heart, and the hand."

55. Lickona, *Educating for Character,* 53.

56. McClellan, *Moral Education,* 104; Thomas Lickona, personal communication, fall 2006.

57. The CEP, for example, has a more inclusive and eclectic approach— "multifaceted," as its research director, Merle Schwartz, told me when I interviewed her, and also as clearly stated in Merle Schwartz and Marvin W. Berkowitz, "Character Education," in *Children's Needs III: Development, Prevention, and Intervention,* 3rd ed., ed. George G. Bear and Kathleen M. Minke (Washington, DC: National Association of School Psychologists, 2006), 13–23.

58. In "The Politics of Character Education," 193, Howard, Berkowitz, and Schaeffer comment on the Christian sensibility of character education and the current convergence of Protestant and Catholic approaches:

> In public schools, the traditional character education approach and the religious approach have largely become one and the same and embrace the traditional character education approach. The direct instruction of virtues and socialization of the young are a point of common ground between Protestant and Catholics. The historical tension between Protestant character education and Catholicism has largely vanished, as is evident in the large number of Catholic educators and philosophers who are prominent advocates of traditional character education, among them Bennett . . . , Kilpatrick . . . , Lickona . . . , and Ryan and Bohlin.

59. Thomas Lickona, "Religion and Character Education," *Phi Delta Kappan* 81, no. 1 (September 1, 1999): 21.

60. Ibid., abstract. Lickona's book *Educating for Character* received a Christopher Award in 1992 for "affirming the highest values of the human spirit." The Christopher organization describes itself on its website (www.christophers.org) as a nonprofit organization that was founded in 1945 by Fr. James Keller and "uses print and

electronic media to spread a message of hope and understanding to people of all faiths and of no particular faith." The mission of the Christopher Awards is, however, explicitly Christian and Christ-centered: "We believe that each person has a God-given mission to fulfill, a particular job to do that has been given to no one else. Love and truth come to us through God, but these gifts are not ours to keep. By sharing them with others each of us becomes a Christ-bearer, a 'Christopher' in the most fundamental sense of that word." Lickona's Center for the 4th and 5th Rs at SUNY Cortland received a grant of nearly $2.7 million from the Templeton Foundation to "instill character education in the nation's high schools," according to a July 16, 2007, news release at www.cortland.edu/news/article.asp? ID=403 (page discontinued). The Templeton Foundation is particularly known for promoting work and funding projects that connect religion and science, and especially those whose aim is to confirm and bolster religion through science.

61. McClellan, *Moral Education*, 1.

62. Ibid., 2.

63. Ibid.

64. Cotton Mather, February 1706, quoted in McClellan, *Moral Education*, 2.

65. McClellan, *Moral Education*, 3.

66. Quoted in ibid., 4.

67. Ibid., 6.

68. Ibid., 7.

69. Ibid., 7–8.

70. Ibid., 8–9.

71. Ibid., 10.

72. Ibid., 11.

73. Ibid., 11–12.

74. Ibid., 12–13.

75. Ibid., 13.

76. Ibid.

77. Ibid., 15.

78. Ibid., 16.

79. Ibid.

80. Ibid.

81. Ibid., 17.

82. Emerson E. White, quoted in McClellan, *Moral Education*, 19. Emerson Elbridge White (1829–1902) was president of Purdue University from 1876 to 1883 and a prolific writer on education.

83. Samuel G. Goodrich, quoted in McClellan, *Moral Education*, 19.

84. McClellan, *Moral Education*, 19–20.

85. Ibid., 21.

86. Ibid., 23.

87. Ibid., 24.

88. Ibid., 22.
89. Ibid., 23.
90. Ibid., 25.
91. Ibid.
92. Ibid., 29.
93. Ibid., 29–30.
94. Ibid., 23.
95. Ibid., 31, 33.
96. Ibid., 45.
97. Ibid., 31, 45.
98. Ibid., 32.
99. Ibid., 34.
100. Ibid., 43–44.
101. Ibid., 44.
102. Ibid., 48.
103. Ibid.
104. Ibid., 49.
105. Ibid.
106. Ibid., 50.
107. Ibid., 51.
108. Ibid.
109. Ibid., 53.
110. Ibid., 54.
111. Ibid., 55.

112. Quoted in James S. Leming, "Research and Practice in Character Education: A Historical Perspective," in Molnar and Rehage, *Construction of Children's Character*, 32.

113. Quoted in Leming, "Research and Practice," 32.

114. Leming, "Research and Practice," 32.

115. Ibid., 32–33.

116. Ibid., 34.

117. Ibid.

118. Hugh Hartshorne, quoted in McClellan, *Moral Education*, 56.

119. Hartshorne and May, quoted in Leming, "Research and Practice," 34–35.

120. Daniel K. Lapsley and Darcia Narvaez, "Character Education," in *Handbook of Child Psychology*, ed. William Damon and Richard M. Lerner, vol. 4, *Child Psychology in Practice*, 7th ed., ed. K. Ann Renninger and I. Siegel (New York: Wiley, forthcoming). They add: "In these studies traits associated with moral character showed scant cross-situational stability and very pronounced situational variability, which is precisely the findings that later personality researchers would report for other traits."

121. McClellan, *Moral Education*, 55

122. Richard J. Bernstein, "John Dewey," in *Encyclopedia of Philosophy*, ed. Paul Edwards (New York: Macmillan, 1967), 2:384.

123. Ibid., 387.

124. McClellan, *Moral Education*, 71.

125. Ibid.

126. Ibid., 72.

127. Ibid.

128. Ibid.

129. Ibid., 73.

130. Ibid., 74.

131. Ibid., 77.

132. Ibid., 79.

133. Ibid. *Values and Teaching* was later modified by Howard Kirschebaum and others.

134. McClellan, *Moral Education*, 80.

135. Ibid.

136. Ibid., 81.

137. See Lapsley and Narvaez, "Character Education." Lapsley and Narvaez point out that moral relativism is the standard charge of conservative critics against liberal models of moral education. They argue, however, that character virtues are as relativist as anything in the liberal handbasket, since the application of virtues to specific situations can be done in all kinds of ways—"one man's integrity is another man's stubbornness."

138. McClellan, *Moral Education*, 82.

139. Ibid.

140. Ibid., 83. I concur to a certain extent but applaud the introduction of empirical data into the formulation of philosophical concepts. See below where I suggest that the problem isn't the rethinking of philosophical questions as a result of empirical research discoveries—something that I hope to do in this book and even in this chapter—but rather the problem arises with the dogmatic use of philosophical presuppositions to frame the questions that empirical research is to investigate.

141. Ibid., 87. That there are hardwired differences in male and female brains, differences that would include how male brains and female brains approach ethics, has been strongly challenged and discredited. See Anne Fausto-Sterling, *Myths of Gender: Biological Theories About Women and Men*, rev. ed. (New York: Basic Books, 1992), and her *Sexing the Body: Gender Politics and the Construction of Sexuality* (New York: Basic Books, 2000); Cordelia Fine, *Delusions of Gender: How Our Minds, Society, and Neurosexism Create Difference* (New York: W.W. Norton, 2010); and Rebecca M. Jordan-Young, *Brain Storm: The Flaws in the Science of Sex Differences* (Cambridge, MA: Harvard University Press, 2010). Fine's exposure of the ways that neuroplasticity enables cultural assumptions and practices of gender differences to structure the brain accords with my focus on the shaping of neuroplasticity by the

cultural presupposition of free will in our standard practices and institutions. Both beliefs, however, are false and create illusions that seem real, obvious, and natural but are instead artifacts of the assumptions and practices themselves.

142. Ibid., 88.

143. Nel Noddings, quoted in McClellan, *Moral Education*, 88–89.

144. Nel Noddings, *Educating Moral People: A Caring Alternative to Character Education* (New York: Teachers College Press, 2002), 2.

145. Ibid.

146. Ibid., 4.

147. Ibid., 5.

148. Ibid., 79.

149. "White Rose," *Wikipedia*, en.wikipedia.org/wiki/White_Rose. See also information about the film *The White Rose* at www.imdb.com and www.rottentoma toes.com.

150. Stephanie Coontz, *The Way We Never Were* (New York: Basic Books, 1992), 101.

151. Ibid., 97.

152. Ibid., 107.

153. Ibid., 97.

154. Ibid., 99.

155. Ibid., 97.

156. Ibid., 108.

157. Ibid., 102.

158. Ibid., 115.

159. Ibid., 104.

160. Ibid., 106.

161. Ibid., 105.

162. A utilitarian view is similar to the Kantian view, but instead of focusing on universal principles that define the right thing to do, utilitarians derive their principles from a general principle of the maximization of goodness or happiness.

163. Lapsley and Narvaez, "Character Education," 14.

164. Professor Max Malikow of LeMoyne College in Syracuse, New York, has transferred the character education model to the college classroom as well in his courses at LeMoyne and in his *Profiles in Character: Twenty-Six Stories That Will Instruct and Inspire Teenagers* (Lanham, MD: University Press of America, 2007).

165. Lapsley and Narvaez, "Character Education," 9.

166. Ibid. In this passage, Lapsley and Narvaez cite Ryan and Bohlin, *Building Character in Schools*, 7, and Edward A. Wynne, "Character and Academics in the Elementary School," in *Moral Character and Civic Education in the Elementary School*, ed. J. Benniga (New York: Teachers College Press, 1991), 145.

167. Lapsley and Narvaez, "Character Education," 10.

168. Leming, "Research and Practice," 36.

169. Lapsley and Narvaez, "Character Education," 5.

170. Ibid.

171. Ibid., 8.

172. Ibid., 23.

173. Ibid., 24.

174. Ibid.

175. Ibid., 25.

176. Ibid.

177. Ibid.

178. Ibid., 26.

179. Ibid., 28.

180. Ibid., 29.

181. Ibid., 22.

182. Ibid.

183. Ibid., 40. I have omitted their embedded citations to the research on these specific points. They discuss a project that is attempting to rethink ethical education in the light of recent research in the cognitive and other mind sciences. It is the Integrative Ethical Education conceptual framework, which is "a conceptual framework that attempts to incorporate insights of developmental theory and psychological science into character education." Narvaez is engaged in this project and has publications on it that the authors cite in their work. Integrative Ethical Education

> attempts to understand character and its development in terms of cognitive science literatures on expertise and the novice-to-expert mechanisms of best practice instruction. It attempts to keep faith with classical Greek notions of *eudaimonia* (human flourishing), *arête* (excellence), *phronesis* (practical wisdom) and *techne* (expertise) with developmental and cognitive science....
>
> In delineating the elemental skills of good character, IEE addresses character education by integrating the findings from developmental psychology, prevention science, and positive psychology. In proposing the best approach to instruction, IEE addresses character education by integrating contemporary findings from research in learning and cognition. (Ibid., 32–33.)

184. Relevant to the teaching of morals through social structure and practice are the experiments on eleven-year-old children conducted by Ralph K. White and Ronald Lippitt, both students of the social psychologist Kurt Lewin, to compare the psychological and moral attitudes and types of social interaction that emerged respectively from authoritarian and democratic social arrangements. See Ralph K. White and Ronald Lippitt, *Autocracy and Democracy: An Experimental Inquiry* (Westport, CT: Greenwood Press Publishers, 1960).

185. Some communitarian social critics propose that liberalism is the origin of the fragmentation of the social domain. For example, see Amitai Etzioni, ed., *The Essential Communitarian Reader* (Lanham, MD: Rowman & Littlefield, 1998). In

the introduction, Etzioni writes, "In the 1890s, a group of political philosophers—Charles Taylor, Michael J. Sandel, and Michael Walzer—challenged individualist liberal opposition to the concept of a common good, although all have been uncomfortable with the label 'communitarian'" (ix).

186. John McCumber, *Time in the Ditch: American Philosophy and the McCarthy Era* (Evanston, IL: Northwestern University Press, 2001).

187. There is a large literature on the loss of civic involvement in America in the last decades of the twentieth century. For a particularly interesting account, see Theda Skocpol, *Diminished Democracy: From Membership to Management in American Civic Life* (Norman: University of Oklahoma Press, 2004); see also Richard Sennett, *The Fall of Public Man* (London: Penguin, 2002), a source frequently cited by Coontz.

188. Randy Cohen, *The Good, the Bad, and the Difference: How to Tell Right from Wrong in Everyday Situations* (New York: Doubleday, 2002), 20–21. Nevertheless, Cohen defines ethics in rational free will terms: "Ethics is the rational determination of right conduct, an attempt to answer the question 'How should I act now?' Ethics is not just knowing; it is doing. And so it is necessarily a civic virtue, concerned with how we are to live in society; it demands an understanding of how our actions affect other people" (10). Here Cohen makes the social claim, but his model is independent personal decision making toward other people, not the social construction of person and action.

2. Moral Lessons of the Holocaust About Good and Evil, Perpetrators and Rescuers

1. Philip G. Zimbardo, *The Lucifer Effect* (New York: Random House, 2007), 436.

2. Theodor W. Adorno et al., *The Authoritarian Personality*, ed. Max Horkheimer and Samuel H. Flowerman (New York: W.W. Norton, 1969).

3. Robert Jay Lifton, *Nazi Doctors: Medical Killing and the Psychology of Genocide* (New York: Basic Books, 2000), 46.

4. Ibid., 41.

5. Ibid., 31. In a recent volume of essays, *Understanding Genocide: The Social Psychology of the Holocaust*, ed. Leonard S. Newman and Ralph Erber (Oxford: Oxford University Press, 2002), Christopher Browning remarks in the introduction that in the 1990s scholarship, especially in German, began to expose and reemphasize the importance of Nazi ideology and of fanatically anti-Semitic SS men in the perpetration of the Holocaust. Certainly the pseudo-biological social Darwinist medical ideology, as described by Lifton, bears out this new turn.

6. Lifton, *Nazi Doctors*, 33.

7. Ibid., 423. Lifton proposes an elaborate theory of the "doubling" of the self, a Nazi self and a prior normal self that continues in other contexts, to explain the psychological state of the Nazi doctors. The evidence from social psychology that I will introduce below suggests that our actions are highly contextual and situation

dependent, so an elaborate theory of a divided self acting in different contexts may not be necessary or, what amounts to the same thing, may be the norm rather than the exception, the Nazi case being but an extreme example that highlights and puts in relief the phenomenon.

8. Christopher Browning, *Ordinary Men: Reserve Police Battalion 101 and the Final Solution in Poland* (New York: HarperCollins, 1992), 58.

9. Ibid., 59.

10. Ibid., 60.

11. Ibid., 66.

12. Samuel P. Oliner and Pearl M. Oliner, *The Altruistic Personality: Rescuers of Jews in Nazi Europe* (New York: Free Press, 1988), 34. Although the Oliners' title suggests an individual character psychology approach, they actually offer what they say is a study "rooted in a social psychological orientation, which assumes that behavior is best explained as the result of an interaction between personal and external social, or situational, factors" (10).

13. Nechama Tec, *When Light Pierced the Darkness: Christian Rescue of Jews in Nazi-Occupied Poland* (New York: Oxford University Press, 1986), 9.

14. Oliner and Oliner, *Altruistic Personality*, 38.

15. Ibid., 43.

16. Ibid., 86.

17. Ibid., 79, 80.

18. Tec, *When Light Pierced the Darkness*, 160.

19. Oliner and Oliner, *Altruistic Personality*, 222.

20. Susan Zucotti, *The Italians and the Holocaust: Persecution, Rescue, and Survival* (Lincoln: University of Nebraska Press, 1996).

21. Tec, *When Light Pierced the Darkness*, 180, 189.

3. The Overwhelming Power of the Group and the Situation

1. Philip G. Zimbardo, Christina Maslach, and Craig Haney, "Reflections on the Stanford Prison Experiment: Genesis, Transformations, Consequences," in *Obedience to Authority: Current Perspectives on the Milgram Paradigm*, ed. Thomas Blass (Mahwah, NJ: Lawrence Erlbaum Associates, 2000).

2. Quoted in Philip Zimbardo, *The Lucifer Effect* (New York: Random House, 2000), 267.

3. Ibid., 275.

4. Zimbardo, Maslach, and Haney, "Reflections."

5. Lee Ross and Richard E. Nisbett, *The Person and the Situation: Perspectives of Social Psychology* (New York: McGraw-Hill, 1991), argue that 1968 was a watershed year because psychologists Walter Mischel and Donald Petersen in separate reviews of the experimental literature came to the conclusion that behavior showed little consistency within individuals across different kinds of situations, so cross-situational

consistency might be the exception and situational specificity might be the rule—a conclusion anticipated by Hartshorne and May in 1929 in their examination of honesty in kids across situations.

6. Ibid., 101, and more generally 90–118 and 145–68, esp. 150 regarding the constancy produced by social role.

The philosopher John M. Doris has devoted an entire book, *Lack of Character: Personality and Moral Behavior* (Cambridge: Cambridge University Press, 2002), to situationism and its implications for revising ethical theory. He talks of character traits as local rather than global, so claims that someone is a compassionate person or a courageous person ought to be redefined as, for example, "dime-finding-dropped-paper compassionate" and "sailing-in-rough-weather-with-one's-friends courageous" (115). Yet his conclusion, as he himself says, is "conservative," namely, that now that we know the power of the situation we can come to resist it. But that seems hardly likely; for whom, or how many, could this in fact be the case, and would it be the case across situations or, as the research that he covers implies, only in some situations for each of those capable of such independence? In fact, studies in which subjects are asked to read about experiments that show the power of situational factors and are then asked to predict what people will do in an analogous situation show that the described experiments have no effect on the subjects' predictions. The subjects still base their predictions of people's behavior on dispositional factors (Ross and Nisbett, *Person and the Situation*, 133). Yet Zimbardo, like Doris, draws back from the edge when faced with the implications for ethics of his own research and theoretical insights, and even his own personal experience of being drawn into perpetration. The risk to the legitimacy of the assignment of full individual responsibility for wrongdoing seems to induce a failure of intellectual courage.

7. Ross and Nisbett, *Person and the Situation*, 92ff. The cognitive motivational determinants of behavior turn out to be much stronger than the presumed characterological ones. See below.

8. Note that Fred's pointed comment (quoted above) that it was "wrong" to continue to shock (in the 2006 iteration of the Milgram experiments) was an after-the-fact reflection upon his reasons for not continuing, one not articulated to the experimenter at the time of his departure.

9. Christopher Browning, in *Understanding Genocide: The Social Psychology of the Holocaust*, ed. Leonard S. Newman and Ralph Erber (Oxford: Oxford University Press, 2002), 4, points out that by the early 2000s most historians would no longer characterize Eichmann as Arendt did, as a faceless bureaucrat, but instead as having a deep ideological identification with the Nazi regime and as extremely ambitious. Nevertheless, there were numerous lower-level bureaucrats who fit Arendt's description well.

10. Quoted in John Sabini and Maury Silver, *Moralities of Everyday Life* (Oxford: Oxford University Press, 1982), 57.

11. Zimbardo, *Lucifer Effect*, 237.

12. Ibid., 445.

13. Ibid., 277.

14. Ross and Nisbett, *Person and the Situation*, 227–30.

15. Ibid., 230–32

16. Ibid., 234.

17. Staub, quoted in Zimbardo, *Lucifer Effect*, 286. See also Ervin Staub, *The Roots of Evil: The Origins of Genocide and Other Group Violence* (Cambridge: Cambridge University Press, 1989), and Ervin Staub, *The Psychology of Good and Evil: Why Children, Adults, and Groups Help and Harm Each Other* (Cambridge: Cambridge University Press, 2003).

18. Zimbardo, *Lucifer Effect*, 287. Zimbardo reports these findings and the analysis of John Steiner, a survivor and sociologist at Stanford University, who conducted decades of interviews of Nazi SS, from privates to generals.

19. Ross and Nisbett, *Person and the Situation*, 164.

20. Zimbardo, *Lucifer Effect*, 303.

21. Christopher Browning, in reflecting on the social forces that were operative in Reserve Police Battalion 101, cites both Milgram's and Zimbardo's research and identifies as particularly important: deference to authority, adaptation to role expectation, the intoxicating effect of exercising unrestrained power over others, pressure for conformity, and isolation from all countervailing cultural and institutional influences while in a foreign and alien territory (*Understanding Genocide*, 6).

22. Zimbardo, *Lucifer Effect*, 312.

23. See Dieter Frey and Helmut Rez, "Population and Predators: Preconditions for the Holocaust from a Control-Theoretical Perspective," in Newman and Erber, *Understanding Genocide*, and also Leonard S. Newman and Ralph Erber, "Epilogue: Social Psychologists Confront the Holocaust," in the same volume, 330.

24. Zimbardo, *Lucifer Effect*, 316.

25. Recall also the experiments of Kurt Lewin's students Ralph K. White and Ronald Lippitt (*Autocracy and Democracy: An Experimental Inquiry* [Westport, CT: Greenwood Press Publishers, 1960]) on the effects on values and relationships of authoritarian versus democratic social arrangements.

26. Elliot Aronson, *Nobody Left to Hate: Teaching Compassion After Columbine* (New York: Henry Holt, 2000), 10.

27. Ibid., 13.

28. Ibid., 15.

29. Ibid., 171.

30. Ross and Nisbett, *Person and the Situation*, 45.

31. Doris, *Lack of Character*, 30–31.

32. Ross and Nisbett, *Person and the Situation*, 28–30.

33. Ibid., 46.

34. Ibid., 70.

35. Irving L. Janis, *Groupthink: Psychological Studies of Policy Decisions and Fiascoes*, rev. ed. (Boston: Houghton Mifflin, 1982), vii.

36. Ibid., 13.

37. Ibid., 174–75.

38. John Rawls's vision (*A Theory of Justice* [Cambridge, MA: Belknap Press, 1971]) is of a polity made up of individuals whose individual free choices and rational decisions are aggregated for governance. But this is living in a dream world—a point anticipated by Spinoza in his political works.

39. Ross and Nisbett, *Person and the Situation*, 192. Staub (*Understanding Genocide*, 16) also sees the dangers of cultural isolation and the need for pluralism in predisposing a society to violence: "A monolithic society, in contrast to a pluralistic society, with a small range of predominant values and/or limitations on the free flow of ideas, adds to the predisposition for group violence."

40. These essays are collected in Heinz Kohut, *Self Psychology and the Humanities: Reflections on a New Psychoanalytic Approach*, ed. Charles B. Strozier (New York: W.W. Norton, 1985).

41. Strozier, "Introduction," in Kohut, *Self Psychology and the Humanities*, xvii–xviii.

42. Kohut, *Self Psychology and the Humanities*, 251–53.

43. Ibid., 63.

4. What Happened to Ethics: The Augustinian Legacy of Free Will

1. I wish to thank Joe Keith Green of Eastern Tennessee University for his acute reading and helpful comments on this chapter.

2. Joe Keith Green reminds me that *akrasia*, as an account of moral failure, has undergone just the historical process that I point to in this chapter: from the Platonic three-part divided and conflicted mind to an Augustinian conflict between (free) will and intellect (reason).

3. See also Nomy Arpaly, *Unprincipled Virtue: An Inquiry into Moral Agency* (New York: Oxford University Press, 2004), and her *Merit, Meaning, and Human Bondage: An Essay on Free Will* (Princeton, NJ: Princeton University Press, 2006).

4. A determinist position that includes consciousness as well as material causes is referred to by philosophers as "anomalous nomism." Joe Keith Green comments (personal communication, August 2011): "There remains a question about what efficient causal explanations of actions (as opposed to events) may cite. One way to define determinism is that there is a true efficient causal explanation for every action (as for every event in nature). But you can still disagree about what will be cited in those explanations as causes. 'Materialists' from whom you distinguish your view, hold that only matter and states of matter can ultimately be cited. Your view is that the distinction between actions and events is not that there are sufficient explanations in terms of efficient causes for one (events) but not the other (actions), for

which 'will' has to be cited as a sort of 'uncaused cause,' but that actions and events are distinguished by the sorts of causes that can be cited. I think you can show that this is also Spinoza's view." See also Charles Jarrett, "Spinoza's Denial of Mind-Body Interaction and the Explanation of Human Action," *Southern Journal of Philosophy* 29, no. 4 (1991): 465ff.

5. Timothy O'Connor, "Free Will," *Stanford Encyclopedia of Philosophy*, plato. stanford.edu/entries/freewill.

6. Michael McKenna, "Compatibilism," *Stanford Encyclopedia of Philosophy*, plato. stanford.edu/entries/compatibilism. McKenna identifies five claims we generally hold:

> Westerners are, it seems, committed in our thought and talk to a set of concepts which, under scrutiny, appear to comprise a mutually inconsistent set. . . .
> Call it the Classical Formulation:
> 1. Some person (qua agent), at some time, could have acted otherwise than she did.
> 2. Actions are events.
> 3. Every event has a cause.
> 4. If an event is caused, then it is causally determined.

If an event is an act that is causally determined, then the agent of the act could not have acted otherwise than in the way that she did. McKenna goes on to point out that the "Classical Formulation involves principles governing six different concepts: a person as an agent, action, could have done otherwise, event, cause, and causal determination."

7. "The spirit of the hard determinist position," McKenna writes, "is sustained by hard incompatibilists, who hold that there is no free will if determinism is true, but also, that there is no free will if determinism is false." For indeterminism "due to quantum indeterminacies . . . poses just as much of a threat to the presumption of free will as determinism would." McKenna, "Compatibilism."

8. Compatibilism has undergone a fairly extensive line of development from Hume and Hobbes to the present. McKenna, "Compatibilism," writes that it is now in its third stage or iteration:

> A useful manner of thinking about compatibilism's place in contemporary philosophy is in terms of at least three stages. The first stage involves the classical form defended in the modern era by the empiricists Hobbes and Hume, and reinvigorated in the early part of the twentieth century. The second stage involves three distinct contributions in the 1960s, contributions that challenged many of the dialectical presuppositions driving classical compatibilism. The third stage involves various contemporary forms of compatibilism, forms that diverge from the classical variety and that emerged out of, or resonate with, at least one of the three contributions found in the second transitional stage.

9. McKenna, in "Compatibilism," defines "classical compatibilism" in this way:

According to the classical compatibilist account of free will, so long as one's action arises from one's unencumbered [that is, uncompelled] desires, she is a genuine source of her action. Surely she is not an ultimate source, only a mediated one. But she is a source all the same, and this sort of source of action, the classical compatibilist will argue, is sufficient to satisfy the kind of freedom required for free will and moral responsibility.

"What the classical compatibilists attempted to do . . . was deny the truth of the . . . premise: If determinism is true, no one can do otherwise." He adds, "The burden of proof rests squarely on the compatibilists."

10. The classical version is now regarded as having failed. McKenna writes in "Compatibilism":

The classical compatibilist attempt to answer the incompatibilist objection failed. Even if an unencumbered agent does what she wants, if she is determined, at least as the incompatibilist maintains, she could not have done otherwise. Since, as the objection goes, freedom of will requires freedom involving alternative possibilities, classical compatibilist freedom falls.

11. On this model, "an agent's control consists in her playing a crucial role in the production of her actions. . . . Control [according to this model] is understood as one's being the source whence her actions emanate." McKenna, "Compatibilism." This is a version of compatibilism that argues for "a source model of control," rather than a model that depends upon alternative choices.

The main versions of contemporary compatibilism have their origins in the 1960s: one eschews the need for choice among alternatives for freedom and responsibility and grants the "Classical Incompatibilist Argument: If determinism is true, no one can do otherwise." This version of compatibilism denies that "a person is morally responsible for what she does only if she can do otherwise." Compatibilists of this stripe argue for ways that human beings have control over their actions without the need for choice among alternatives. Harry Frankfurt argued that the freedom that underlies responsibility does not depend upon the possibility to choose among alternatives.

McKenna points out that more recently, John Martin Fischer, a dominant figure in the contemporary debate since the 1980s, has introduced a relevant distinction between what he calls "regulative control" and "guidance control." Regulative control "regulates" between different alternatives, while guidance control brings about an action even when there are not alternatives. So guidance control is all one needs for moral responsibility since it marks actions as one's own. McKenna points out that, "[o]n a view like Fischer's or Frankfurt's, it is only guidance control that is necessary for moral responsibility." Some contemporary compatibilists focus solely on guidance control as necessary for freely willed action, while others argue (as in classical compatibilism) that regulative control is also necessary, a position developed by some in a new dispositionalism, which attempts to develop a new account of the choice among alternatives.

12. McKenna, "Compatibilism."

13. Modern versions of compatibilism often appeal to biology. For example, the philosopher P.F. Strawson appeals to morally reactive attitudes. He proposes that moral reactions are deeply natural to the human species and interwoven in the human way of life and hence should not be given up. He believes that moral attitudes are psychologically impossible to dispense with and that the relinquishing of practices of moral responsibility—even if true to the facts of natural determinism—would impoverish human life. And such moral attitudes, he claims, presuppose an allegiance to some form of free will.

Another version of contemporary compatibilism appeals to a complex internal harmony within the psyche ("mesh" and "hierarchical theories"). Harry Frankfurt proposed a hierarchical mesh theory in 1971. In it he explains that freely willed action "must suitably mesh within hierarchically ordered elements of a person's psychology. The key idea," McKenna says, "is that a person who acts of her own free will acts from desires that are nested within more encompassing elements of her self." Frankfurt argues that desires can be of two different kinds: first-order and second-order. The former are simple desires: to eat a piece of cake, to see a certain film, and the like. The latter, however, are "desires about desires. They have as their objects, desires of the first-order," for examples the desire to keep fit and exercise and eat healthy foods. Frankfurt argues that only persons have second-order desires, that is, desires distinct from the immediate impulsive ones. "Persons care about which desires lead them to action." This theory is about the nature of guidance control rather than regulative control (the choice between alternatives). It has been characterized as a "real self theory."

> According to Frankfurt, the sort of freedom needed for assessments of moral responsibility turns crucially on whether or not the agent reveals herself in acting as she does, or if instead her conduct is in some way alien to her. By desiring at a hierarchically higher-order level of reflection that one's will be a certain way (or not be a certain way), one reveals her deeper self, not merely at the surface of her conduct, but in terms of how she herself regards her very own motives issuing in her conduct. When she acts of her own free will, those motives are hers, are *of* her. She owns them. Hence, they reflect her true self. When she acts, but does not act of her own free will, she disavows her motives. They do *not* reflect her true self.

Frankfurt argues that "the willing addict possesses the sort of freedom required for moral responsibility because the will leading to her action is the one that she wishes it to be; she acts with guidance control." The willing addict both wants the drug and also wants to want the drug; she wants to be an addict. They reflect her true self and hence she has guidance control and moral responsibility. But does this version of compatibilism succeed? One objection is that a conflict can occur not only between one's desires and the desires one desires to have but also between desires and any given level. One can want to be an addict and a pursuer of health, for example. So the problem of free will, on this model, occurs at any level and can recur: "It needs supplementing so as to avoid the problem of a spiraling reoccurrence of challenges to an agent's freedom," McKenna points out. Yet there are deeper objections. We recall that, "for Frankfurt, an agent that acts of her own free will does not merely reveal her desires in action, she reveals how she wishes herself to be as a person." The objection has been raised that the kind of person one wishes oneself to be can be manipulated—for the drug addict, for example, by the drug itself or by some other manipulation. In that case the person is not free. Moreover, this objection can be expanded to determinism itself—isn't "a deterministic history simply a more elaborate form of manipulation that happened to take a very long time to achieve the same sort of result"? "If one sort of causal history giving rise to a Frankfurtian mesh [that is, internal consistency] can undermine an agent's freedom and moral responsibility, then why not a deterministic history?" McKenna concludes, "Frankfurt must either show why manipulation cases fail, or instead, bite the bullet and accept that, on his theory, agents so manipulated can still be free and morally responsible persons." In my view Frankfurt's position ought to play out in the end as a version of incompatibilism, rather than compatibilism. It seems intuitive that the person who is an addict who affirms her addiction as true to herself is less free, is acting more compulsively, than the person who is in personal conflict. The conflicted person has a chance of recovery whereas the happy addict seems completely unreachable and destitute.

Daniel Dennett, of Tufts University and a major player in the contemporary philosophical response to the cognitive and neurosciences, also offers a version of true-self compatibilism. He proposes what McKenna calls a multiple viewpoint compatibilism. According to Dennett, it is legitimate to use folk psychology in explaining intentional actions because we can have multiple viewpoints and take varying stances toward a system. They are valid insofar as each contributes to understanding it, making predictions about it, and interacting with it. Dennett proposes that "even a thermostat can be interpreted as a very limited intentional system since its behavior can usefully be predicted by attributing to it adequate beliefs and desires to display it as acting rationally within some limited domain." So we can legitimately say, for example, that "the thermostat *desires* that the room's temperature (or the engine's internal temperature) not go above or below a certain range. If it *believes* that it is out of the requisite range, the thermostat will respond appropriately to *achieve* its desired results." "For Dennett, the propriety of adopting the intentional stance towards a system is settled pragmatically in terms of the utility of its application in interacting with the system." Dennett further argues that we don't need guidance control but only regulative control to underlie the kind of freedom that supports moral responsibility. McKenna concludes that despite disparaging incompatibilists and their arguments, Dennett takes the source model of free will seriously. "By appealing to views on intentionality, rational action, agency, and personhood, Dennett offers a suggestive account of how it is that an agent can be an authentic source of her action." McKenna, "Compatibilism."

14. Spinoza does not distinguish between bodily events and mental reasons as a difference in the kind of cause. "For Spinoza (by contrast), actions are a subset of 'events,' and not distinguished from other events by virtue of the fact that no efficient causes explain actions, but are sufficient explanations of events," Joe Keith Green says (personal communication, August 2011). What that means is that actions, like events, according to Spinoza are deeply embedded within causal contexts and not separable from them. Nor are mental and bodily causes (ontologically, as the philosophers say) of different realms of being. According to Spinoza, mind and body, consciousness and matter, are two different ways of describing everything, from each individual to the universe as a whole. So far we agree. But Green proposes that Spinoza holds the following version of compatibilism:

> One major participant in current debates (John Martin Fisher) echoes Spinoza, and articulates a view very much like Spinoza's, in claiming that we can talk about actions not as free, but "free"—or identify actions for which we are "accountable"—where we can explain an agent's action by correctly citing the reasons to which she responded. And this is, of course, entirely compatible with determinism in the sense in which Spinoza is a determinist. It is a view that Spinoza shares, at least when reasons are "clear and distinct." Here, the true and sufficient explanation also constitutes a justification of it. But he also arguably holds this view where the reasons upon which an agent acts are grounded in inadequate ideas. Though here, they do not constitute a *sufficient* explanation—a true explanation in this case does not also constitute a rational justification for it. Other causes "outside" the agent act upon the agent, bringing about a movement of his appetites as an effect, and his "desire" is his awareness of this effect.
>
> So: in terms of the contemporary debate, Spinoza is a compatibilist of a particular kind. All the "participatory emotional responses" that are forms of taking joy in actual "strengths" of mind and body are entirely good so long as they are responses to qualities or strengths that persons actually have (and we are not self-deceived about our own). Our regarding these "strengths" as "meritorious" are entirely consistent with there being sufficient explanations for our having them in terms of efficient causes, of which we are unaware. Blame and any related forms of hatred are *never* merited as a response to any action, from any causes, since we are constituted to feel hatred only when we do not adequately understand the causes that bring about even the hatred of others. But just punishment, and other social conventions are appropriate, sans hatred, not because we are "free," but because, as agents who can mostly and largely have only an inadequate idea of the causes working upon us and others, and in acting in

terms of these conventions, we are "taking the intentional stance." So, if this representation of Spinoza's view is correct, Spinoza is *not* a "classical compatibilist." But if we take into account all the forms of interaction (and all the social conventions) that are grounded in having and attributing "merit," they are not rendered somehow nonrational by virtue of the fact that we are "determined." This is what McKeon and others call "multiple-viewpoints compatibilism."

I disagree that Spinoza holds this version of compatibilism, for the reasons we do an act (in any normal sense of the term *reasons*) do not in fact encompass the entire system of causes of the action in Spinoza's sense of their infinity. There are no background conditions that would not be included in the causes of the act for Spinoza—to infinity. The reasons we do an act, even if we are pretty aware of our unconscious, are proximate causes, and those have no special status for Spinoza—in fact, quite the contrary. For Spinoza to be free (which includes responsibility) is to own the self *in God*, which is to say in a kind of ex post facto way, as Oedipus did in Sophocles's portrayal. What's wrongheaded in the above quotations is the (still Augustinian) framing of the question in terms of discrete willed actions of an atomic agent. But our actions are the actions of the universe, Spinoza tells us. In fact, Green cites Spinoza's invoking of God in the moral context but fails to grasp what Spinoza is actually saying:

> Spinoza immediately comments that he "said so much" on Descartes' view because "in this matter he is perfectly consistent," then distinguishes his own view only by asserting what he perceives to be an advantage in it: That everything is ultimately "explainable" in terms of God's willing it, but "free from all superstition," which, in the context, seems to mean perceiving God to be an agent in an anthropomorphic sense.

Spinoza's point is that actions are the necessary playing out of the whole of nature (or God). They are not actions of individuals as atomic at all but only of the universe from a particular standpoint and starting point. So, *pace* Joe Keith Green, for Spinoza merit is as vacuous as blame. For God can be neither merited nor blamed— God is not "good" for either Spinoza or Maimonides. Nor, in a sense, are we. The moral distinction is between the benighted and the enlightened: the enlightened, as including consciously more of the causes of the universe giving rise to them, thus act from a wider swath of the universe, rather than from the narrowest of perspectives or selves. Moreover, praise, although pleasurable, is corruptive in Spinoza's estimation—it appeals to ambition and hence is a weakening of the conatus in terms of a false and partial pleasure. It fosters a type of dependency on others rather than understanding, which is the source of independence of mind and of all virtue, Spinoza tells us.

So Spinoza's position is a version of incompatibilism according to which, while there can be no true praise or blame, reward or punishment, except as a means of inducing future behavior, he does develop an account of enhanced agency—via the conatus. It is the basis for the pursuit of one's own psychic health and cannot function as a basis for legitimate punishment or reward. For we are all determined by our biology and histories. Those of us who are lucky enough to encounter Spinoza's philosophy have the great luck of a path to joy and contentment. And compassion—it is compassion that this kind of determinism undergirds, not moral judgment of others.

15. Peter Brown, *Augustine of Hippo* (Berkeley: University of California Press, 1967), 502–12.

16. Ibid.

17. Ibid.

18. Augustine, *The City of God Against the Pagans*, ed. and trans. R.W. Dyson (Cambridge: Cambridge University Press, 1998), 15.3.585.

19. Augustine, *The Literal Meaning of Genesis*, trans. and annotated John Hammond Taylor (New York: Newman Press, 1982), 10.12.110.

20. Ibid. 10.17.118.

21. Harry A. Wolfson, "Extradeical and Intradeical Interpretations of Platonic Ideas," *Journal of the History of Ideas* 22, no. 1 (January–March 1961): 3–32.

22. Stephen Menn, *Augustine and Descartes* (Cambridge: Cambridge University Press, 1998), 202–3.

23. But the divine wisdom is not yet reduced to the divine will by Augustine himself, whereas it will be so reduced later in the Latin West as Augustinianisms become more radicalized by Descartes and others. As Stephen Menn remarks (*Augustine and Descartes*, 350), "For Augustine the eternal truths are not made: so with regard to these Augustine is not saying that 'in God to see and to will are one and the same thing' [as Descartes says]. . . . So also . . . the rules of number exist originally in the divine wisdom, and thence proceed to creatures." While "the rules themselves are not established at God's command, but exist in God and are coeternal with him," nevertheless, "these rules proceed to creatures according to God's will."

24. Saint Augustine, *Confessions*, trans. Henry Chadwick (Oxford: Oxford University Press, 1992), 7.3.5.114 and 7.16.22.126.

25. Bonnie Kent, *Virtues of the Will: The Transformation of Ethics in the Late Thirteenth Century* (Washington, DC: Catholic University of America Press, 1995), 111. Kent cites as illustrative Augustine's *De Trinitate* 10.

26. Augustine, *Literal Meaning of Genesis* 8.14.53–55 and 11.10.142.

27. Augustine, *City of God* 14.7.592. Augustine further proposes that "a righteous will then is a good love; and a perverted will is an evil love." Logically, given the claim that "emotions . . . are all no more than acts of the will," what Augustine means here must be that "a good love is a righteous will" and "an evil love is a perverted will."

28. As mentioned in the first chapter, it was Aristotle who introduced character as the shaper of moral action. Aristotle defined the virtues of character in terms of the

cognitive grasp (interpretation) of situations, not in terms of individuals' will to conform to virtues. See below and also following chapter.

Scholarly support for my reading of Augustine as having reduced emotions to acts of will comes from the historian of philosophy Richard Sorabji in his account of the subtle shift from the Stoic moral psychology to the Christian, *Emotion and Peace of Mind: From Stoic Agitation to Christian Temptation* (Oxford: Oxford University Press, 2000). Sorabji proposes that Christianity introduced a new twist into the Stoic moral psychology: for the Stoics, the moral problem was how to quell the painful emotions that agitate the mind, but Christianity ends up instead with the seven deadly sins, which must be voluntarily resisted. The linchpin, Sorabji argues, is Augustine. The turning point in Western Christianity was Augustine's interpretation or revision of the Stoic doctrine of the "first movements" of emotions. For the Stoics, not only were these involuntary, but they were also *pre*-passions, rather than full-blown emotions. The Stoics' moral aim was to avoid agitating emotions. Sorabji proposes that Augustine misread or misunderstood an ancient text that described the ideal Stoic sage as growing jittery in a proverbial storm at sea, and hence just as afraid as a non-Stoic or someone who was not a sage, the Stoic's virtue being in his refusal to "consent" to the painful emotion, not in his experience of the emotion. Bonnie Kent, in her review of the book, suggests that this was not so much a failure of philological understanding (as Sorabji proposes) as Augustine's deliberate recruitment of an ancient text for his own philosophical agenda. In any case, according to Sorabji, the fateful revision, a new notion of emotions as objects of consent or resistance, was introduced into normative Christianity with the regrettable consequences (Sorabji opines) of a refocusing of the moral aim upon sin and temptation, both matters of the will, and away from the Hellenistic moral psychology that called for a kind of therapy of the emotions through understanding, as Martha Nussbaum has argued in *The Therapy of Desire: Theory and Practice in Hellenistic Ethics*, 3rd ed. (Princeton, NJ: Princeton University Press, 2009).

29. Stephen Menn, in *Augustine and Descartes*, 174, describes Augustine's notion of God's wisdom or truth working in the material arena as operating as follows:

> As bodies receive numerical form from the divine Truth, and are corrupted by the extent that they depart from this numerical form, so rational souls receive the rules of Wisdom from the same divine Truth, and so they too are corrupted when they depart from these rules of wisdom. God is Wisdom itself and immutable form, but the human soul is merely a mutable thing informed by the divine wisdom capable of turning [i.e., via free will] from wisdom to foolishness, and thus giving rise to moral evil.

30. Augustine, *Literal Meaning of Genesis* 8.9.45–46.

31. Ibid. 8.21.62.

32. The "interior, natural movement," in Augustine's view, however, turns out to be neither strictly internal, as we think of nature as operating, nor necessary and

automatic. For it is carried out by the direct divine intervention of the good angels in each instance and at all times, or nature would come to a standstill. "If [God] were to withdraw His creative power, so to speak, from things," Augustine warns us, "they would no more exist than they did before they were created" (*City of God*, 12.26.538). It is the angels who carry out the natural type of providence, at the divine direction. In *The Literal Meaning of Genesis* 8.24–25, Augustine describes in more detail the role of the angels in the natural working of providence. The angels are said to have under their direct control "every corporeal being, every irrational life, every weak or wayward will." They "see in God immutable truth, and according to it they direct their wills. . . . They are moved . . . by His command even in time, although He is not moved in time." "They carry out [God's] commands in their subjects, moving themselves through time and moving bodies through time and space in accordance with what is proper to their activity." The incorporeal angels are "spiritual creatures," whom God speaks to from within, "interiorly in a mysterious and indescribable manner." They perform their divine functions by operating on things both from the outside and also from within them, "intrinsically," Augustine says. He explains here in chapter 24 that this is what he meant by his earlier reference in chapter 9 to the twofold working of the divine providence.

33. Augustine, *Literal Meaning of Genesis* 9.10.80–82.

34. Ibid. 7.13.194.

35. Augustine, *On Free Will*, quoted in *Free Will*, ed. Sidney Morgenbesser and James J. Walsh (Englewood Cliffs, NJ: Prentice-Hall, 1962), 14–17.

36. Augustine, *City of God* 12.8.508–9.

37. Augustine, *Literal Meaning of Genesis* 8.11.49–50 and 8.13.52–53.

38. Augustine, *On Free Will*, quoted in Morgenbesser and Walsh, *Free Will*, 14–17.

39. Brown, *Augustine of Hippo*, 35–59.

40. Ibid., 128, 139.

41. Augustine first takes recourse in a Platonizing strategy also prevalent in the Jewish Platonist Philo of Alexandria: the first creation he says refers not to the creation of the actual things themselves but to an earlier stage of creation, a primordial one, in which God (by His Wisdom-Logos-Christ) creates the intellectual "seeds" of the things mentioned in the Bible, which can then be brought into being at a later time. (In Wolfson's "Intradeical and Extradeical," Christ is identified with the Wisdom-Logos dimension of God.) A heavenly domain, a highest and most intelligible created reality, which Augustine refers to as the heaven of heavens, is created first, and it is the locus of this intellectual creation. It is the world of the unfallen angels and the home and future abode of holy people. All those who reside there enjoy an unbroken communion with God. Augustine frequently refers to these first intellectual creations in the Stoic-Neoplatonic language of "seeds." Things have their "rational causes" or "seminal reasons," potentials to be what they will become, planted into a first spiritual order of creation. Augustine attributes to these rational

causes a mathematical meaning concerning measure and weight and number, conceptions that will introduce into actual created embodied things their stability, harmony, and endurance. It is these rational causes that are first (yet atemporally) brought into being by God during the first six days of creation of Genesis 1.

Nevertheless, we are not to regard these seminal reasons or causes as functioning as mere potencies, natural essences or forms whose realization follows a *necessary* unfolding in time, as we might expect, and as Plotinus, following the Stoics, proposed. Augustine scholar Rowan Williams warns us that "this should not be understood to mean that things are created simply with immanent capacities for growth, and that subsequent history does no more than unfold what has been there from the beginning in terms of natural processes," as Plotinus held, as well as Aristotle, with whom the idea originates. Instead, the seminal reasons amount to a range of potentialities, only some of which will be realized by the direct intervention of God's will. "It is the will of God that makes things happen." For "God's continuing providence or 'administration' of the world is what determines events, not any internal principles of necessity." "God's will," Williams further points out, "is not a cause among others" for Augustine, "but the power that activates a particular set of causes at the appropriate time." Rowan Williams, "Creation," in *Augustine Through the Ages*, ed. Allan Fitzgerald and John C. Cavadini (Grand Rapids, MI: Eerdmans, 1999), 252.

42. Augustine, *City of God* 13.14.555–56.

43. Ibid. 14.13.608, 14.13.609.

44. Augustine, *Literal Meaning of Genesis* 8.13.52–53. Zeev Harvey (personal communication, September 2011) points out that "in fact, God had not commanded this. Eve said this when she paraphrased the command to the Serpent. See Rashi on the verse."

45. Augustine, *Literal Meaning of Genesis* 8.23.64.

46. The philosopher Eleonore Stump points out in her essay "Augustine on Free Will," in *The Cambridge Companion to Augustine*, ed. Eleonore Stump and Norman Kretzmann (Cambridge: Cambridge University Press, 2001), 126, the Augustinian notion of will is both a "first-order" will and a "second-order" will in the sense that the philosopher Harry Frankfurt distinguishes. "A first-order volition is the will's directing some faculty or bodily power to do something. A second-order volition, by contrast, is a will to will something." Stump argues that Augustine's concept is that "the will can command itself . . . and [also] resist its own commands." Augustine is not a "compatibilist," which is to say, someone who holds the position that our human actions are determined but nevertheless we know we have free will, Stump points out. (Joe Keith Green in a personal communication [August 2011] points out that Augustine can be thought of as a compatibilist from the standpoint of holding that God is all-powerful but at the same time human beings have free will. I agree with this point but for my purposes the overriding issue is the magical character of both the divine free will and the human free will—neither is the working of the

necessity of nature. Both human and divine are outside nature and natural causality.) For Augustine insists that "a will determined by nature or causal necessity is not a free will (in fact it is not a will at all, properly speaking)," Stump says. He embraces the position that our will is the sole ultimate cause of action. "Nothing else," Stump quotes Augustine as saying, "makes the mind the ally of evil desire except its own will and free choice." This fundamental position does not undergo any significant change from the early articulation just quoted from *De Libero Arbitrio*, Stump argues (131). As Bonnie Kent, the reigning expert on free will in the history of Christian theology and its secularization within the dominant Western philosophical tradition, puts it, for Augustine "either the will is the first cause of sin, not merely one more link in a chain of natural efficient causes, or there is no sin" (Bonnie Kent, "Augustine's Ethics," in Stump and Kretzmann, *Cambridge Companion to Augustine*, 222). Sidney Morgenbesser and James Walsh (*Free Will*, 11) point out that whatever Augustine proposes is lost to human nature by the divine punishment meted out to Adam and his descendants for the original sin, the capacity for free will as the very definition of what it means to be human, nevertheless, remains. They point to a distinction between human *free will*, as the very definition of the human, and *freedom*, the capacity to use that free will wisely:

> It has seemed to many that St. Augustine shifted from his earlier insistence on the reality of free will to strict theological determinism. He himself denied this, and there is . . . the distinction which has been much emphasized as the key to his consistency. This is the distinction between free will (*liberum arbitrium*), which belongs to man by nature and which he can never lose, and freedom (*libertas*), the ability to make good use of free will. It is freedom which is lost with original sin and recovered with grace.

47. Augustine, *Literal Meaning of Genesis* 7.26.28.
48. Ibid. 11.7.140.
49. Ibid. 8.12.50–52.
50. Ibid. 8.11.49–50.
51. Augustine, *Literal Meaning of Genesis* 7.26, 7.28.
52. Augustine, *City of God* 12.3.501–2.
53. Augustine, *Literal Meaning of Genesis* 11.9.142.
54. Ibid. 11.4.137.
55. Ibid. 9.10.80–82. Although Adam and Eve's bodies were not "spiritual," like those of the angels, but natural, they were nevertheless not "this body of death" from which and with which we have been born.
56. Augustine interprets Paul's phrase in Romans 7:24, "Who will rescue me from this body that is subject to death?" (according to the translation of the New International Version) or (in the King James translation), "Who shall deliver me from the body of this death?" as "Who will deliver me from *this body of death*?" Augustine's understanding of the phrase is that the post-Fall human body is a body

whose nature in life—and not just in death—has been changed by God as a punishment for Adam's sin.

57. Our bodies even from birth are "in a condition that is a punishment for sin," for in "this body . . . all men are by nature 'children of wrath,'" Augustine writes, interpolating a phrase from Ephesians 2:3.

58. Augustine, *Literal Meaning of Genesis* 9.11.82–83.

59. Augustine, *City of God* 14.19.618–19.

60. Ibid. 13.23.572.

61. Augustine, *Literal Meaning of Genesis* 9.10.80–82.

62. Augustine, *City of God* 13.3.544.

63. Ibid. 13.13.555.

64. Ibid. 13.12.555: "From the very beginning of our existence in this dying body there is never a moment when death is not at work in us." Throughout "life—if indeed it is to be called life," the body "leads us toward death."

65. Ibid. 14.10.603.

66. The death of the soul is only obliquely referred to here in Genesis, Augustine proposes, "because God wished to keep it hidden until the dispensation of the New Testament, where it is declared most plainly." Augustine writes, "When God said, 'Adam, where art thou?' He signified by this the death of the soul, which comes about when He forsakes it; and when He said, 'Dust thou art, and unto dust shalt thou return,' He signified the death of the body, which comes about when the soul departs from it."

67. Augustine continues: "By the grace of God, through a Mediator, He has redeemed from the second death [that is, of the soul] those who were 'Called according to His purpose,' as the apostle says."

68. Augustine, *City of God* 13.23.571–72.

69. As Zeev Harvey points out (personal communication, September 2011), the dearth of Aristotle in Latin Christendom was because "Aristotelian philosophy had for centuries been successfully stifled by Augustine, the Platonist."

70. Kent, *Virtues of the Will*, 40.

71. Ibid., 44.

72. Tobias Hoffman, "Intellectualism and Voluntarism," in *The Cambridge History of Medieval Philosophy*, ed. Robert Pasnau (Cambridge: Cambridge University Press, 2010), 1:414–27. Historian Bonnie Kent calls Henry "an ardent voluntarist" and "a dominant figure" in the faculty at Paris in the thirteenth century, when voluntarism was receiving a much more precise form and various articulations in the wake of the influx of the radical naturalist Aristotelianism from the Iberian Peninsula.

73. Hoffman, "Intellectualism and Voluntarism," cites and comments on *Quodlibet* 9.6, *Opera* 13.142–43.

74. Kent, *Virtues of the Will*, 77.

75. Aristotle, *Nicomachean Ethics* 3.3.1112b14, in *The Basic Works of Aristotle*, ed. Richard McKeon (New York: Random House, 1970).

76. Ibid. 3.3.1113a11.

77. Ibid. 2.1.1103b15ff: "By doing the acts that we do in our transactions with other men we become just or unjust . . . and by doing the acts we do in the presence of danger, and being habituated to feel fear or confidence, we become brave or cowardly. The same is true of appetites and feelings of anger. . . . Thus, in one word, states of character arise out of like activities. . . . It makes no small difference, then, whether we form habits of one kind or another from our very youth; it makes a very great difference, or rather *all* the difference."

78. Ibid. 3.3.1114a15ff.

79. Ibid. 2.1.1103b3ff.

80. Kent, *Virtues of the Will*, 17.

81. Ibid., 93.

82. Zeev Harvey comments (private communication, September 2011): "One might say: Aquinas follows Aristotle but tries to accommodate Augustine, while Bonaventure follows Augustine but tries to accommodate Aristotle."

83. Bonnie Kent, "Reinventing Augustine's Ethics: the Afterlife of *City of God*," in *Augustine's "City of God": A Critical Guide*, ed. James Wetzel (Cambridge: Cambridge University Press, 2012). Joe Keith Green comments (personal communication, August 2011): "You might note that Kent's very choice of words emphasizes how Aquinas 'shrank' the will from Augustine's seemingly broader view to 'electio'—a faculty of the intellectual appetite that chooses—choosing is its 'end.' Aquinas's view is, however, a more or less inevitable outcome of Augustine's language."

84. Bonnie Kent, "Aquinas and Weakness of Will," *Philosophy and Phenomenological Research* 75, no. 1 (2007): 70–91, cites *Summa Theologica* 2–3.77.a.3, 2–3.77.ad 2.

85. Kent, *Virtues of the Will*, 85. Kent suggests that it is generally the straddling positions that recent attempts at reviving virtue ethics have actually articulated and advocated in the name of Aristotle. It is the Christian compromises of Aquinas, for example, rather than Aristotle's actual moral psychology, that is at the heart of Alasdair MacIntyre's and Bill Bennett's virtue ethics. Hence my discovery, mentioned in the first chapter, that teaching virtues in American schools seems to cash out as a free choice enterprise rather than as an attempt to revive and implement an actual training in habits of virtue, has a long Christian historical trail behind it. As a medievalist colleague once put it to me, "Remember . . . what people might now regard as 'common-sense moral intuitions' owe a great deal to the historical influence of Christianity." Disguised as simply a commonsense, universal, and secular moral education, current versions of character education instead (we can now realize) infuse the American classroom and school with disguised normative Christian ideologies and conceptions of the human; and hence they both implicitly missionize and indoctrinate, infusing our children with unconscious Augustinian moral reflexes. Also, they just, plain and simple, get it wrong—as the second half of this book exposes.

86. Ibid., 96.

87. Ibid., 148–49.

88. *Wikipedia*'s article "René Descartes" (en.wikipedia.org/wiki/Descartes) points out that "the Cartesian coordinate system . . . allow[s] algebraic equations to be expressed as geometric shapes in a two-dimensional coordinate system" and that Descartes "is credited as the father of analytical geometry, the bridge between algebra and geometry, crucial to the discovery of infinitesimal calculus and analysis."

89. Kurt Smith, "Descartes' Life and Works," *Stanford Encyclopedia of Philosophy*, plato.stanford.edu/entries/descartes-works.

90. Menn, *Augustine and Descartes*, 18–19.

91. Ibid., 22–23.

92. Ibid., ix, 21.

93. Ibid., 4.

94. Ibid., 22–23.

95. Ibid., 25.

96. Ibid., 43–44.

97. Ibid., ix.

98. Ibid., 46–49.

99. Ibid., 51–53.

100. Ibid., 54.

101. Ibid., 5.

102. Ibid., 55.

103. Ibid., 5–6.

104. See Susan James, *Passion and Action: The Emotions in Seventeenth Century Philosophy* (Oxford: Oxford University Press, 1997); and Heidi M. Ravven, "Spinoza's Intermediate Ethics for Society and the Family," *Animus* 6 (Winter 2001). *Passivity* and *activity*—the terms of the debate about emotions and moral psychology in the seventeenth century—identify the direction of causal agency, either from soul to body or from body to soul. So the human moral problem is the struggle between body and soul for Descartes (but not, we will see, for Spinoza).

5. Another Modernity: The Moral Naturalism of Maimonides and Spinoza

1. I wish to thank Zeev Harvey of Hebrew University in Jerusalem for his astute comments and reflections on this chapter.

2. Jon McGinnis, "Arabic and Islamic Natural Philosophy and Natural Science," *Stanford Encyclopedia of Philosophy*, plato.stanford.edu/entries/arabic-islamic-natural.

3. "Justinian," *Wikipedia*, en.wikipedia.org/wiki/Justinian.

4. "Neoplatonism," *Wikipedia*, en.wikipedia.org/wiki/Neoplatonism.

5. Christina D'Ancona, "Greek into Arabic: Neoplatonism in Translation," chap. 2 in *The Cambridge Companion to Arabic Philosophy*, ed. Peter Adamson and Richard C. Taylor (Cambridge: Cambridge University Press, 2005), 18–19.

6. Ibid.

7. Ibid., 19–20.

8. Peter Adamson and Richard C. Taylor, introduction to *Cambridge Companion to Arabic Philosophy*, 1–9.

9. On Alexandria versus Rome, see also Maren Niehoff, *Jewish Exegesis and Homeric Scholarship in Alexandria* (Cambridge: Cambridge University Press, 2011). Zeev Harvey comments (personal communication, September 2011): "However, in the conflict . . . of Alexandria vs. Rome, Alexandria represents the old Greek philosophic tradition, while Rome is basically Augustine. The exceptional or revolutionary tradition here is not the Alexandrian one, but the Augustinian one."

10. Christina D'Ancona, "Greek Sources in Arabic and Islamic Philosophy," *Stanford Encyclopedia of Philosophy*, plato.stanford.edu/entries/arabic-islamic-greek.

11. "Al-Farabi," *Wikipedia*, en.wikipedia.org/wiki/Alfarabi.

12. David C. Reisman, "Al-Farabi and the Philosophical Curriculum," in Adamson and Taylor, *Cambridge Companion to Arabic Philosophy*, 52–71. Reisman's extensive quotation (*Cambridge Companion*, 55) is from Al-Farabi's *Appearance of Philosophy*.

13. Thérèse-Anne Druart, "Metaphysics," in Adamson and Taylor, *Cambridge Companion to Arabic Philosophy*, 327–28, 346.

14. Alexander Altmann, "Maimonides and Thomas Aquinas: Natural or Divine Prophecy?" *AJS Review* 3 (April 1978): 1–19, doi:10.1017/S0364009400000295.

15. Peter Adamson, "The Arabic Tradition," in *The Routledge Companion to Ethics*, ed. John Skorupski (London: Routledge, 2010), 65.

16. See also Muhsin Mahdi, *Alfarabi and the Foundation of Islamic Political Philosophy* (Chicago: University of Chicago Press, 2010). On the philosophic religion of al-Farabi, see Carlos Fraenkel, "From Maimonides to Samuel ibn Tibbon: Interpreting Judaism as a Philosophical Religion," in *Traditions of Maimonideanism*, ed. Carlos Fraenkel (Leiden: Brill, 2009), 177–212.

17. Adamson, "Arabic Tradition," 65.

18. There were ascetic and moderate versions of the moral intellectualism that ascribed virtue strictly to intellectual endeavors for their own sake. Both versions focused on coming to know "intelligible objects," that is, the bodies of knowledge of the theoretical sciences and philosophy. The moral ascetic believed that intellectual aims alone are worth pursuing, while the moral moderate believed that practical and social goods could have instrumental benefit or were indifferent. The two positions could also be reconciled in a "two-level ethic" that ascribed to the elite alone the goal of the pursuit of intellectual virtue and the intellectual life, whereas the rest of humanity was relegated to having their proper moral aim be the mundane life of moderation and practical virtues. Only the philosophically and scientifically educated had the opportunity to pursue the superior moral life of intellectual virtue, which developed the (theoretical rational) soul, rather than having to settle for the pursuit of the lesser virtue of moderation, a virtue focusing on developing the (psychic abilities associated with the) body and the social body.

19. Adamson, "Arabic Tradition," 66–69.

20. Ibid., 65.

21. Harry A. Wolfson, "Extradeical and Intradeical Interpretations of Platonic Ideas," *Journal of the History of Ideas* 22, no. 1 (January–March 1961): 3–32.

22. Aristotle had, in any case, proposed that the practical reason involved in moral deliberation was only about determining the *means* toward the good and not about identifying and choosing the good. He had held that the end or goal, that is, the identification of what *is* good, in contrast with the means, was set by desire and not by practical rational deliberation. Desire in turn was determined by how a person's character shaped his or her cognitive grasp or understanding. A bad person was someone who mistakes bad ends for good ones, while a good person had moral cognitive clarity, knowing and recognizing the true good. Hence for Aristotle moral agency was all in all, in both its expressions, as practical wisdom, as well as, as theoretical intellect, a cognitive capacity. Moral agency was about moral discernment and broad understanding of the world, an understanding that transformed one's sense of what was desirable and thereby identified the true source of ultimate human fulfillment and joy. Choice, as dependent on having the right desires, was a mere necessary outgrowth of such knowledge rather than the crucial moral step it was deemed to be in the Latin West. Hence failure to act morally was a form of cognitive error, a failure of understanding and desire rather than of will. Moral agency, in both its forms, practical and theoretical, was about clearly recognizing, knowing the good, from which right desire would follow and, thereupon, action.

23. Shlomo Pines, "Spinoza's *Tractatus Theologico-politicus*, Maimonides, and Kant," in *Further Studies in Philosophy*, ed. Ora Segal (Jerusalem: Magnes, 1968).

24. Harry A. Wolfson, *The Philosophy of Spinoza: Unfolding the Latent Processes of His Reasoning* (Cambridge, MA: Harvard University Press, 1934).

25. The interpretation of Maimonides offered here summarizes and extends my arguments in three published papers: Heidi M. Ravven, "Some Thoughts on What Spinoza Learned from Maimonides About the Prophetic Imagination," *Journal of the History of Philosophy* 29, no. 3 (April 2001) and no. 4 (July 2001); Heidi M. Ravven, "The Garden of Eden: Spinoza's Maimonidean Account of the Genealogy of Morals and the Origin of Society," *Philosophy and Theology* 13, no. 1 (2001): 3–47; Heidi M. Ravven, "Maimonides' Non-Kantian Moral Psychology: Maimonides and Kant on the Garden of Eden and the Genealogy of Morals," *Journal of Jewish Thought and Philosophy* 20, no. 2 (2012): 199–216.

26. Shlomo Pines, "Truth and Falsehood Versus Good and Evil: A Study in Jewish and General Philosophy in Connection with the *Guide of the Perplexed*," in *Studies in Maimonides*, ed. Isadore Twersky (Cambridge: Cambridge University Press, 1990).

27. In *The Guide of the Perplexed*, Maimonides does away entirely with the practical intellect. Rather than merely asserting that it is of comparatively less importance,

he absorbs practical reason into the imagination, a subordinate, nonrational, bodily capacity, and hence a capacity that has no play in the human ascent to the divine. (The status of the practical intellect in both Maimonides and Spinoza is a matter of some controversy. What we can say is that in the thinking of both, the contemplative stance of the theoretical intellect is ultimately both salvific and the source of the only true moral and spiritual transformation.) If we had any doubt about the implications of such a move, Maimonides does not hesitate to draw them: he is regarded by scholars as the philosopher who in the most extreme way in the history of philosophy set the true in opposition to the good. True and false were real and discoverable by reason; good and bad, by contrast, he said, were mere products of the imagination and hence did not capture anything real about the natural order in its divine origins and expressions in causal principles. Instead, Maimonides regarded good and bad as products of the bodily imagination and, as we would put it today, capturing well Maimonides's point, of the social imaginary. He regarded the social imagination as the source of social conventions that could not be determined apart from social definition and political enforcement—a position that Spinoza later adopted from Maimonides.

28. Maimonides, *The Guide of the Perplexed*, trans. Shlomo Pines (Chicago: University of Chicago Press, 1963), pt. 3, chap. 13.

29. The notion of ethics as bringing the human person into harmony with the natural cosmos is of Stoic origin and influence.

30. Maimonides, *Guide*, pt. 3, chap. 53. So, too, Spinoza.

31. Ibid., pt. 2, chap. 48 (Pines, 410).

32. In his commentary on the Mishnah's Tractate Avot, Maimonides adopts and adapts Aristotle's doctrine of moral virtue as the mean between two extremes as the underlying practical rational purpose of much rabbinic law. However, this is not his last word on the subject of moral psychology. In the *Guide* we learn about the imaginative status of moral virtue in contrast to intellectual virtue. See Ravven, "Some Thoughts," and Ravven, "The Garden of Eden."

Spinoza adopted Maimonides's doctrine of the conventionality of morals and articulated it succinctly in the *Ethics*: "In a state of nature nothing can be said to be just or unjust: this is so only in a civil state, where it is decided by common agreement what belongs to this or that man. From this it is clear that justice and injustice, wrong-doing and merit, are extrinsic notions, not attributes that explicate the nature of the mind." Baruch Spinoza, EIVP37S2, in *Baruch Spinoza: The Ethics and Selected Letters*, trans. Samuel Shirley and ed. Seymour Feldman (Indianapolis: Hackett, 1982), 176–77.

33. Michael Rohlf, "Immanuel Kant," *Stanford Encyclopedia of Philosophy*, plato.stanford.edu/entries/kant.

34. The description of Herder's account of Eden from his *Ideas for the Philosophy of History of Humanity* is adapted from Allen W. Wood's account of it in *Kant's Ethical Theory* (Cambridge: Cambridge University Press, 1999), 228.

35. All quotations from Kant's "Conjectural Beginning of Human History" are taken from Emil Fackenheim's English translation, in *On History: Immanuel Kant*, ed. Lewis White Beck (Indianapolis: Bobbs-Merrill, 1963).

36. Kant, "Conjectural Beginning," 55.

37. Ibid.

38. Ibid.

39. Ibid., 56.

40. Ibid.

41. Ibid., 57.

42. Ibid.

43. Ibid., 57–58.

44. Ibid., 58.

45. Ibid., 68.

46. Ibid., 58. This, we recall, is in sharp contrast with Maimonides's insistence that man is not the goal of nature but nature is its own end.

47. Ibid.

48. Ibid., 59.

49. Ibid., 60.

50. Ibid.

51. Ibid., 65.

52. Maimonides, *Guide*, pt. 1, chap. 2, and pt. 2, chap. 30.

53. Ibid., pt. 2, chap. 30.

54. Maimonides's immediate source is the midrash in *Genesis Rabbah* 8:1, which alludes to Plato's *Symposium*.

55. Maimonides, *Guide*, pt. 1, chap. 2. Zeev Harvey and Sara Klein-Braslavy also maintain that Adam had a perfected imaginative capacity before the Fall. But nothing in the text suggests that, and much precludes it. First, Adam was seduced by the body. Second, he did not do what the perfected imagination is essentially characterized by, according to Maimonides—namely, to govern and institute morals. In fact, that's exactly what Maimonides explicitly claims Adam could not do till after his punishment. So I do not think that the claim of Adam's perfected imagination in pre-Fall Eden is tenable.

56. Lawrence Berman, in "Maimonides on the Fall of Man," *AJS Review* 5 (1980): 9, points out that for Maimonides "Paradise (*gan eden, pardes*) is identical with theoretical speculation."

57. Maimonides, *Guide*, pt. 1, chap. 2.

58. Here in this chapter and systematically throughout the *Guide* Maimonides avoids mention of the practical intellect, and instead relegates its practical and deliberative moral functions to the bodily faculty of the imagination; in his interpretation of Eden he draws a stark contrast between theoretical reason and the practical concerns of the imagination. (See Pines, "Truth and Falsehood." See also my critique of Pines on this matter, below.) So Maimonides downgrades moral knowledge

by relegating it not to the practical intellect, as the Aristotelian tradition (and Maimonides's own *Eight Chapters*) had done, but to a bodily faculty outside of reason—namely, to the imagination. (Lawrence Berman, in "Maimonides on the Fall of Man," 10, shares the view that Maimonides is arguing that "the fall story symbolizes man's coming under the power of the imagination from an original idyllic state in which he was completely devoted to truth and falsehood, not being concerned with good and evil actions." For an extended treatment of the claim that according to Maimonides Adam becomes endowed with a full-fledged imagination as a result of his disobedience, see my "The Garden of Eden.") So the moral of the tale is that the imagination is what all human beings have by nature (in addition to intellect or reason), and, by unfortunate situation, it is what all human beings need. Maimonides follows al-Farabi in extending the imagination to include the capacity for political leadership. So Adam in the story gets an enhanced imagination, which improves on the merely primitive imagination that took him initially by surprise when he followed impulse. When this newly enhanced imagination is *perfected* (through knowledge and practice), Maimonides says, its vital function will be in instituting social accord and conventional morals, and generally in governance. Maimonides interprets the biblical claim that Adam, after eating the fruit, has become like *elohim* to mean he has become like a ruler: he has gained the capacity for defining and instituting morals and moral leadership. Like an excellent ruler, Adam has acquired the capacity to institute and enforce conventional regulations that create harmony in the passions of self and society.

59. Maimonides, *Guide*, pt. 1, chap. 2.

60. Ibid.

61. In "Truth and Falsehood," Shlomo Pines argues that Maimonides, in an unprecedented interpretation of Adam's sin, radically opposes the true to the good. It is this radical opposition that Spinoza will later both adopt as the cornerstone of his ethical theory and explain theoretically in terms of his own elaborate psychological doctrine. See also Berman, "Maimonides on the Fall of Man." Berman argues (8–9) that, according to Maimonides,

> the fall of man consisted in a change of priorities, from an interest in the things of the mind to becoming interested in the things of the body; from being a philosopher, a master of his passions, to becoming a beast in human form, mastered by his passions; from being a solitary thinker, to becoming a ruler of cities, being informed by imagination only. . . .
>
> Thus previous to the fall, Adam was not concerned with matters relating to values but only with the truth.

Berman goes on to suggest that since "imagination did not enter in Moses' prophecy . . . thus Adam and Moses were identical, the difference being that Adam, before the fall, represents the ideal for man, not living in society, while Moses represents the ideal for man living in society" (8n22).

Maimonides asserts of Adam in Eden that "when man was in his most perfect and excellent state, in accordance with his inborn disposition and possessed of intellectual cognition, he had no faculty that was engaged in any way in the consideration of generally accepted things and he did not apprehend them." There was no need for practical intellect in Eden. Instead, Adam could focus on theoretical contemplation alone. Hence, morals and the moral capacity are compromises necessary for living in a post-Eden world, a world of bodily needs and social necessities.

62. Aristotle expressed ambivalence about exactly how human theoretical intellection really is. For in *Nicomachean Ethics* Book 10, in his paean to the theoretical life as alone providing supreme happiness to humanity, he also suggests that it reaches beyond the human: contemplation is the activity par excellence of the gods. It makes human beings immortal and is far more like the divine life than the normal human concerns that engage practical thinking and demand moral virtues. The latter Aristotle (like Maimonides in the *Guide*) associates with the body. We have discovered in *Nicomachean Ethics* Book 10 the immortal, contemplative, divine-like Adam before his disobedience and punishment. He is living a divine-like life with no need to devote himself to practical thinking about the body and social virtues. Aristotle even says of his gods, who are engaged exclusively in contemplation, exactly what Maimonides says of Adam: that they have no need to concern themselves with matters of justice or social policy or moral deliberation. For Aristotle divine beings had no need for the practical intellect and that's what distinguished gods from the rest of us. The transition to life outside the Garden will be marked by Adam no longer living by contemplation alone. In falling to earth, so to speak, Adam must engage his practical intellect—for Maimonides, the perfected imagination—in the pursuit of normal human individual and social virtues. As in Aristotle, he must acquire practical wisdom embedded in a body, which for Maimonides is the imagination, while still yearning for the contemplation and the contemplative life and stealing moments of transcendence whenever possible. Maimonides has radicalized and systematized Aristotle's theology of the philosopher, assimilating it into Judaism. He resolves Aristotle's ambivalence about the relative merit of intellectual versus moral virtue in the same intellectual direction that other philosophers in the Arabic tradition did, and then takes it one step further, actually *opposing* the good to the true, as Shlomo Pines points out—we, poor post-Eden Adams, get so caught up in the good we forget the true and have a hard time finding time for the pleasures of the mind. (One wonders at how autobiographical a tale for Maimonides himself his Adam has turned out to be.)

63. See Maimonides, *Guide*, esp. pt. 2, chap. 40.

64. Mosaic law, Maimonides maintains, gives the community governed by justice and convention a taste of the ultimate joy of the intellectual life. For Mosaic law, unlike the strictly imaginative law of the non-prophetic statesman, is divine, Maimonides says, because the prophets, unlike mere political leaders (i.e., the founder

and followers of the *nomos*), are those whose intellect, as well as imagination, is operative in governance. They are thus able to apply the imagination to *conceptual* knowledge, thereby rendering it in imaginative form precisely geared to the understanding of the masses. In this way, a glimpse of intellectual paradise opens to the law's adherents (*Guide*, pt. 2, chap. 40; *Moreh* 2:84b). For an excellent treatment of this chapter of the *Guide* and of the prophetic versus the nomoitic constitution, see Miriam Galston, "The Purpose of the Law According to Maimonides," in *Maimonides: A Collection of Critical Essays*, ed. Joseph A. Buijs (Notre Dame, IN: Notre Dame University Press, 1988). See also Shlomo Pines, "Translator's Introduction," xci–xcii, in Maimonides, *Guide* (Pines, lvii–cxxxiv).

65. Aristotle, *Nicomachean Ethics* 6.8.1142a.

66. Ibid. 6.7.1141a.

67. Ibid. 6.12.1144a.

68. Maimonides, *Guide*, pt. 3, chap. 51. On God's attributes of action, see, e.g., pt. 1, chap. 51.

69. Ibid., pt. 3, chap. 51.

70. Ibid.

71. We also discover throughout the Kantian critical philosophy the notion of reason as will. The human universal subjectivity is an imposition upon, and the human shaping of, nature (for example, in the contribution of the human intersubjective mental categories of causality and time as the prism through which the world is grasped). These are acts of *will* rather than reason thought of as the discovery of what's already there. (We see a nod to Descartes's voluntarist account of cognition in Kant's theory here, too, of course.) The ubiquity of reason functioning as will in this way is, for Kant, a presupposition of our experiencing anything at all; human will thus replaces the divine will as constitutive of (the explanation of) nature in this secularized account. Kant picks up on the Cartesian legacy of created eternal truths but attributes them to the human knower: these are ideas of nature that God wills and could have created differently. God knows the world because He knows His own will—and so, too, human beings know the world because we see it through the categories we impose upon it.

72. Maimonides, *Guide*, pt. 3, chap. 51ff.

73. Spinoza, *Ethics*, pt. 5, chap. 32 (Shirley, 219).

74. In this Spinoza owes a debt to the Stoic approach to ethics as well as to the Maimonidean. The Stoics thought of ethics as a therapeutic praxis and also understood it as accomplished through the individual coming into harmony with nature as a whole. Spinoza, however, replaces Stoic calm as the aim of the moral life with Maimonidean "love," which is to say rapturous joy. This is a point he makes clear in both the first part of his *Short Treatise* and in Part V of the *Ethics*. See Heidi M. Ravven, "Spinoza's Individualism Reconsidered: Some Lessons from the *Short Treatise on God, Man, and His Well-Being*," in *Iyyun: Jerusalem Philosophical Quarterly* 47 (July 1998): 265–92.

75. See, e.g., Zeev Harvey's article on Spinoza and Ibn Ezra in Y.Y. Melamed and M.A. Rosenthal, *Spinoza's TPT* (Cambridge: Cambridge University Press, 2010), 41–55.

76. Spinoza's determinism was influenced not only by Maimonides but also by the fourteenth-century Jewish philosopher Hasdai Crescas, whose overall approach was to out-Aristotle Aristotle.

77. For a longer list and discussion of what Spinoza owed Maimonides, see Ravven, "Some Thoughts." See also Zeev Harvey, "A Portrait of Spinoza as a Maimonidean," *Journal of the History of Philosophy* 19, no. 2 (April 1981): 151–72.

78. In *Spinoza, Descartes and Maimonides* (New York: Russell and Russell, 1963), first published in 1924, Leon Roth proposed that Spinoza raised Cartesian questions to which he provided Maimonidean answers. Since then a number of scholars have addressed the relationship between Spinoza and Maimonides, including Shlomo Pines, Zeev Harvey, and myself. For a fuller bibliography on the topic, see Ravven, "Some Thoughts."

79. Spinoza held that there were infinite ways to conceive the person and the universe. These were what he termed, in medieval philosophical nomenclature, the "divine attributes." But human beings had access to only two of these, namely, thought and extension (or mind and body).

80. According to Aristotle, and to Maimonides as an Aristotelian, rational knowledge could be only of species and of universal scientific principles and could not extend to and grasp in a scientific way the uniqueness of particular individuals. All human beings, according to Aristotle, had in common an identical human nature, a nature that was particularized into unique human beings only through their embodiment in matter. But body or matter, unlike Aristotle's rational "form" that all human beings had in common, was in principle unknowable, according to the Aristotelian tradition. Hence it was via the universal and identical reason inherent in all human minds that, according to the Aristotelian Maimonides, the discovery could be made of the universal rational principles embedded and embodied in all natural things.

To come to know the underlying scientific causal principles of the particular material phenomena surrounding us, an Aristotelian believed, one had to "abstract" each from its matter and see the underlying universal rational causes at work, causes that were the same in similar phenomena and also identical with the general principles inherent in the rationality of the mind. To come to know was thus to shed one's own bodily particularity in favor of one's own inherent rational universality of mind when the mind engaged in abstracting the universal rational causes of natural phenomena. Thus, our universal human reason, through its theoretical endeavors, could come to unite the mind with the divine reason embedded in, and now abstracted from, nature.

81. Spinoza, *Ethics*, pt. 3, Definition of the Emotions #1 (Shirley, 142).

82. Spinoza's notion of the conatus was partly adapted from Hobbes's materialist conception of a basic human striving for power, an urge that included the power to maintain bodily stability in the face of external onslaught.

83. Even though Aristotle defined human beings as rational animals, he was ambiguous about how reason functioned in human action, for human aims (ends) were set by desire, not in any way independently by reason. Reason followed desire when it came to human action, yet the truly noble desire was to live the life of theoretical rational endeavor and engagement. In setting the desire for theoretical rational understanding and self-understanding as the proper goal and fulfillment of a human life, Spinoza is standing on solid Aristotelian ground.

84. The neuroscientist Antonio Damasio has called Spinoza a "proto-biologist." Spinoza based his quasi-biological science of the emotions on his claim of the identity of mind and body. Mind and body, he argued against Descartes, are not two substances but one whole, one thing, described in two ways. Spinoza defined mind as the consciousness of the body, the body made conscious. Mind is the awareness (an awareness that is both cognitive and emotionally rich) of the changes in the body-mind as it is affected by, and affects, the environment. A consequence of the bodily rootedness of cognition was that immediate awareness is of self and context and of self in context. So perception and feeling tone, which is to say thought and affect, he argued, always begin with the self; as they are built upon self-reflectively, with reference to bodies of scientific explanation, they remain emotionally charged, no matter how theoretical and broad the explanations become, because they are about the self and redefine the experiences of the self.

Spinoza writes:

> The mind does not know itself save in so far as it perceives ideas of the affections of [i.e., changes in] the body (Pr. 23, II). Now it does not perceive its own body (Pr. 19, II) except through ideas of [awareness of and reflections upon] affections of the body, and it is only through these affections that it perceives external bodies (Pr. 26, II). (*Ethics*, pt. 2, prop. 29, corollary [Shirley, 84])

Proposition 26 in Part II of *Ethics* reads: "The human mind does not perceive any external body as actually existing except through the ideas of its own body." See also my "Spinoza's Individualism Reconsidered."

85. Spinoza believed that every interaction, every experience, every impingement, whether interior (like a stomachache) or exterior (like a traffic jam), either contributed to or detracted from the proper functioning of one's conatus, and thus had a feeling tone—up or down. Each interaction and experience, he believed, makes us feel better or worse, and hence the impingements of the objects, situations, and people occasioning such experiences came to be colored with positive or negative affects (moods and emotions) by association. No object could be neutral; always associated with it was an "up" or "down" assessment (pleasure or pain) in

terms of how it affected the self. According to Spinoza, anything could become the source of pleasure (or pain) by mental association.

Spinoza argued that sometimes, or perhaps even often, the feeling tone of our perceptions does not accurately express whether the functioning of our conatus has actually been enhanced or debilitated by our experiences. Our emotional assessments are thus subject to error—for example, there are many pleasures that feel good but work against us and cause pain and weakening of our capacities in the long run—and also to self-deception. We err because we take isolated or partial pleasures to be beneficial and we are seduced by pleasure to suppress or repress the awareness of associated pain (addictions fall into this category). If pleasure always accurately signaled the enhancement of our overall well-being and pain its detraction, ethics would be easy and straightforward: all one would have to do is listen to the feelings of pleasure or pain generated by the conatus and simply follow the pleasure in order to promote one's own well-being and the enhanced agency of the well-functioning conatus. But Spinoza argued that, unfortunately, ignorance and self-deception are the common human fate.

86. The Aristotelian tradition located the emotions in the "sensitive" part of the soul, the part devoted to perception and appetite. Perception and appetite represent two types of receptive posture toward the external world. In the Aristotelian tradition, emotions span body and soul; they are states of body and soul as a composite. They are bodily changes with their accompanying feelings, and so they are both physical and psychological states. This psychophysical aspect of Aristotelian theory was retained by seventeenth-century mechanical theorists of the emotions along with the distinction between activity and passivity, divested of many of its Aristotelian metaphysical assumptions. Activity and passivity were reinterpreted as aspects of the new mechanical account of causality, in its reduction of all causes to what in the Aristotelian taxonomy were efficient causes of motion (like billiard balls).

87. See Ravven, "Spinoza's Intermediate Ethics"; Heidi M. Ravven, "Spinoza's Ethic of the Liberation of Desire," in *Women and Gender in Jewish Philosophy*, ed. Hava Tirosh-Samuelson (Bloomington: Indiana University Press, 2004); and Heidi M. Ravven, "Spinoza's Materialist Ethics: The Education of Desire," in *Spinoza: Critical Assessments*, vol. 2, ed. Genevieve Lloyd (London: Routledge, 2001).

88. In *Ethics*, pt. 3, chap. 59 (Shirley, 140), Spinoza writes: "Among all the emotions related to the mind in so far as it is active, there are none that are not related to pleasure or desire."

89. *Ethics*, pt. 3, chap. 58 and dem. (Shirley, 140):

P58: Besides the pleasure and desire that are passive emotions, there are other emotions of pleasure and desire that are related to us in so far as we are active.

Dem.: When the mind conceives itself and its power to act, it feels pleasure (Pr. 53, III). Now the mind necessarily regards itself when it conceives a true, that is, adequate idea (Pr. 43, II). . . . Therefore, it feels pleasure, too, in so far as it conceives adequate ideas, that is (Pr. 1, III), in so far as it is active. . . .

Therefore desire is also related to us is so far as we understand, i.e., in so far as we act (Pr. 1, III).

90. Spinoza thus substitutes the internalization of one's particular constitutive external causes as self for the Aristotelian formal essence, an internal genetic-like program playing itself out in human development and particularly in rational understanding and action.

91. See my articles on Spinoza's systems theory of ethics: Heidi M. Ravven, "The Self Beyond Itself: Further Reflections on Spinoza's Systems Theory of Ethics," in *Proceedings of the 3rd International Multi-Conference on Complexity, Informatics and Cybernetics, March 25th–28th, 2012—Orlando, Florida, USA*, ed. Nagib Callaos, Belkis Sánchez, Michael J. Savoie, Mohammad Siddique, Andrés Tremante, and C. Dale Zinn, 133–38; Heidi M. Ravven, "Spinoza's Systems Theory of Ethics," in *Cognitive, Emotive, and Ethical Aspects of Decision Making in Humans and in Artificial Intelligence*, ed. George E. Lasker, vol. 3, ed. Iva Smith and Wendell Wallach (Windsor, ON: International Institute for Advanced Studies in Systems Research & Cybernetics, 2004); Heidi M. Ravven, "Notes on Spinoza's Critique of Aristotle's Ethics: From Teleology to Process Theory," *Philosophy and Theology* 4, no. 1 (Fall 1989): 3–32; Heidi M. Ravven, "What Can Spinoza Teach Us Today About Naturalizing Ethics? Provincializing Philosophical Ethics and Freedom Without Free Will," in *Cognitive, Emotive, and Ethical Aspects of Decision Making in Humans and in Artificial Intelligence*, vol. 3, 99–104.

92. As scholars have noted, Spinoza's vision of the bringing of the self into harmony with nature and universe is profoundly Stoic in tenor. Spinoza combines a Maimonidean ecstatic intellectualism with a Stoic moral naturalism. The echoes of that Stoic moral naturalism were, albeit more faintly, to be heard in Maimonides himself, however, as well.

93. Spinoza, *Ethics*, pt. 1, def. 7 (Shirley, 31).

94. Ibid., pt. 2, chap. 48 (Shirley, 96).

95. "Very roughly, non-naturalism in meta-ethics is the idea that moral philosophy is fundamentally autonomous from the natural sciences. . . . Most often, 'non-naturalism' denotes the metaphysical thesis that moral properties exist and are not identical with or reducible to any natural property or properties in some interesting sense of 'natural'. . . . Understood in this way, non-naturalism is a form of moral realism." Michael Ridge, "Moral Non-Naturalism," *Stanford Encyclopedia of Philosophy*, plato.stanford.edu/entries/moral-non-naturalism/.

Broad argued for a form of free will agency that he defined in terms he called "categorical substitutability": "Categorical substitutability involves a negative and a positive condition. These are necessary and jointly sufficient for obligability. The *negative condition* says that the agent's willing of a certain action was not completely determined by the nomic and singular conditions which existed at the time of willing." Not surprisingly, Broad regards an action that merely fulfils this condition as "an *accident,* lucky or unlucky as the case may be." Neither the agent nor "anything else in the universe can properly be praised or blamed for it" (1934, 212, Broad's italics; quoted in Kent Gustavsson, "Charles Dunbar Broad," *Stanford Encyclopedia of Philosophy,* plato.stanford.edu/entries/broad). "The *positive condition* (which must also be fulfilled) says that the willing of the action is "literally determined by the agent or self, considered as a substance or continuant, and not by a total cause which contains as factors *events in* and *dispositions of* the agent" (1934, 214–5; Broad's italics). Thus a free action is an action that is caused by the *agent.* And "In writings up to and including *Five Types of Ethical Theory* Broad assumes that ethical sentences express judgements. He favours an intuitionist or non-naturalist analysis." Gustavsson, "Charlie Dunbar Broad."

6. Surveying the Field: How the New Brain Sciences Are Exploring How and Why We Are (and Are Not) Ethical

1. Walter Sinnott-Armstrong, ed., *Moral Psychology,* 3 vols. (Cambridge, MA: MIT Press, 2008).

2. Owen Flanagan, Hagop Sarkissian, and David Wong, "Naturalizing Ethics," in ibid., 1:2, 7.

3. Ibid., 1:8.

4. Ibid., 1:9.

5. Ibid., 1:7.

6. Ibid., 1:8–9.

7. See, for example, the discussion later in this chapter of Jonathan Haidt and Frederik Bjorklund, "Social Intuitionists Answer Six Questions About Moral Psychology," in Sinnott-Armstrong, *Moral Psychology,* 2:181–217.

8. This characterization of the Khmer notion of karma is taken from Nancy J. Smith-Hefner's ethnographic study of the Khmer immigrant community in Boston in the 1980s and 1990s, *Khmer American: Identity and Education in a Diasporic Community* (Berkeley: University of California Press, 1999).

9. Flanagan, Sarkissian, and Wong, "Naturalizing Ethics," 1:18.

10. Ibid., 1:19.

11. William Casebeer, "Comment on Flanagan et al.," in Sinnott-Armstrong, *Moral Psychology,* 1:29.

12. Debra Lieberman, "Moral Sentiments Relating to Incest: Discerning Adaptations from By-Products," in Sinnott-Armstrong, *Moral Psychology,* 1:165–90. See also Walter Sinnott-Armstrong, introduction to *Moral Psychology,* 2:xv.

13. Sinnott-Armstrong, "Introduction," 2:xiii.

14. Marc Hauser, Liane Young, and Fiery Cushman, "Misreading the Linguistic Analogy: Response to Jesse Prinz and Ron Mallon," in Sinnott-Armstrong, *Moral Psychology,* 2:172–73.

15. Haidt and Bjorklund, "Social Intuitionists."

16. In "Does Social Intuitionism Flatter Morality or Challenge It: Comment on Haidt and Bjorklund," in Sinnott-Armstrong, *Moral Psychology*, 2:219–32, Daniel Jacobson draws the distinction between philosophical moral intuitionism and Haidt and Bjorklund's moral intuitionism in cognitive science. The philosophical theory is that moral sentiments or insights are basic and can have no further explanation or justification than intuition. The scientific claim, on the other hand, is of the explanation of the *origins* of morals. The justification for moral insights derived from intuition in the scientific account could come from either reason or intuition itself. Jacobson points out (2:219–20) that Haidt and Bjorklund have not distinguished between these two meanings and practices of moral intuitionism.

17. Haidt and Bjorklund, "Social Intuitionists," 2:186.

18. Ibid., 2:204, 205.

19. Ibid., 2:190.

20. Ibid.

21. Ibid., 2:181.

22. Ibid., 2:193.

23. Ibid., 2:186.

24. Ibid., 2:193.

25. Daniel Jacobson, in "Does Social Intuitionism Flatter Morality," 2:220–21, points out that Hume's actual position was more nuanced than Haidt and Bjorklund suggest, for in Hume's view, a moral sentiment could presuppose much prior reasoning about its object, the moral situation. The moral sentiment could be mistaken and a given object might not in fact warrant a given sentiment.

26. Haidt and Bjorklund, "Social Intuitionists," 2:189.

27. Ibid., 2:204.

28. Dan Sperber, quoted in Haidt and Bjorklund, "Social Intuitionists," 2:205.

29. Haidt and Bjorklund, "Social Intuitionists," 2:206–7.

30. Haidt and Bjorklund, ibid., 2:207, say that they borrow the term and concept from Paul Churchland.

31. Ibid., 2:209.

32. Ibid., 2:210.

33. Ibid., 2:211.

34. Jacobson, "Does Social Intuitionism Flatter Morality," 2:224, 226. How the social enculturation dimension fits with the intuitionism seems to be ad hoc, Jacobson suggests (2:230).

35. Ibid., 2:219–20.

36. Walter Sinnott-Armstrong, "Framing Moral Intuitions," in *Moral Psychology*.

37. I wish to thank political scientist Edwin Winckler for pointing me to this essay and for suggesting that I look at the work of Terrence W. Deacon on the evolution of language as well.

38. Ursula Goodenough and Terrence W. Deacon, "From Biology to Consciousness to Morality," *Zygon* 38, no. 4 (December 2003).

39. Ibid., 809.

40. John Gray reviews Frans de Waal's *Primates and Philosophers: How Morality Evolved* (Princeton, NJ: Princeton University Press, 2006) in the *New York Review of Books*, May 10, 2007, 26–28, and presents de Waal's arguments against a common assumption of philosophers that Hobbesian competitive striving is borne out by evolution and that morality is a mere "veneer." (Gray also offers an excellent critique of the provincially Western Christian religio-cultural presuppositions of Marc Hauser's *Moral Minds: How Nature Designed Our Universal Sense of Right and Wrong.*)

41. De Waal, *Primates and Philosophers.*

42. So argues Ron Mallon in "Reviving Rawls's Linguistic Analogy Inside and Out," in Sinnott-Armstrong, *Moral Psychology*, 2:145–55, as does Jesse Prinz in "Resisting the Linguistic Analogy: A Commentary on Hauser, Young, and Cushman," in Sinnott-Armstrong, *Moral Psychology*, 2:157–70.

43. Jaak Panksepp and Jules B. Panksepp, "The Seven Sins of Evolutionary Psychology," *Evolution and Cognition* 6, no. 2 (2000): 108–31 (discussed at length in the next chapter). In addition, others who have argued and provided evidence against a moral faculty include Darcia Narvaez, "The Social Intuitionist Model: Some Counter-Intuitions," in Sinnott-Armstrong, *Moral Psychology*, 2:233–40; Haidt and Bjorklund, "Social Intuitionists"; and Joshua D. Greene, "The Secret Joke of Kant's Soul," in Sinnott-Armstrong, *Moral Psychology*, 3:35–80.

44. Jesse Prinz, "Is Morality Innate?" in Sinnott-Armstrong, *Moral Psychology*, 1:387.

45. Ibid., 1:391.

46. Ibid., 1:390.

47. Ibid., 1:391.

48. Jesse Prinz, "Reply to Dwyer and Tiberius," in Sinnott-Armstrong, *Moral Psychology*, 1:427.

49. Prinz, "Is Morality Innate?" 1:368.

50. Gerd Gigerenzer, "Moral Intuition = Fast and Frugal Heuristics?" in Sinnott-Armstrong, *Moral Psychology*, 2:1–26.

51. Ibid., 2:5.

52. Ibid.

53. Ibid., 2:9.

54. Ibid., 2:19.

55. Sinnott-Armstrong, "Framing Moral Intuitions," 2:54ff. Sinnott-Armstrong also cites the cognitive framing experiments described in Amos Tversky and Daniel Kahneman, "The Framing of Decisions and the Psychology of Choice," *Science* 211 (1981): 453–58; Lewis Petrinovich and Patricia O'Neill, "Influence of Wording and Framing Effects on Moral Intuitions," *Ethology and Sociobiology* 17 (1996): 145–71; and Jonathan Haidt and Jonathan Baron, "Social Roles and the Moral Judgments of Acts and Omissions," *European Journal of Social Psychology* 26 (1996): 201–18.

56. Sinnott-Armstrong, "Framing Moral Intuitions," 2:67.

57. Sinnott-Armstrong, "Introduction," 2:xiv.

58. Sinnott-Armstrong, "Framing Moral Intuitions," 2:67.

59. Daniel Kahneman, quoted in Haidt and Bjorklund, "Social Intuitionists," 2:189.

60. Narvaez, "Social Intuitionist Model," 2:236–37.

61. Ibid., 2:239.

62. Ibid.

63. John M. Doris and Alexandra Plakias, "How to Argue About Disagreement: Evaluative Diversity and Moral Realism," in Sinnott-Armstrong, *Moral Psychology*, 2:303.

64. J.L. Mackie, cited in Doris and Plakias, "How to Argue." See also Stephen Stich, "The Persistence of Moral Disagreement," Second Leverhulme Lecture on Moral Psychology, University of Sheffield, May 2009.

65. Moody-Adams, cited in Doris and Plakias, "How to Argue," 2:314.

66. Richard E. Nisbett and Dov Cohen, cited in Doris and Plakias, "How to Argue," 2:316.

67. Doris and Plakias, "How to Argue," 2:318.

68. Ibid.

69. Brian Leiter, "Against Convergent Moral Realism: The Respective Roles of Philosophical Argument and Empirical Evidence," in Sinnott-Armstrong, *Moral Psychology*, 2:335.

70. Paul Bloomfield, "Disagreement About Disagreement," in Sinnott-Armstrong, *Moral Psychology*, 2:343.

71. While Driver provides much evidence for this surprising discovery, she then paradoxically turns to develop a new defense of the free-cause-entails-responsibility principle.

72. Julia Driver, "Attributions of Causation and Moral Responsibility," in Sinnott-Armstrong, *Moral Psychology*, 2:425–26. Driver cites here cases presented in the work of H.L.A. Hart and Tony Honoré, *Causation in the Law* (Oxford: Clarendon Press, 1959).

73. M.D. Alicke, "Culpable Causation," *Journal of Personality and Social Psychology* 63 (1992): 368–78, cited in Driver, "Attributions," 2:427.

74. Joshua Knobe, "Cognitive Processes Shaped by the Impulse to Blame," *Brooklyn Law Review* 71 (2005): 929–37, paraphrased in Driver, "Attributions," 2:429.

75. Driver, "Attributions," 2:430.

76. Joshua Knobe and Ben Fraser, "Causal Judgment and Moral Judgment: Two Experiments," in Sinnott-Armstrong, *Moral Psychology*, 2:441–447.

77. Ibid., 2:446.

78. John Deigh, "Can You Be Morally Responsible for Someone's Death if Nothing You Did Caused It?" in Sinnott-Armstrong, *Moral Psychology*, 2:450–58.

7. Beginning Again: The Blessing and Curse of Neuroplasticity: Interpretation (Almost) All the Way Down

1. Inbal Ben-Ami Bartal, Jean Decety, and Peggy Mason, "Empathy and Pro-Social Behavior in Rats," *Science* 334, no. 6061 (December 9, 2011): 1427–30, doi:10.1126/science.1210789.

2. "Rats Display Human-like Empathy and Will Help Rodents in Distress," *Daily Telegraph*, December 10, 2011.

3. Jaak Panksepp, "Empathy and the Laws of Affect," *Science* 334, no. 6061 (December 9, 2011): 1358–59, doi:10.1126/science.1216480.

4. In *Baboon Metaphysics: The Evolution of a Social Mind* (Chicago: University of Chicago Press, 2007), Dorothy L. Cheney and Robert M. Seyfarth report on the studies of empathy in baboons and other monkeys and compare those findings with the evidence in human beings. Although human beings are capable from a young age, according to them, of taking the perspective of others and monkeys are not (it is not clear whether apes are or are not), shared emotional feeling, if not shared cognition, is present in monkeys. See esp. 146–98.

5. Marc Bekoff and Jessica Pierce, *Wild Justice: The Moral Lives of Animals* (Chicago: University of Chicago Press, 2010).

6. See also Marc Bekoff, *The Emotional Lives of Animals* (New York: New World Library, 2008).

7. Bekoff and Pierce, *Wild Justice*, 41.

8. Ibid., 59.

9. Ibid., 13. I thank Edwin Winckler, political scientist, China expert, and colleague in the exploration of the new neuro- and cognitive sciences, for this summary of Bekoff and Pierce's argument.

10. William Hornaday, quoted in Bekoff and Pierce, *Wild Justice*, ix.

11. Bekoff and Pierce, *Wild Justice*, 20.

12. "The Neurobiology of We: Exploring the Neurobiology of Empathy, Compassion, and Relatedness," August 5–9, 2009, Upaya Institute and Zen Center, Santa Fe, NM.

13. I thank departmental colleague and Buddhism scholar Richard Hughes Seager and my friend and China scholar Edwin Winckler for bringing my attention to the relevance of Buddhism to my social neuroscience concerns. Seager has also brought my attention to what is going on in the interaction between Buddhism scholars and neuroscientists. I also thank Buddhism and science and psychology expert Ann Klein, fellow recipient of Ford Foundation grants, for first bringing my attention to this exciting encounter.

14. Quoted in Norman Doidge, *The Brain That Changes Itself: Stories of Personal Triumph from the Frontiers of Brain Science* (New York: Viking, 2007), iv.

15. Quoted in ibid.

16. Ibid., xviii.

17. Ibid., xviii–xix.

18. Ibid., 12.

19. Ibid., 9.

20. Ibid., 14.

21. Ibid., 23.

22. Ibid., 16.

23. Ibid., 11–12.

24. Ibid., 15.

25. Ibid., 18.

26. Susan Hurley, "Neural Plasticity and Consciousness," *Biology and Philosophy* 18 (2003): 131–68; also available in manuscript form on Hurley's website, www.bristol.ac.uk/philosophy/hurley/index.html.

27. Doidge, *Brain*, 42–43.

28. Ibid., 27–44.

29. Jaak Panksepp and Jules B. Panksepp, "The Seven Sins of Evolutionary Psychology," *Evolution and Cognition* 6, no. 2 (2000): 108–30.

30. Ibid., 109.

31. Ibid., 117.

32. Ibid., 108.

33. Ibid., 114–17.

34. Ibid., 111–12.

35. Ibid., 114.

36. Ibid., 125.

37. Ibid., 124.

38. Doidge, *Brain*, 45.

39. Ibid., 48–49.

40. The nervous system has two kinds of cells: neurons and glial cells that help the neurons function. Neurons have three parts: (1) the cell body, which contains its DNA; (2) the dendrites, which are branches projecting from the neuron that receive messages from other neurons; and (3) the axons, nerve fibers that convey electrical impulses away from the cell body to the dendrites of other neurons. Neurons receive signals that either excite them or inhibit them. When excited, a neuron fires its own signal. The axons don't actually touch the neurons but are separated from them by spaces known as synapses. When an electrical signal reaches the end of the dendrite, it triggers the release into the synapse of a chemical messenger called a neurotransmitter, which goes over to the nearby dendrites of another neuron, causing it to be excited or inhibited. "When we say that neurons 'rewire' themselves," Doidge says, "we mean that alterations occur at the synapse, strengthening and increasing, or weakening and decreasing, the number of connections between the neurons." Doidge, *Brain*, 55.

41. Ibid., 59–60.

42. Ibid., 61.

43. Ibid., 56.

44. Ibid., 51–52.

45. Ibid., 91.

46. "Neuroplasticity," *Wikipedia*, en.wikipedia.org/wiki/Neuroplasticity.

47. Doidge, *Brain*, 196–214.

48. Ibid., 199.

49. Malcolm Gladwell, *Outliers: The Story of Success* (Boston: Little, Brown, 2008).

50. Doidge, *Brain*, 206–7.

51. Ibid., 208.

52. Ibid., 209.

53. Ibid., 220.

54. Ibid., 293–94, 399n.

55. See also "Frame Semantics (Linguistics)," *Wikipedia*, en.wikipedia.org/wiki/Frame_semantics_(linguistics).

56. George Lakoff, *The Political Mind: Why You Can't Understand 21st-Century American Politics with an 18th-Century Brain* (New York: Viking, 2008), 14.

57. Ibid., 15.

58. Ibid., 22.

59. Ibid.

60. Ibid., 26.

61. Ibid., 27.

62. Ibid.

63. Ibid., 28–29.

64. Ibid., 23.

65. Ibid., 33.

66. For contemporary Neo-Aristotelians that part of the moral equation has fallen out.

8. The Self in Itself: What We Can Learn from the New Brain Sciences About Our Sense of Self, Self-Protection, and Self-Furthering

1. Cordelia Fine, *A Mind of Its Own: How Your Brain Distorts and Deceives* (New York: W.W. Norton, 2007), 6.

2. Ibid., 2.

3. An encyclopedic list of the brain's problematic cognitive biases can be found in the article "List of Cognitive Biases," *Wikipedia*, en.wikipedia.org/wiki/Cognitive_biases.

4. Fine, *Mind of Its Own*, 78.

5. Ibid., 58.

6. Ibid., 61.

7. Ibid., 70.

8. Ibid., 73.

9. Ibid., 74.

10. Ibid., 26.

11. Carol Tavris and Elliot Aronson, *Mistakes Were Made (but Not by Me): Why We Justify Foolish Beliefs, Bad Decisions, and Hurtful Acts* (Orlando, FL: Harcourt, 2007), 10.

12. Ibid., 14.

13. Ibid., 17.

14. Ibid., 19.

15. Ibid., 27.

16. Ibid., 37.

17. Ibid., 34.

18. Ibid., 200.

19. Ibid., 201.

20. Ibid., 205.

21. Sandra Blakeslee and Matthew Blakeslee, *The Body Has a Mind of Its Own: How Body Maps in Your Brain Help You Do (Almost) Everything* (New York: Random House, 2007), 190.

22. Ibid., 188.

23. Robert C. Solomon, *True to Our Feelings: What Our Emotions Are Really Telling Us* (New York: Oxford University Press, 2008), 218–19, 224, 217.

24. Blakeslee and Blakeslee, *The Body Has a Mind of Its Own*, 189.

25. Ibid.

26. J. Christopher Perry's research has provided neuroscientific evidence for the mind's defense mechanisms. See, for example, his "Defense Mechanisms Rating Scales in Psychotherapy," 136:165–94, in *Defense Mechanisms: Theoretical, Research, and Clinical Perspectives*, ed. Uwe Hentschel, Gudmnd Smith, Juris G. Draquns, and Wolfram Ehlers (North Holland: Elsevier, 2004).

27. Drew Westen, *The Political Brain: The Role of Emotion in Deciding the Fate of the Nation* (New York: Public Affairs, 2007).

28. Ibid., ix.

29. Ibid., xi.

30. Ibid., xiv.

31. Ibid., xiii.

32. Ibid., xiv.

33. Ibid., xv.

34. Ibid., 70.

35. "Somatic Marker Hypothesis," *Wikipedia*, en.wikipedia.org/wiki/Somatic_marker_hypothesis.

36. Fine, *Mind of Its Own*, 37.

37. Heidi M. Ravven, "Spinoza's Anticipation of Affective Neuroscience," *Consciousness and Emotion*, November 2, 2003.

38. Antonio Damasio, *Descartes' Error: Emotion, Reason, and the Human Brain* (New York: Putnam, 1994), xi.

39. Ibid., xii.

40. Ibid.

41. Ibid., xiii.

42. "Amygdala," *Wikipedia*, en.wikipedia.org/wiki/Amygdala.

43. Westen, *Political Brain*, 57.

44. Ibid., 61.

45. Ibid.

46. Ibid., 100.

47. Ibid., 81, 82.

48. Ibid., 120.

49. Stephen Colbert, quoted in ibid., 103–4.

50. Westen, *Political Brain*, 103.

51. Ibid., 112.

52. Ibid., 112–13.

53. Joshua Greene, "The Secret Joke of Kant's Soul," in *Moral Psychology*, ed. Walter Sinnott-Armstrong (Cambridge, MA: MIT Press, 2008), 3:35–79.

54. Ibid., 36–37. Greene tends to follow the modular hardwired intuitionism of Jonathan Haidt.

55. Ibid., 40. Cognition here refers to reasoning, planning, grasping, and assessing information, working memory, and the like. Emotion can refer to mood but in this context refers to discrete motivating urges and feelings.

56. Ibid., 43, 44.

57. Ibid., 51–54.

58. Ibid., 55–57.

59. Jaak Panksepp and Heidi M. Ravven, "Review of Damasio, *Looking for Spinoza*," *Neuropsychoanalysis* 5, no. 2 (2003): 3.

60. Greene, "Secret Joke," 61.

61. Fine, *Mind of Its Own*, 39.

62. Michael Gazzaniga and Joseph LeDoux are famous for having done much of the split-brain research. Mark Solms and Oliver Turnbull in *The Brain and the Inner World: An Introduction to the Neuroscience of Subjective Experience* (New York: Other Press, 2001), who studied the psychological coping of neurologically impaired adults, also focus on the right-brain/left-brain tendencies to confabulate, deny, and rationalize why they could no longer perform certain kinds of actions, such as move the left side of the body.

63. Greene, "Secret Joke," 62–63. Hence deontology claims for our standard gut moral responses an undeserved elevated (universal and rational) status and justification. So much for Kantian impartiality, universality, and detached moral rationality.

64. Jaak Panksepp, personal communication.

65. Jorge Moll, Mirella L.M.F. Paiva, Roland Zahn, and Jordan Grafman, "Response to Casebeer and Hynes," in Sinnott-Armstrong, *Moral Psychology*, 3:32.

66. "Homology (Biology)," *Wikipedia*, en.wikipedia.org/wiki/Homology _(biology).

67. Paul Sheldon Davies, "Ancestral Voices in the Mammalian Mind: Philosophical Implications of Jaak Panksepp's Affective Neuroscience," *Neuroscience and Biobehavioral Reviews* 35, no. 9 (October 2011), doi:10.1016/j.neurobiorev.2010.10,010. Jaak Panksepp in his writings capitalizes the names of the seven basic emotional systems to indicate that these are scientific terms rather than the informal use of these words in normal speech (private communication, January 2, 2013).

68. Stephen T. Asma and Thomas Greif, "Affective Neuroscience and the Philosophy of Self," *Journal of Consciousness Studies*, forthcoming, available at www. academia.edu/1312559/Affective_Neuroscience_and_the_Philosophy_of_Self.

69. Ibid.

70. This section summarized Jaak Panksepp, "The Basic Emotional Circuits of Mammalian Brains: Do Animals Have Affective Lives?" *Neuroscience and Biobehavioral Reviews* 35, no. 9 (October 2011), doi10.1016/j.neubiorev.2011.08.003.

71. Asma and Greif, "Affective Neuroscience."

72. Ibid.

73. Davies, "Ancestral Voices."

74. Ibid.

75. Asma and Greif, "Affective Neuroscience."

76. Ibid., n4, citing Thomas Metzinger, "Phenomenal Transparency and Cognitive Self-Reference,"*Phenomenology and the Cognitive Sciences* 2 (2003): 353–93. See also Thomas Metzinger, "Self Models," *Scholarpedia*, www.scholarpedia.org/article /Self_models.

77. Asma and Greif, "Affective Neuroscience" (6), quoting David J. Chalmers, *The Conscious Mind: In Search of a Fundamental Theory* (New York: Oxford University Press, 1996).

78. Asma and Greif, "Affective Neuroscience" (8), quoting Daniel Dennett, *Kinds of Minds: Toward an Understanding of Consciousness* (New York: Basic Books, 1996).

79. Asma and Greif, "Affective Neuroscience."

80. Ibid., 7.

81. However, not only is it not good for the cardiovascular system to experience so much stress, but also numbing awareness as a response to psychological threats, in contrast with physical ones, may often be more counterproductive than self-protective.

82. The neuroscientist of the emotions Antonio Damasio has proposed a speculative account of how the symbolic self emerges from a basic self-map of the feeling of the body in *The Feeling of What Happens: Body and Emotion in the Making of Consciousness* (New York: Harcourt Brace, 1999).

83. Martha Stout, *The Myth of Sanity: Divided Consciousness and the Promise of Awareness* (New York: Viking, 2001).

84. Ibid., 8.

85. The definition of dissociation is from Philip Bromberg, *Standing in the Space: Essays on Clinical Process, Trauma, and Dissociation* (Hillsdale, NJ: Analytic Press, 1998), and is quoted in David W. Mann, "The Mirror Crack'd: Dissociation and Reflexivity in Self and Group Phenomena," *Contemporary Psychoanalysis* 44, no. 2 (2008): 237.

86. Stout, *Myth of Sanity*, 19–20.

87. Ibid., 17–18.

88. Ibid., 19–20.

89. Ibid., 27.

90. Ibid., 56–58.

91. Ibid., 212.

92. Ibid.

93. Ibid., 212, 231–32.

94. Jonathan Lear, *Love and Its Place in Nature: A Philosophical Interpretation of Freudian Psychoanalysis* (New Haven: Yale University Press, 1998), 170–71. "Where it was, there I shall become" is the exact translation from the German of Freud's famous description of his psychotherapeutic aim. This phrase was translated into English by James Strachey as "Where id was, there ego shall be" in the standard twenty-four-volume English edition.

95. Ibid., 172.

96. The cognitive neuroscientist Michael Gazzaniga has argued in his Gifford Lectures (*Who's in Charge?: Free Will and the Science of the Brain*, Ecco, 2011) that the brain constrains the mind, but the mind still has free will. My response is as follows. The problem with Gazzaniga's argument is that it presumes that all scientific causes are reductively material causes, so he does not take into account neuroplasticity. Neuroplasticity means that our brains are partly structured by their experience and that the pathways of meaning are made by experience. So he, unlike the affective neuroscientists, sees the science of the brain as only about the material, corporeal causal system and believes that the way that culture, meaning, and language influence us is voluntary. This is a dated view, in my estimation. Contemporary neuroscientists believe that culture, meaning, and understood experience also write themselves into the neocortical pathways of the brain; they are not voluntary but constitutive, along with physical causes, such as genes, chemicals, and the like.

Gazzaniga thinks of the brain as a computer with an internal program running. For Gazzaniga (and Pinker), that program can be prompted to run by events in the environment but is basically internally set and inherited and has a unidirectional causality from set internal patterns toward the world rather than from the neocortex offering a general-use cognitive capacity that gets programmed, shaped, and put to use by experience. The latter theory is an account that has two-way causality between world and high-level cognitive functions, so culture shapes the actual brain, what it does, and how it works, as well as vice versa. (This is Panksepp's view, Damasio's view, and the view of the neuroscientists that Doidge cites.) According to this theory, the cultural and experiential shaping of the neuroplastic pathways is not voluntary but flexible. It can be changed by more experience and different experience, but it cannot be changed by an act of free will. We cannot undo our mental associations, for example, by an act of will but only by different training and re-training.

97. Ran R. Hassin, James S. Uleman, and John A. Bargh, eds., *The New Unconscious* (Oxford: Oxford University Press, 2005), 29.

98. Ibid., 28.

99. Ibid., 20.

100. Ibid.

101. Blakeslee and Blakeslee, *Body Has a Mind of Its Own*, 208.

102. Hassin, Uleman, and Bargh, *New Unconscious*, 21.

103. Ibid., 22.

104. Ibid., 32.

105. Ibid., 29.

106. Ibid., 23.

107. Ibid., 25.

108. Ibid., 33.

109. Timothy D. Wilson, *Strangers to Ourselves: Discovering the Adaptive Unconscious* (Cambridge, MA: Belknap Press, 2002), 5. For the sake of clarity I will draw a distinction between unconscious and nonconscious processes. The philosopher of mind John Searle makes a distinction between unconscious and nonconscious *mental* processes, maintaining that the unconscious ones demarcate those that are capable of being brought to conscious awareness, whereas the nonconscious ones are not. Although I think this distinction is important, current evidence suggests that it may not be that easy to draw that distinction and we may not yet know which are capable of being brought to consciousness. So I will fall back on my distinction. I think Wilson could have used the distinction to good effect rather than muddying the waters by using *unconscious* and *nonconscious* as synonyms.

110. Hassin, Uleman, and Bargh, *New Unconscious*, 3.

111. Wilson, *Strangers to Ourselves*, 6–7.

112. Ibid., 19.

113. Ibid., 21.

114. Ibid., 50, 51.

115. Ibid., 38.

116. Ibid., 68.

117. Ibid., 73.

118. Ibid., 84–86.

119. Ibid., 90.

120. Ibid., 106.

121. Ibid., 125.

122. Ibid., 130–32.

123. Ibid., 112.

124. Wilson, in *Strangers to Ourselves*, tells a joke about two behaviorists who have just had sex, and one says to the other, "I know it was good for you, but was it good for me?"

125. Ibid., 91.

126. Asma and Greif, "Affective Neuroscience," 5.

127. Jaak Panksepp, quoted in Asma and Greif, "Affective Neuroscience," n. 11.

128. Damasio, *Feeling of What Happens*, 10.

129. Ibid., 30.

130. Ibid., 127.

131. Antonio Damasio, *Self Comes to Mind: Constructing the Conscious Brain* (New York: Pantheon, 2010).

132. Asma and Greif, "Affective Neuroscience."

133. Damasio, *Descartes' Error*, xiv.

134. Ibid., xv.

135. Ibid.

136. Ibid., 159.

137. Olaf Blanke and Thomas Metzinger, "Full-Body Illusions and Minimal Phenomenal Selfhood," *Trends in Cognitive Science* 13, no. 1 (2008): 7.

138. Damasio, *Feeling of What Happens*, 23.

139. Ibid., 24.

140. Ibid., 24–25.

141. Ibid., 10.

142. Ibid., 11.

143. Metzinger, "Self Models."

144. Blanke and Metzinger, "Full-Body Illusions," 8.

145. Damasio, *Descartes' Error*, xvi.

146. Ibid.

147. Ibid., 145.

148. Douglass F. Watt, "Emotion and Consciousness: Part II (Review of Anthony Damasio, *The Feeling of What Happens*)," *Journal of Consciousness Studies* 7, no. 3 (March 2000): 73.

149. Damasio, *Feeling of What Happens*, 25.

150. Damasio, *Descartes' Error*, xv.

151. Ibid., 30.

152. Ibid., 127.

9. The Self Beyond Itself: The "We That Is I" and the "I That Is We"

1. See Heidi M. Ravven, "Spinoza's Anticipation of Contemporary Affective Neuroscience," *Consciousness and Emotion* 4, no. 2 (2003); and Heidi M. Ravven, "Spinoza and the Education of Desire," *Neuropsychoanalysis* 5, no. 2 (2003): 218–29.

2. Sandra Blakeslee and Matthew Blakeslee, *The Body Has a Mind of Its Own: How Body Maps in Your Brain Help You Do (Almost) Everything* (New York: Random House, 2007), 142–43, 117–18.

3. Andy Clark, *Supersizing the Mind: Embodiment, Action, and Cognitive Extension* (Oxford: Oxford University Press, 2008).

4. Ibid., 82.

5. Ibid., xvi.

6. Ibid., 30–31.

7. Philippe Rochat, *Others in Mind: Social Origins of Self-Consciousness* (Cambridge: Cambridge University Press, 2009), 56, 57.

8. This is the conclusion of Susan M. Andersen, Inga Reznik, and Noah S. Glassman in their essay "The Unconscious Relational Self," in *The New Unconscious*, ed. Ran R. Hassin, James S. Uleman, and John A. Bargh (Oxford: Oxford University Press, 2005). The authors call what emerged from their research a "contextual model based in social-cognitive processes that are known to transpire outside of awareness and without effort." Their research exposed that people have what they call "an overall repertoire of selves, each of which stems from a relationship with a significant other."

9. See, e.g., Heinz Kohut, *The Restoration of the Self* (New York: International Universities Press, 1977); Ernest S. Wolf, *Treating the Self: Elements of Clinical Self Psychology* (New York: Guilford Press, 1988).

10. Hassin, Uleman, and Bargh, *New Unconscious*, 421–81.

11. Heinz Kohut, *The Kohut Seminars on Self Psychology and Psychotherapy with Adolescents and Young Adults*, ed. Miriam Elson (New York: W.W. Norton, 1987), 277.

12. Olaf Blanke and Thomas Metzinger, "Full-Body Illusions and Minimal Phenomenal Selfhood," *Trends in Cognitive Science* 13, no. 1 (2008): 7–13.

13. Ibid.

14. Blakeslee and Blakeslee, *Body Has a Mind of Its Own*, 103–4.

15. Ibid., 122.

16. Donald W. Pfaff, *The Neuroscience of Fair Play: Why We (Usually) Follow the Golden Rule* (New York: Dana, 2007).

17. Ibid., 61, 62.

18. Ibid., 62.

19. Ibid., 70.

20. Ibid., 74.

21. Ibid., 79.

22. Ibid., chap. 6, 80–98.

23. Ibid., 99–120.

24. Ibid., 119.

25. Ibid., 123.

26. Ibid., 147–48.

27. Ibid., 202–3.

28. Blakeslee and Blakeslee, *Body Has a Mind of Its Own*, 133–37.

29. Greg J. Stephens, Laruen J. Silbert, and Uri Hasson, "Speaker-Listener Neural Coupling Underlies Successful Communication," *Proceedings of the National Academy of Sciences of the United States of America* 107, no. 32 (August 10, 2010): 14425–30.

30. Vittorio Gallese, "The 'Shared Manifold' Hypothesis: From Mirror Neurons to Empathy," *Journal of Consciousness Studies* 8 (2001): 35.

31. G. Buccino et al., "Short Communication: Action Observation Activates Premotor and Parietal Areas in a Somatotopic Manner: An fMRI Study," *European Journal of Neuroscience* 13 (2001): 400.

32. Giacomo Rizzolatti, Leonardo Fogassi, and Vittorio Gallese, "Neurophysiological Mechanisms Underlying the Understanding and Imitation of Action," *Nature Reviews/Neuroscience* 2 (September 2001): 662, 667.

33. Buccino et al., "Short Communication," 403.

34. Rizzolatti, Fogassi, and Gallese, "Neurophysiological Mechanisms," 661.

35. Giacomo Rizzolatti, *Mirrors in the Brain: How Our Minds Share Actions and Emotions* (Oxford: Oxford University Press, 2008), xi.

36. Gallese, "'Shared Manifold,'" 38.

37. Ibid., 33–34.

38. Rizzolatti, Fogassi, and Gallese, "Neurophysiological Mechanisms," 667.

39. Marco Iacoboni, *Mirroring People: The New Science of How We Connect with Others* (New York: Farrar, Straus and Giroux, 2008), 210.

40. Ibid., 108–9.

41. Ibid., 121–24.

42. Gallese, "'Shared Manifold,'" 42–43.

43. Ibid., 43.

44. Ibid., 44.

45. Ibid., 46.

46. Iacoboni, *Mirroring People*, 212–13.

47. Ibid., 214ff.

48. Ibid., 133ff.

49. Malcolm Gladwell, *Outliers: The Story of Success* (Boston: Little, Brown, 2008), 19.

50. Ibid., 55.

51. Ibid., 49.

52. Norman is quoted in ibid., 50.

53. Gladwell, *Outliers*, 50.

54. Ibid., 53.

55. Ibid., 55.

56. Ibid., 64–67.

57. Ibid., 67.

58. Richard E. Nisbett, *Intelligence and How to Get It: Why Schools and Cultures Count* (New York: W.W. Norton, 2009).

59. Ibid., 154.

60. Ibid., 3.

61. Ibid., 171–72.

62. Ibid., 180–81.

63. Alva Noë, *Out of Our Heads: Why You Are Not Your Brain, and Other Lessons from the Biology of Consciousness* (New York: Hill and Wang, 2009), 5.

64. Ibid., 169.

65. Ibid., 180.

66. Philip Robbins and Murat Aydede, "A Short Primer on Situated Cognition," in *The Cambridge Handbook of Situated Cognition*, ed. Philip Robbins and Murat Aydede (Cambridge: Cambridge University Press, 2009), 3.

67. Noë, *Out of Our Heads*, 48–49.

68. Robbins and Aydede, "Short Primer," 5.

69. Blakeslee and Blakeslee, *Body Has a Mind of Its Own*, 106.

70. Ibid.

71. Noë, *Out of Our Heads*, 183, 184.

72. I wish to thank my friend and psychologist-coach Michael Jay Sullivan of Cambridge, Massachusetts, for the insight that the type of identification we have with our discrete thoughts, feelings, and actions can be distanced while still owning them as our own (acceptance) and instead having a stronger identification with the observer-self.

73. Jaak Panksepp (in an unpublished manuscript) calls the complex the "emotion action systems." He maintains that the evidence shows that "emotional feelings . . . are fundamentally experienced action systems." Panksepp goes on, "The coherence of our bodily existence, and I think mental also, is based upon action coordinates; and in the realm of emotions, the data overwhelmingly indicates that the Unconditioned Emotional Response Systems are exactly where *emotional* reward and punishments arise." Here is the evidence for Hurley's prescient account.

74. Susan Hurley, *Consciousness in Action* (Cambridge, MA: Harvard University Press, 1998).

75. Noë, *Out of Our Heads*, 82, 83.

76. The perception/action cut has the virtue of exposing to view the "constitutive interdependence" of the two, insofar as "both depend on complex dynamic feedback systems" so that the "significance of boundaries like the skin or the skull fades."

Environments are very complex combinations of what's outside the person per se and how human beings have influenced, constructed, and interpreted their worlds, that is, outputs, Hurley argues. Conversely, the internal conceptual structures via which we perceive the world are not just hardwired internal mental contributions but are substantially constituted by the way the world is itself structured—and structured by us, and hence taken in by us already so structured. The connecting link is action and interaction, which inform both mind and world.

77. Hurley, *Consciousness in Action*, 35.

78. "So long as we preserve autonomy [from the world] on the side of the mind, we'll need intrinsic aboutness [a content that comes from the mind itself without the contribution of the world] and still be unable to make sense of it" (Hurley, *Consciousness in Action*, 261). In fact, the world will be "radically indeterminate" in content if we think it has no content to contribute to what we perceive (ibid., 263). Hurley argues further that the Kantian autonomy of the mind is mutually implicative of the claim of free will. The cognitive autonomy of the mind makes knowing the world impossible, while the claim of free will makes acting incapable of being actually about the world. She writes: "[I]f experience can't be about the world at all, it can't be wrong about the world. . . . Three parallel points can be made about a conception of the inner aspect or agency . . . as autonomous. First, . . . the world is inscrutable with the presence or absence of true agency; . . . Second, . . . the possibility that our tryings may be utterly futile in relation to the world is open and threatening. . . . Third, . . . most importantly, this double dissociability of mind and world makes the aboutness of our intentions and tryings, their content or directedness, mysterious. . . . At whatever point we stop, it's still mysterious how an autonomous will manages to be intrinsically about the world at all" (ibid., 262).

79. Ibid., 263, 264. Hurley says that the externalism she advocates is a version of contextualism.

80. Ibid., 15.

81. Ibid., 22, 23.

82. Ibid., 250.

83. I have written on Spinoza's anticipation of systems theory and his rethinking of ethics in terms of it. See Heidi M. Ravven, "What Can Spinoza Teach Us Today About Naturalizing Ethics? Provincializing Philosophical Ethics and Freedom Without Free Will," in *Cognitive, Emotive, and Ethical Aspects of Decision Making in Humans and in Artificial Intelligence*, ed. Iva Smith and Wendell Wallach (Windsor, ON: International Institute for Advanced Studies in Systems Research and Cybernetics, 2005), 3:99–104. See also the new study by Rainer E. Zimmermann, *New Ethics Proved in Geometrical Order: Spinozist Reflexions on Evolutionary Systems* (Litchfield Park, AZ: Emergent Publications, 2010).

84. A brilliant treatment of the social hierarchical origins of normativity is Dorothy L. Cheney and Robert M. Seyfarth, *Baboon Metaphysics: The Evolution of a Social Mind* (Chicago: University of Chicago Press, 2008). Jaak Panksepp doubts that the

preference for hierarchy is hardwired in the brain. He proposes that all we need to produce hierarchy is the seeking system and a context of scarcity.

85. Gladwell, *Outliers*, chap. 7, "The Ethnic Theory of Plane Crashes," 177–223.

86. Ibid., 200, 294, which cites Ho-min Sohn, "Intercultural Communication in Cognitive Values: Americans and Koreans," *Language and Linguistics* 9 (1993): 93–136.

87. William J. Clancey, "Scientific Antecedents of Situated Cognition," in *Cambridge Handbook of Situated Cognition*, 14, identifies the following as the characteristic features of complex systems: emergence; feedback loops; open, observer-defined boundaries; having a history; and compositional networks. Emergence refers to patterns of behavior that result from interactions among the components. Feedback loops in the context of situated cognition can be conscious or unconscious, and even nonlinguistic and nonconceptual, and often "highlight conceptual aspects that pertain to identity and social relations." Open systems are "observer-defined" because they depend on the interests and concerns and questions of the knower. And complex systems have a history because past interactions "have changed both the parts and what constitutes their system environment." Finally, "the components of complex systems are often themselves complex adaptive systems."

88. Ibid., 24.

89. Ibid., 28.

90. Merlin Donald, "How Culture and Brain Mechanisms Interact in Decision Making," in *Better than Consciousness? Decision Making, the Human Mind, and Implications for Institutions*, ed. Christoph Engel and Wolf Singer (Cambridge, MA: MIT Press, 2008), 192.

91. Ibid., 202.

92. Ibid., 192.

93. Christoph Engel and Wolf Singer, "Better than Conscious? The Brain, the Psyche, Behavior, and Institutions," in *Better than Consciousness?*, 8–9.

94. Murray Gell-Mann, *The Quark and the Jaguar: Adventures in the Simple and the Complex* (New York: W.H. Freeman, 1994), 9.

95. "Complex Adaptive Systems," *Wikipedia*, en.wikipedia.org/wiki /Complex_adaptive_system.

96. This is known as the recycling effect.

97. John H. Holland, *Hidden Order: How Adaptation Builds Complexity* (Reading, MA: Addison-Wesley, 1995), 29.

98. Ibid., 27. The recycling effect and diversity are addressed on 25–31.

99. Gell-Mann, *Quark and the Jaguar*, 9.

100. This is known as edge of chaos theory.

101. See Ravven, "What Can Spinoza Teach Us"; Heidi M. Ravven, "Spinoza's Systems Theory of Ethics," in Smith and Wallach, *Cognitive, Emotive, and Ethical Aspects of Decision Making*; Heidi M. Ravven, "Notes on Spinoza's Critique of

Aristotle's Ethics: From Teleology to Process Theory," *Philosophy and Theology* 4, no. 1 (Fall 1989): 3–32.

102. Martin A. Nowak, Corina E. Tarnita, and Edward O. Wilson, "The Evolution of Eusociality," *Nature* 466 (August 26, 2010): 1057–62.

103. Ibid., 1058.

104. Ibid.

105. Ibid.

106. Ibid., 1060.

107. Ibid.

10. What Is Ethics? How Does Moral Agency Work?

1. Spinoza uses *affect* very broadly to indicate how we are affected by our interactions, of which emotions are an expression in awareness. The cognitive scientist Art Markman (in his essay "Disgust and Morality," *Psychology Today*, December 8, 2009, www.psychologytoday.com/blog/ulterior-motives/200912/disgust-and-morality) points out that the current usage defines emotion as cognized, interpreted "affect."

2. See, e.g., Antoine Lutz et al., "Long-Term Meditators Self-Induce High-Amplitude Gamma Synchrony During Mental Practice," *Proceedings of the National Academy of Sciences* 101, no. 46 (November 8, 2004), doi:10.1073/pnas.0407401101.

3. For a related treatment of Spinoza's moral agency as rooted in embodied relations in the world, see Andrew Collier, "The Materiality of Morals: Mind, Body and Interests in Spinoza's *Ethics*," *Studia Spinozana* 7 (1991): 69–93.

4. Baruch Spinoza, E V, prop. 15 and dem., in *Baruch Spinoza: The Ethics and Selected Letters,* trans. Samuel Shirley and ed. Seymour Feldman (Indianapolis: Hackett, 1982), 212: "*He who clearly and distinctly understands himself and his emotions loves God, and the more so the more he understands himself and his emotions.* Dem. He who clearly and distinctly understands himself and his emotions feels pleasure (Pr. 53, III) accompanied by the idea of God."

5. The philosopher Hegel, perhaps significantly inspired by his reading of Spinoza, envisioned the dialectical and progressive interplay and mutual interpenetration of mind and world in *The Phenomenology of Mind* (1807).

6. I wish to thank my student Sorina Seeley for her excellent senior thesis at Hamilton College, "Can Spinoza Help Us? Cultivating a New Ecological Consciousness Through Spinoza's Philosophy and Deep Ecology," developing a more detailed and stronger Spinozist foundation and argument for deep ecology. For further exploration of Spinoza and the deep ecology movement, see the works of Arne Naess, the Norwegian Spinoza scholar who developed deep ecology; Hasana Sharp's *The Politics of Renaturalization* (Chicago: University of Chicago Press, 2011); Essy De Jonge's *Spinoza and Deep Ecology: Living as if Nature Mattered* (Salt Lake City, UT: G.M. Smith, 2007); and George Sessions, ed., *Deep Ecology for the*

Twenty-First Century (Boston: Shambhala, 1995). My recruitment of Spinoza's vision to a contemporary theory of moral agency parallels and may broaden the scope of the deep ecology movement, a movement advocating the preservation of the natural environment as a whole and for itself, not just as a means to human survival. The deep ecology movement was significantly inspired by Spinoza and some of its important founders were scholars of Spinoza. My Spinozist theory of moral agency adds to the biological and ecological vision of the deep ecology movement Spinoza's account of the stages of personal awareness of oneself as ecologically constituted. I have attempted a resurrection of Spinoza's individual and social psychology of moral agency that is similar to the moral cosmological vision of Spinoza that deep ecology has revived and recruited. The two revivals are not only parallel but also complementary, the one offering the rationale in individual fulfillment for the cosmological normative ideal of the other. One offers the global vision and the other its individual rationale. I hope that each of us, as many as possible of us, come to a self-reflective understanding of the eco-biological reality as it affects oneself, integrating that self-reflection into the self—and thereby transforming the self as within its eco-world and universe.

7. Spinoza's philosophy has been designated a coherence theory of knowledge, but we might recast it more precisely as a coherence account, a systems theory of the mind. It is the urge of the biological self-organizing energy to maintain itself at the heart of the human organism as expressed in mind, in its *self*-understandings. It is the mind minding the body; it is the embodied mind in its environments.

8. If we recall how cognitive frames shape our understanding of self and world, we can think of the initial unreflective self as emerging out of the interpretive frames about self and environment that we unconsciously take in and enact. The initial pathways, the neural networks written into our brains, are pruned early on and are both highly emotionally charged and also implicit directors of meaning and action. These pathways and connections are instantiated through a process of pruning as well as ingrained through repetition. Perhaps it is this early stage of the unreflective and largely prelinguistic child's swallowing of self in its world as set in the mind that drives a primitive sense of self and belonging. It is also perhaps this initial implicit neuro-network framing that lends a feeling of normativity, of rightness and necessity, of ought, to that early environmentally and socially marked self. Perhaps it is this neural net incising that makes it in a sense authoritative for a person, often throughout life. Thus the narrow localism and fanaticism of Spinoza's descriptions of an immediate unreflective kind of agency—his account of the imagination or what I would call "the imaginative mind" both in the *Ethics* and the *Theological Political Treatise*—would seem to be supported by some of the recent discoveries about the brain's neuroplasticity that Norman Doidge, for example, outlined.

Index

Publishing in the Public Interest

Thank you for reading this book published by The New Press. The New Press is a nonprofit, public interest publisher. New Press books and authors play a crucial role in sparking conversations about the key political and social issues of our day.

We hope you enjoyed this book and that you will stay in touch with The New Press. Here are a few ways to stay up to date with our books, events, and the issues we cover:

- Sign up at www.thenewpress.com/subscribe to receive updates on New Press authors and issues and to be notified about local events
- Like us on Facebook: www.facebook.com/newpressbooks
- Follow us on Twitter: www.twitter.com/thenewpress

Please consider buying New Press books for yourself; for friends and family; or to donate to schools, libraries, community centers, prison libraries, and other organizations involved with the issues our authors write about.

The New Press is a 501(c)(3) nonprofit organization. You can also support our work with a tax-deductible gift by visiting www.thenewpress.com/donate.